Ecology, Environment and Conservation

Ecology, Environment and Conservation

Edited by **Anne Offit**

SYRAWOOD
PUBLISHING HOUSE
New York

Published by Syrawood Publishing House,
750 Third Avenue, 9th Floor,
New York, NY 10017, USA
www.syrawoodpublishinghouse.com

Ecology, Environment and Conservation
Edited by Anne Offit

International Standard Book Number: 978-1-68286-061-8 (Hardback)

Contents

Preface

Every book is a source of knowledge and this one is no exception. The idea that led to the conceptualization of this book was the fact that the world is advancing rapidly; which makes it crucial to document the progress in every field. I am aware that a lot of data is already available, yet, there is a lot more to learn. Hence, I accepted the responsibility of editing this book and contributing my knowledge to the community.

This book integrates some of the key issues and concepts pertaining to ecology, environment and conservation. The rapid degradation of natural resources, diminishing reserves of conventional fuel and mineral sources, and the impoverished state of environmental health has necessitated the re-evaluation of damage caused by various industrial and human activities. The topics covered in this extensive book deal with some of the crucial aspects such as emerging trends in recycling and waste management, strategies to improve sustainability and productivity, diverse branches of ecology, population dynamics and utilization of natural resources, green house effects, etc. From theories to researches to practical applications, case studies related to all contemporary topics of relevance to this field have been included in this book. This book is a resource guide for experts as well as students.

While editing this book, I had multiple visions for it. Then I finally narrowed down to make every chapter a sole standing text explaining a particular topic, so that they can be used independently. However, the umbrella subject sinews them into a common theme. This makes the book a unique platform of knowledge.

I would like to give the major credit of this book to the experts from every corner of the world, who took the time to share their expertise with us. Also, I owe the completion of this book to the never-ending support of my family, who supported me throughout the project.

Editor

Estimation of conidial concentration of freshwater Hyphomycetes in two streams flowing at different altitudes of Kumaun Himalaya

Pratibha Arya[1]* and S. C. Sati[1]

[1]Botany Department, Government P. G. College Augustyamuni, Rudraprayag, Garhwal, Uttarakahand, India.
[2]Botany Department, Kumaun University, Nainital, Uttarakhand, India.

Freshwater Hyphomycetes commonly occurs in all types of natural freshwater streams and form one of the most important components of freshwater ecosystem as decomposers. They produce tremendous conidia in submerged condition. In the present study, concentration of these fungi in per unit volume of water in two freshwater streams situated at different altitudes viz., Ratighat (1200 m asl) and Vinayak (1500 m asl) was determined by using the Millipore filter paper technique following Iqbal and Webster (1973b). The results demonstrate that in Ratighat stream, 21 species of freshwater Hyphomycetes belonging to 15 genera were found. Maximum concentration of conidia was observed in the month of December (41.05×10^3 conidia/litre) and the minimum number of conidia was observed during May (12.83×10^3 conidia/litre). In Vinayak stream 26 species of freshwater Hyphomycetes belonging to 18 genera were found. Maximum concentration of conidia per litre of stream water sample was observed during the month of January (74.40×10^3 conidia/litre) and the minimum concentration of conidia was observed during June (30.79×10^3 conidia/litre).

Key words: Freshwater Hyphomycetes, conidial concentration, millipore filter.

INTRODUCTION

Freshwater Hyphomycetes frequently colonize the submerged decaying leaf litter of trees and shrubs in well aerated streams. The branched septate mycelium of these fungi ramifies the dead leaf tissues and their conidiophores are projected into the water to release the spores and taken away by the water currents (Iqbal and Webster, 1973a). These spores (conidia) are often collected in accumulated water foam and scum at the barriers. Biologically these fungi are very important as they provide inoculum potential, responsible for decomposition of organic matter in running freshwater bodies (Barlocher, 1985; Gessner et al., 1997; Suberkropp, 1992, 1998; Graca, 2001; Gessner and vanRyckegem, 2003).

A vast literature has been accumulated about their morphology, development, occurrence and ecology from different parts of the world (Ingold, 1965, 1976; Iqbal et al., 1980; Descals and Webster, 1982; Nawawi, 1985; Sridhar et al., 1992; Sati and Tiwari, 1997; Marvanova, 1997). However, there is a paucity of knowledge on the study of conidial concentration of these fungi.

The objective of present investigation was to determine the conidial concentration of freshwater Hyphomycetes per unit volume of water in two freshwater streams flowing at different altitudes viz., Ratighat (1200 m asl) and Vinayak (1500 m asl) of Kumaun Himalaya.

MATERIALS AND METHODS

In order to quantify the conidial concentration of Freshwater Hyphomycetes in unit volume of stream water, Iqbal and Webster (1973b) was followed. Water samples were taken at an interval of 30 days for the period of twelve months from each stream situated at different altitudes viz., Ratighat (1200 m asl, 29° 25′ N Latitude and 79° 32′ E Longitude) and Vinayak (1500 m asl, 29° 22′ N Latitude and 79° 33′ E Longitude). 10 samples of water

*Corresponding author. E-mail: arya.pratibha_82@yahoo.co.in.

Table 1. Freshwater Hyphomycetes recorded by the Millipore filtration of the unit volume of water from Ratighat and Vinayak streams.

S. No.	Fungi	Ratighat stream	Vinayak stream
1.	*Acaulopage tetraceros* Drechsler	♦♦	–
2.	*Alatospora acuminata* Ingold	♦♦♦	♦♦♦
3.	*A. pulchella* Marvanova	♦	♦
4.	*Anguillospora crassa* Ingold	♦	–
5.	*A. longissima* (Sacc. & Therry) Ingold	♦♦	♦♦
6.	*Beltrania rhombica* Penzing	♦	–
7.	*Camposporium pellucidum* (Grove) Hughes	–	♦
8.	*Campylospora chaetocladia* Ranzoni	♦	♦♦♦
9.	*C. parvula* Kuzaha	♦♦♦	♦
10.	*Clavariopsis aquatica* de Wildeman	♦♦	♦♦♦
11.	*Diplocladiella longibrachiata* Nawawi & Kuthub.	–	♦
12.	*Helicomyces roseus* Link	♦	–
13.	*Heliscella stellatacula* Marvanova	–	♦
14.	*Heliscus lugdunensis* Sacc. & Therry	♦♦	–
15.	*Lemonniera terrestris* Tubaki	–	♦
16.	*L. cornuta* Ranzoni	–	♦♦♦
17.	*L. pseudofloscula* Dyko	–	♦
18.	*Lunulospora curvula* Ingold	♦♦♦	♦♦♦
19.	*L. cymbiformis* Miura	♦♦♦	♦♦♦
20.	*Pestalotiopsis submersus* Sati & Tiwari	♦	♦♦
21.	*Pleurophragmium sonam* Sati & Tiwari	–	♦
22.	*Setosynnema isthmosporum* Shaw & Sutton	–	♦♦♦
23.	*Speiropsis scopiformis* Kuthub.& Nawawi	♦♦	♦♦
24.	*Tetrachaetum elegans* Ingold	♦♦	♦♦♦
25.	*Tetracladium apiense* Sinclair & Eicker	–	♦♦
26.	*T. marchalianum* de Wildeman	♦♦♦	♦♦♦
27.	*T. setigerum* (Grove) Ingold	♦♦♦	♦♦
28.	*Tricladium chaetocladium* Ingold	–	♦
29.	*Tricladium* sp.	♦	–
30.	*Tripospermum myrti* (Lind)Hughes	–	♦
31.	*Triscelophorus acuminatus* Nawawi	♦♦♦	♦♦♦
32.	*T. monosporus* Ingold	♦♦♦	♦♦♦

♦♦♦ = Commonly frequent species; ♦♦ = frequent species; ♦ = rare species; □ = species absent.

(approximately 100 ml water) were collected from different substations in both the streams and a composite sample of one litre water was made for each stream separately. This was brought to the laboratory and filtered with a Millipore filter (5 μm pore size). Millipore filters were then treated with cotton blue and lactophenol to stain the conidia. Treated filter papers were then heated in lactic acid at 50 to 60°C to make them transparent for low power microscope examination. This allows the recognition and identification of various kinds of conidia on the filter paper. The conidia were identified with the help of relevant monographs and papers (Ingold, 1975; Marvanova, 1997). Data was recorded for each month and each stream separately.

RESULTS

The composite water samples of unit litre volume each of

two different studied streams were filtered at monthly intervals and data are presented in Tables 1 and 2. The filtration process causes many conidial forms overlapping to each other; therefore, conidial counting was made very carefully at each vision field of microscope to scan the treated millipore filter.

In the Ratighat stream 21 species of Freshwater Hyphomycetes belonging to 15 genera viz., *Acaulopage*, *Alatospora*, *Anguillospora*, *Beltrania*, *Campylospora*, *Clavariopsis*, *Helicomyces*, *Heliscus*, *Lunulospora*, *Pestalotiopsis*, *Speiropsis*, *Tetrachaetum*, *Tetracladium*, *Tricladium* and *Triscelophorus* were reported (Table 1) Of these, *Alatospora acuminata*, *Campylospora parvula*, *Lunulospora curvula*, *L. cymbiformis*, *Tetracladium marchalianum*, *T. setigerum*, *Triscelophorus acuminatus*

Table 2. Estimation of conidial concentration of Freshwater Hyphomycetes on unit litre of composite water sample from Ratighat and Vinayak streams.

Month	Ratighat			Vinayak		
	(No. of conidia × 10³ /litre)	Temperature (°C)	pH	(No. of conidia × 10³ /litre)	Temperature (°C)	pH
March-07	25.65	10.5	7.2	35.92	13.0	7.6
April	20.52	15.8	7.9	41.05	17.9	7.9
May	12.83	22.5	9.0	35.92	19.3	8.8
June	23.09	21.8	7.6	30.79	24.3	8.2
July	17.96	22.9	7.6	38.48	23.7	8.0
August	15.39	23.6	7.9	48.74	23.0	8.2
September	30.79	21.0	7.7	43.61	19.5	8.6
October	33.35	21.5	7.9	46.18	19.5	8.7
November	35.92	19.5	7.3	51.31	15.5	8.3
December	41.05	15.5	7.2	61.57	12.0	8.2
January-08	30.79	9.0	7.5	74.40	9.0	8.2
February	28.22	8.0	7.2	69.27	8.0	8.2

and *T. monosporus* were found as frequently occurring species in each month of observation.

As evident from Table 2 the maximum number of conidia was found in the month of December (41.05 × 10³ conidia/litre) and the minimum number of conidia was observed during May (12.83 × 10³ conidia/litre). The seasonal fluctuation of Freshwater Hyphomycetes was found negatively correlated with water temperature and pH (for temperature r = -0.370456, for pH r = -0.635664).

In the Vinayak stream 26 species of freshwater Hyphomycetes belonging to 18 genera viz., *Alatospora*, *Anguillospora*, *Camposporium*, *Campylospora*, *Clavariopsis*, *Diplocladiella*, *Heliscella*, *Lemonniera*, *Lunulospora*, *Pestalotiopsis*, *Pleurophragmium*, *Setosynnema*, *Speiropsis*, *Tetrachaetum*, *Tetracladium*, *Tricladium*, *Tripospermum* and *Triscelophorus* were reported (Table 1). Of these 10 species viz., *Alatospora acuminata*, *Campylospora chaetocladia*, *Clavariopsis aquatica*, *Lemonniera cornuta*, *Lunulospora curvula*, *L. cymbiformis*, *Setosynnema isthmosporum*, *Tetrachaetum elegans*, *Tetracladium marchalianum*, *Triscelophorus acuminatus* and *T. monosporus* were found frequently in all the months.

Maximum concentration of conidia per litre of stream water sample was observed during the month of January (74.40 × 10³ conidia/litre) and the minimum concentration of conidia was observed during June (30.79 × 10³ conidia/litre). The seasonal fluctuation of Freshwater Hyphomycetes was found negatively correlated with water temperature and pH (for temperature r = -0.783035) and positively correlated with water pH (r = 0.031326).

DISCUSSION

Foam collected at well-aerated streams contains a dense accumulation of spores and it is an effective trap for studying the spores of Freshwater Hyphomycetes (Ingold, 1961, 1967; Nilsson, 1964; Iqbal and Webster, 1973a). Observation of foam is a very quick way of getting an idea of the Freshwater Hyphomycetous fungus flora of that stream. In the present study the stream water was filtered through a Millipore filter to get conidial concentration of water borne conidial fungi in a unit volume of stream water. Though the present method is the best to measure quickly the concentration of spores of water borne conidial fungi in any water body but filtration process causes overlapping of some conidial forms to one other. Therefore, it requires a careful counting of conidia under the microscope. During this study Vinayak (higher altitudinal stream) was found with high concentration of conidia in per unit volume of stream water ranging from- 30.79 - 74.40 ×10³ conidia/litre than the Ratighat stream (low altitudinal stream) ranging from 12.83 – 41.05 ×10³ conidia/litre.

As evident from Table 2 the stream water of two sites had considerable monthly fluctuations in the conidial concentration per unit volume of water. Ingold (1976) has also reported a sharp seasonal variation in the monthly conidial concentration. Maximum conidial concentration in per unit volume of composite water sample was recorded during low temperature months i.e., December and January. This indicates that the winter months support greater conidial concentration than the other months. These findings are almost similar to Iqbal and Webster (1973 b), Muller – Haeckel and Marvanova (1979).

During this study it was interesting to note that higher altitudinal stream supports greater species richness (26 species) than the lower altitudinal stream (21 species). Raviraja et al., (1998), Chauvet (1991) and Fabre (1996) also found greater number of species in higher altitudinal streams as compared to lower altitudinal streams.

Conclusion

The estimation of conidial concentration of Freshwater Hyphomycetes in any water body would be useful to have a quick idea not only for its fungal flora but also to get the conidial inoculum potential of a stream in the particular season responsible for the decomposition of litter in the aquatic ecosystem.

ACKNOWLEDGEMENT

The authors thank the Department of Botany for facilities provided.

REFERENCES

Barlocher F (1985). The role of fungi in the nutrition of stream invertebrates. Bot. J. Linn. Soc., 91: 83-94.

Chauvet E (1991). Aquatic hyphomycetes distribution in South-Western France. J. Biogeogr., 18: 699-706.

Descals E, Webster J (1982). Taxonomic studies on "aquatic hyphomycetes" III some new species and a new combination. Tran. Brit. Mycol. Soc., 78: 405-437.

Fabre E (1996). Relationships between aquatic hyphomycetes communities and riparian vegetation in 3 Pyrenean streams. Comptes Rendus Hebdomadaires des Seances des l' Academie des Sciences, Paris, 319: 107-111.

Gessner MO, vanRyckegem G (2003). Water fungi as decomposers in freshwater ecosystems. In G. Bitton (Ed.), Encyclopedia of Environmental Microbiology. New York: John Wiley and Sons. (corrected online version).

Gessner MO, Suberkropp K, Chauvet E (1997). Decomposition of plant litter by fungi in marine and freshwater ecosystem. In D. T. Wicklow and Soderstrom, B. (Eds.), The Mycota, Environmental and microbial relationships. Berlin: Springer, 4: 303-322.

Graca MAS (2001). The role of invertebrates on leaf litter decomposition in streams- A review. Int. Rev. Hydrobiol., 86: 383-393.

Ingold CT (1961). Another aquatic spore-type with clamp connections. Trans. Brit. Mycol. Soc., 44: 27-30.

Ingold CT (1965). Hyphomycete spores from mountain torrents. Trans. Brit. Mycol. Soc., 48: 453-458.

Ingold CT (1967). Spores from foam. Bull. Brit. Mycol. Soc., pp. 60-63.

Ingold CT (1975). An illustrated guide to aquatic and water borne hyphomycetes (Fungi Imperfecti) with notes on their biology. Freshwater Biol. Assoc. Scient. Publ. No. 30, England, p. 96.

Ingold CT (1976). The morphology and biology of fresh water fungi excluding Phycomycetes. In "Recent advances in aquatic Mycology". (Ed. E.B. Gareth Jones) Publ. Paul Elek . Ltd. 335-357.

Iqbal SH, Webster J (1973a). The trapping of aquatic hyphomycetes spores by air bubbles. Trans. Brit. Mycol. Soc., 60: 37-48.

Iqbal SH, Webster J (1973b). Aquatic Hyphomycetes spora of the River Exe and its tributaries. Trans. Brit. Mycol. Soc., 61: 331-346.

Iqbal SH, Bhatty SF, Malik KS (1980). Fresh water hyphopmycetes of Pakistan. Bull. Mycol., 1: 1-25.

Marvanova L (1997). Freshwater hyphomycetes: A survey with remarks on tropical taxa In Tropical Mycology (eds. K. K. Janardhanan, C. Rajendran, K. Natrajan and D. L. Hawksworth). Sci. Publ. Inc., 169-226.

Muller-Haeckel A, Marvanova L (1979). Periodicity of aquatic hyphomycetes in the Sub arctic. Trans. Brit. Mycol. Soc., 73: 109-116

Nawawi A (1985). Aquatic Hyphomycetes and other water borne fungi from Malaysia. Malay. Nat. J., 39: 75-134.

Nilsson S (1964). Fresh water Hyphomycetes. Taxonomy, Morphology and Ecology. Symb. Bot. Upsal., 18: 1-30.

Raviraja NS, Sridhar KR, Barlocher F (1998). Fungal species richness in Western Ghats streams (southern India): is it related to pH, temperature or altitude? Fungal Divers., 1: 179-191.

Sati SC, Tiwari N (1997). Glimpses of conidial aquatic fungi in Kumaun Himalaya. In "Recent Researches in Ecology, Environment and pollution". Eds. S.C.Sati, J. Saxena and R.C. Dubey. Today and Tomorrow printers and Publishers, New Delhi, 10: 17-37

Sridhar KR, Chandrasekhar KR, Kaveriappa KM (1992). Research on the Indian Subcontinent. In "The Ecology of Aquatic Hyphomycetes (ed. Bärlocher, F.) Springer-Verlag, Heidelberg, pp.182-211.

Suberkropp K (1992). Interactions with Invertebrates. In: The Ecology of Aquatic Hyphomycetes (Ed. F. Barlocher). Springer-Verlag, Berlin, pp. 118-134.

Suberkropp K (1998). Microorganisms and organic matter decomposition. In R. J. Naiman and Bilby, R. E. (Eds), River ecology and management: Lessons from the Pacific Coastal Ecoregion. New York: Springer, pp. 120-143.

Spatial assessment of farmers' cooperative organizations in agricultural development in Gurara area of Niger State, Nigeria

ADEFILA, J. O.

Department of Geography, Ahmadu Bello University, Zaria, Nigeria. E-mail:olufiladr@yahoo.com.

The cooperative organizations among farmers are viewed as contributory forces towards growth and development of Agricultural production in recent times. The cardinal objective was to assess such elements that enhance their performance. Both primary and secondary data were used for the study. The primary data were collected with the aid of structured questionnaire administered among 60 different registered farmers' cooperative organizations in the study area. Data were analysed by employing descriptive statistics such as frequencies, mean, percentages and ordinary least square multiple regression. The study revealed that income generation, duration and years of cooperative experience, type of agricultural activities and quality of leadership were found to be significant at 0.01 alpha value while enrolment in terms of population size was found to be significant at 0.05 alpha value. Based on the findings, it was recommended among other things that agricultural policy makers and economic planners should take into cognizance the morale boosters for the cooperative organizations such as lowering interest rates on granting credit facilities, empowering the farmers' cooperatives to perform through legislative control and adopting cooperatives as an effective strategy towards socio-economic transformation of the less-privileged and disadvantaged lagging areas and by so doing, it will assist in fostering regional balance in our developmental efforts.

Key words: Agriculture, cooperative, farming, organization, development.

INTRODUCTION

The history and importance of agricultural cooperative organizations in Nigeria is a long-standing one. Ihimodu (1998) traced their origin to British administration in 1935 with the enactment of the cooperative society law. Moreover, before the legislative control there had been indigenous attempts to form associations such as cocoa farmers' society and kola-nut planters union. These associations were formed in major cocoa producing areas and they were independent of government support (Ihimodu, 1998: 50). The collapse of the traditional mode of cooperatives was attributed to incapacitation of members to bear risk, expectation of high returns on investment and poor management. Cooperative organizations have undergone changes over the years ranging from traditional, informal to modern and formal institutions (Harris and Stefanson, 2005). The cardinal

objective of introducing agricultural cooperative was to increase crop production and credit facilities to cultivators. They have been deeply involved in activities that have impacted on the livelihood of members in particular and rural people in general.

This opinion was shared by Omotosho (2007) that cooperatives often ploughed back resources in terms of dividend on share capital and distributed proportionally to members as patronage bonus. These voluntary social organizations are found in communities possessing common interests but differ in size and degree of interaction among members (Thompson, 2002). In these societies, members have had the ability to influence ideas and actions of the government through a common bargaining power. Royer (2005) and Chambo (2009) had the belief in principle that agricultural marketing cooperatives were competing favourably with private individuals including multinational companies amidst of various challenges such as price fluctuations, legislative controls and low capital accumulation. In this regard, most community and agricultural development agencies have sought the support of these organizations as effective means of imparting new ideas, techniques, harnessing their resources towards improving agricultural production and this constitutes the significance of farmers' cooperative organizations towards the development of agricultural sector.

The country embarked on many agricultural development strategies such as input subsidization, marketing boards, and institutional reforms geared towards improvement of agricultural production. The failure of many agricultural development programmes in Nigeria could be traced to poor organizational structure and implementation at the grassroots level (Omotosho, 2007: 57). The rural poor farmers are isolated, under-educated and lack the means to win greater access to means of production such as capital, labour and this engendered pulling together financial resources towards a common goal. Donald (2002) remarked that some projects targeted ranged from medium to large-scale producers and supporting them with technology, credit and extension services hoping that improvements will gradually extend to the more backward and disadvantaged rural area but unfortunately none of such projects brought about increases in the yield of crops for participants and non-participants. Indeed, a good number of factors are responsible for this, such as constantly changing technology through education and research, availability of equipment and supplies including the ability of farmers to obtain them on time, poor transportation network, among others (Adefila, 2011).

It is the gap arising from the poor performance of government and other institutions that led to the formation of farmers' organizations as means of achieving goals of common interests (Odigbo, 1998: 213). These agricultural cooperative societies do engage in the production, processing, marketing and distribution of agricultural products. An important form of agricultural cooperative in Nigeria is the Group Farming Societies (GFS). Members of this society engage in the production of a variety of crops while they also arrange for the marketing of the products. Some other agricultural cooperatives are devoted to the cultivation of single crops and such societies are named after the crops such as Tobacco Growers Cooperatives (TGC), Cooperative Credit and Marketing Societies (CCMS). In addition, there are Cooperative Production and Marketing Societies (CPMS) in marketing crops such as cocoa, groundnuts and palm produces.

Moreover, there are modern agricultural processing cooperatives for crops such as oil seeds and groundnuts (Ihimodu, 1998: 50; 2007: 36). Farmers' cooperatives have played far reaching roles in agricultural development.

Certain factors influence the role performance of these organizations. This study therefore evaluates some of the factors influencing role performance of these farmers' cooperative organizations in agricultural development in the study area.

The study area

The study area is located between Latitude 9° 30" north of the equator and Longitude 6° 15" east of the prime meridian. It is bounded with Suleja Local Government Area to the south-west, Federal Capital Territory to the south-east and Paikoro Local Government Area to the north. The area enjoys a tropical climate with distinctive dry and wet seasons. The seasons are governed by two principal air masses namely, the Tropical maritime (Tm) and Tropical continental (Tc) air masses. While the former air mass is moist and originates from the Atlantic Ocean bringing rain to the entire region between the months of April and October, the latter air mass is dry, cold and dusty because it originates from the desert, bringing its dry effects to the area between the months of November and February. The annual rainfall is about 1,230 mm and the mean maximum temperature is 37°C (ABU,2011)) and this may increase to about 40°C, especially during the month of March that precedes the commencement of rain.

The formation of the soil is largely a function of lithology, topography, climate and vegetation cover. The soils have developed from metamorphic and sedimentary parent material through several cycles of tropical wet and dry conditions. The physiographic nature of the area is much related to a gently undulating terrain but supported savanna woodland vegetation. However, anthropogenic activities such as bush burning, agricultural activities, deforestation and animal husbandry have relegated the vegetation to poor savanna grasslands. The major ethnic groups are the Nupes, Gbagyis, Hausas, Kambaris and Kadaras (ABU, 2011). The people are predominantly farmers and cattle rearers. The farmers produce food crops such as guinea-corn, maize and cassava at subsistence level. The farmers are quite enterprising and constitute themselves into cooperative societies. Farmers' cooperatives are prominent in the study area, easily accessible due to its location along Kaduna-Abuja Express Highway, possessing relevant records for the research and thus call for investigating their role performance in the development of agriculture. The list of the farmers' cooperatives in the study area is presented in Table 1.

METHODOLOGY

A reconnaissance survey was carried out in order to familiarize oneself with the activities of the cooperative societies and to make spot assessment of the unions. In addition, a focused group

Table 1. List of registered farmers' cooperatives in Niger State, Nigeria.

S/N	Local Government Area	No. registered	Membership
1	Agaie	519	31,140
2	Agwara	141	5,640
3	Bida	579	34,265
4	Borgu	348	12,180
5	Bosso	577	17,310
6	Chanchaga	2,203	168,120
7	Edati	136	4,080
8	Gboko	339	1,450
9	Gurara*	60	10,170
10	Katcha	379	7,595
11	Kontagora	1,048	36,905
12	Lapai	523	15,690
13	Lavun	541	13,525
14	Magama	419	8,380
15	Mariga	364	9100
16	Mashegun	250	6,250
17	Mokwa	444	13,320
18	Munya	170	4,760
19	Paikoro	474	14,220
20	Rafi	384	20,220
21	Rijau	219	4,385
22	Shiroro	674	22,925
23	Suleja	655	26,200
24	Tafa	33	500
25	Wushishi	236	4,995
	Total	11,715	493,325

Source: Ministry of Investment, Commerce and Cooperatives Minna, Niger State (March, 2011). *: Study area.

discussion (FGD) was conducted, so as to collect relevant data about the operations of the cooperative organizations. In the study area there were 60 registered farmers' cooperatives. The list of farmers' cooperative organizations was compiled from the related ministries in the study area. The questionnaire was a major research instrument for the study which was assessed for content validity by a team of experts in Agricultural economics and cooperative studies. The items of the instrument had a confidence coefficient of 0.95 reliability.

The questionnaire was administered among all the registered farmers' cooperatives represented by their respective Chairmen and Secretaries, thereby constituting the sample size for the study. The secondary sources of data included those collected from the related ministries, government gazettes, agencies, parastatals and published journals. The study employed descriptive statistical techniques involving calculation of the mean, percentages, frequencies and inferential statistics such as multiple regressions. The multiple regression model is stated as follows:

Y is the dependent variable, while X is the independent variable:

$$Y = f(X_1, X_2, X_3, X_4, X_5, e)$$

where; Y= Role performance of the farmers' cooperative organizations; X_1 = Income of the farmers in Naira; X_2 = Experience in farming in terms of years; X_3 = Number of people enrolled (population size); X_4 = Type of agricultural activities - crop = 1, animal husbandry = 2, Fisheries = 3 and agro – allied enterprises = 4; X_5 = Assessment of leadership (good leadership = 1; poor leadership = 0); E = error term.

Null hypothesis

There is no significant relationship between farmers' personal traits – age, gender, marital status, income, level of education and performance in agricultural development.

DATA ANALYSIS AND RESULTS

Socio-economic characteristics of the farmers' cooperatives

A cursory glance at Table 2 revealed that a majority (58.4%) of the farmers' cooperatives collected funds from credit facilities made available by their respective organizations. In addition, 30.0 and 8.3% of the cooperative societies got their income from levies and dues respectively. Indeed, only 3.3% of the income came from launching programmes. This implies that the farmers' cooperatives had little or no access to external financial resources in executing agricultural development activities. Merrett and Walzer (2001) remarked that funding of the cooperatives often came from contributions made by members and rarely did they receive donations from external sources.

It is obvious that if financial resources were to be increased, it would enhance role performance of the cooperatives in agricultural development. Credit is one of the basic pre-requisites to increasing agricultural production. Rotan (2000) had earlier remarked that cooperatives deserved higher income in order to boost agricultural production. The type of agricultural activities being practiced by the farmers' cooperatives showed that majority (41.7%) of the members engaged in cultivation of crops, some (33.3%) of them engaged in animal husbandry. In addition, some 20.0 and 5.0% of them practiced fisheries and agro-allied enterprises respectively. This is an indication that cultivation of crop and animal husbandry dominate other agricultural activities in the study area. USDA (2001, 2005) had reported that livestock production has been the major source of income and food to most people in the world. FOS (2006) reported that livestock contributed about 6.6% and poultry accounted for about 6% of gross domestic products (GDP) in Nigeria.

The annual income of the farmers' cooperatives was presented in Table 2. The level of income was based on the assessment made by the executives from each cooperative organization. It showed that some (46.6%) of

Table 2. Socio-economic traits of the farmers' cooperatives.

Variable	Category	Frequency	Percentage
Sources of fund	Credit facilities from the cooperatives	35	58.4
	Loan from financial institutions	5	8.3
	Levies and dues from members	18	30.0
	Launching of fund	2	3.3
Type of Agric. activity	Raising of crops	25	41.7
	Animal husbandry	20	33.3
	Fisheries	12	20.0
	Agro-allied enterprises	3	5.0
Level of income (p.a)	10,000 – 40,000	15	25.0
	41,000 – 80,000	12	20.0
	81,000 – 120,000	28	46.7
	121,000 – 160,000 +	5	8.3
Farming experience (years)	0 – 10	22	36.7
	11 – 20	18	30.0
	21 – 30	12	20.0
	31 – 40+	8	13.3
Enrolment (size)	0 – 50	7	11.7
	51 – 100	22	36.7
	101 – 200	28	46.6
	201 – 300	3	5.0
Age of establishment	0 – 10	25	41.7
	11 – 20	17	28.3
	21 – 30	10	16.7
	31 and above	8	13.3

Source: Author, 2011.

the farmers' cooperatives organizations had annual income level of 81,000 to N120,000, while 25.0% had income of about 10,000 to N40,000. In addition, 20.0 and 8.3% had annual incomes of 41, 000 to N80, 000 00 and 121, 000 to N160,000 and above respectively. The incomes of the farmers' cooperative organizations range from 10, 000 to N160, 000 per annum. This is an indication that income earnings by the farmers' cooperatives are still meager for meaningful agricultural development. The years of experience of the farmers' cooperatives in agriculture were investigated and it revealed variations in terms of length of experience. For instance, some (46.7%) of them have obtained 0 to 10 years experience in farming while 30.0 and 20.0% of them acquired 11 to 20 years and 21 to 30 years experience in farming respectively.

In addition, some 13.3% of the cooperative organizations attained 31 to 40 years experience in cooperative farming. Indeed, experience goes along with skill acquisition, which is fundamental to efficiency and

effectiveness in any job operation. The result implies that most cooperative societies have acquired reasonable years of experience in cooperative farming which can have spread effects on agricultural development. This result apparently corroborated with Torgerson (1990), and Trechter (1996) that farming experience significantly correlated with adoption of improved soil conservation practices. It is essentially an indication that farmers with more experience would likely adopt innovative ideas and techniques that would enhance increase in agricultural productivity.

The size of membership of the farmers' cooperatives was presented in Table 2. It showed that (46.6%) had large enrolment size of 101 to 200 while 36.7 and 11.7% of them had a membership size of 51 to 100 persons, 0 to 50 persons. This enrolment per cooperative organization indicated an optimal population capable of embarking on agricultural activities geared towards increasing productions. The age of the cooperative organizations revealed that about 41.7% of the farmers'

Table 3. Expected roles of farmers' cooperative organizations.

Expected roles	Ranking
Granting credit facilities to members	1
Enlightenment and educating members	2
Introducing new ideas and techniques of farming	3
Rendering guidance and counseling services	4
Create a strong beginning for marketing products	5
Ensure unity and peace within the society	6
Subsidize agricultural inputs to members	7
Organize agricultural exhibition, seminars and workshops	8
Fund raising for agricultural activities	9

Source: Author, 2011.

cooperatives that involved in agricultural development, fell within the age of 0 to 10 years. This period showed that the cooperatives were established in not quite a long period of time. In addition, 28.3 and 16.7% of them have existed for 11 to 20 years and 21 to 30 years respectively. One could remark that the proliferation of farmers' cooperative organizations was largely attributed to the recent policy of the Federal Government on the Fadama Project towards boosting food security by the year 2020.

Expected roles of farmers' cooperative organizations

The responses of the farmers' cooperatives to the expected roles towards agricultural development in the study area are presented in Table 3. The responses were ranked according to the degree of importance. Indeed, a topmost priority was given to granting of credit facilities to members and then followed by enlightening and educating members. The third ranking was for introducing new ideas and techniques towards improving agricultural productivity. The fourth in the rank concerned offering guidance and counseling services to members. The fifth rank was to create a strong bargaining power for agricultural products. This role is viewed to be instrumental to changing government policies that affect cooperative farmers.

This role is followed by ensuring peace and tranquility within the society. The seventh role was to subsidize the price of agricultural inputs for members. The farming cooperatives were noted for buying the agricultural inputs in bulk from the producers or the wholesalers and retail the items to members in subsidized rates. Of considerable importance was organization of exhibitions, workshops, seminars and film shows to showcase the cooperative products to the public. The last and possibly the least role of the farmers' cooperative in the study area was the raising of funds through launching and it was rated ninth in that order. There is no gainsaying that

farmers' cooperative organizations are indispensable to agricultural development in our rural communities. This is because they have been involved either directly or indirectly in agricultural activities. For instance, Hogeland (2002) had remarked that provision of credit facilities to farmers' cooperatives was geared to helping them increase their production and obtain higher standard of living. In addition, Walzer and Merrett (2002) rightly observed that farmers' cooperative organizations also assisted in spreading new ideas, innovations and incentives to allow the majority of the people to be positively involved in the development of agriculture.

Factors influencing performance in agricultural development

The relationship between the roles of farmers' cooperatives and the stated variables namely – income per annum, experience in farming, population size, type of agricultural activity and quality of leadership was examined in this study. The highest coefficient of multiple regressions is $r = 0.715$ and F-ratio = 56.267. Generally, the variables were found to be significantly correlated at different alpha values. The result revealed that the level of income, years of cooperative farming experience, type of agricultural activity and quality of leadership were found to be significant at 0.01 alpha value while enrolment size was found to be significant at 0.05 alpha value. The hypothesis is hereby rejected hence; there was significant relationship between role performance (Y) and the stated variables namely, income of the farmers (X_i), experience in farming (X_{ii}), membership size (X_{iii}), type of agricultural activity (X_{iv}) and leadership assessment (X_v).

The coefficient of income was significantly correlated to the farmers' cooperative role performance. The positive relationship implies that the richer the farmers' cooperatives, the higher the level of their involvement in agricultural development activities. Increment in income is capable of increasing agricultural production (all things being equal). The period of cooperative experience had a coefficient $r = 2.681$, which was equally correlated to the role performance by farmers' cooperatives. This implies that the more the farming experiences, the more the roles cooperatives are likely to perform. This is equally in agreement with the relationship between age of establishment of the cooperatives and role performance. Enrolment in terms of population size with coefficient $r = 2.423$ was correlated to the role performance of farmers' cooperatives. This implies that the cooperative societies with a large number of membership invariably would have more roles to perform and thus having multiplier effects towards agricultural development. The coefficient of the types of agricultural activity was $r = 3.071$. This was significantly correlated to the role performance of farmers' cooperatives.

This result corroborated with the findings (Bhuyan and Leistritz, 2000) that there was a positive relationship between income, experience in agriculture, agricultural activities and membership size with the adoption of soil conservation techniques. Moreover, quality of leadership was found to be significant at 0.01 alpha value with a coefficient r = 2.874 that was correlated with the role performance. This means that quality of leadership determines the level of involvement in agricultural development. Walzer and Merrett (2002) had remarked that leadership involves close monitoring, organizing, coordinating members in order to attain organizational goals. Leadership is essentially human skill that binds a group together geared towards organizational success.

RECOMMENDATIONS

It is obvious from the study that farmers' cooperative societies have limited financial resources to execute their onerous agricultural activities. This could generate agricultural policy that favours the growth of agricultural cooperatives by putting in place financial institutions that could grant credit facilities to the farmers at low interest rates. Ortmann and King (2007) remarked that increased inaccessibility to credit facilities had contributed immensely to the agricultural development in the country. Indeed, the establishment of micro-finance banks and agricultural banks has genuine interest in granting loans to prospective borrowers but the cooperatives are still finding it difficult to access the loans due to strings and conditions attached to it. It is recommended that interest rates on loans should be reduced to the minimum.

Leadership is paramount to the success of any organization and this explains the failure of many cooperatives since they lack proper coordination, administrative skills and managerial acumen. Government at all levels should be interested in the formulation, administration and accountability through the related agencies such as the ministry of commerce and industry by supervising these farmers' cooperatives with a view to ensuring stability and continuity of the organizations. Fulton and Gray (2006) observed that there ought to be strong alliance among cooperatives especially in the area of marketing agricultural products. The ministries and agencies can organize and sponsor seminars, workshops, and conferences for the farmers' cooperatives with a view to exposing them to new techniques and ideas on cooperative philosophy.

In this regard, all cooperatives in whatever form are seriously viewed as catalyst in the process of rural socio-economic transformation. In this regard, all hands are urged to be on deck to ensure their successful operations not only in Nigeria but in all developing countries of the world. This laudable goal can only be achieved if it is backed with legislative controls. The law should empower the cooperatives to perform certain functions, such as

strengthening the bargaining power as effective agents of socio-economic rural transformation. The cooperatives need proper education and enlightenment which can be achieved through government involvement.

Conclusion

The study has revealed that farmers' cooperative societies are variously involved in agricultural development. Moreover, certain factors are influencing their role performance which includes annual income, experience in farming, leadership training and membership size. In this regard, serious attempts ought to be made to address these issues at stake that are serving as impediments to the growth and development of the farmers' cooperatives. The appeal for the promotion of cooperatives at the grassroots and community levels should be seen as an instrumental strategy towards sustainable rural development now that government cannot be depended upon to meet individual numerous needs.

Meijerink (2007) emphasized the important role of agriculture in sustainable development by evolving appropriate strategy and policy. The attitudes of government and the generality of the people must be changed positively towards cooperative development, since it will be too difficult to achieve a meaningful balanced development without involving and stimulating the under-utilized rural resources which these cooperatives are trying to pool together to develop themselves. The government should create an enabling environment for holding and managing the means for production in the process of developing under-privileged and disadvantaged areas.

REFERENCES

ABU (2011). Ahmadu Bello University, Department of Geography. Occasional paper, pp. 11:6-32.
Adefila JO (2011). An assessment of cooperatives as a rural economic development strategy in Nigeria. Paper presented at the International conference of the research and development institute (IRDI). Held at Ambrose Ali University. Ekpoma, Edo State, Nigeria. 4-6 May.
Bhuyan S, Leistritz FL (2000). Cooperatives in non-agricultural sectors: Examining a potential community development tool. J. Community Dev., 31: 89-109.
Chambo SM (2009). An analysis of the socio-economic impact of cooperatives on Africa and their institutional context. ICA Regional Office for Africa, Nairobi.
Donald AF (2002). Anti-trust status of farmer cooperatives: The story of the copper-Volstead Act. USDA rural business-cooperative services report. 59: 249-252.
FOS (2006). Federal office of Statistics. Minna. Government printers.
Fulton JR, Gray C (2006). Strategic alliance and joint venture agreements in grain marketing cooperatives. J. Coop., 11: 15-28.
Harris A, Stefanson R (2005). New generation cooperatives and cooperative theory. J. Coop., 11: 15-28.
Hogeland JA (2002). The changing federated relationship between local and regional cooperatives. USDA research report, 190: 5.
Ihimodu II (1998). Cooperative economics: A concise analysis in theory

and application. University of Ilorin, pp. 50-55.

Ihimodu II (2007). Reforms in the agricultural sector' In Nigeria's reform programme issues and challenges. Edited by Salihu, H. A. and Amali, E. Ibadan. Vantage publisher, pp. 236-264

Meijerink G (2007). The role of agriculture in development, market chains, and sustainable development strategy. Policy paper, No. 5. Stitching DLO. Wageningen.

Merrett CD, Waltzer N (2001). A cooperative approach to local economic development. Westport Connecticut. Quorum Books.

Odigbo PC (1998). Promoting cooperative effectiveness for rural development in Nigeria. Afr. J. Soc. Policy Stud., 1(2): 213-216.

Omotosho OA (2007). Cooperatives as a vehicle for mobilizing resources for poor farmers in Nigeria. In general reading studies in Nigeria. University of Ilorin Press, pp. 57-62.

Ortmann F, King RP (2007). Agricultural cooperatives: History, theory and problems. J. Agric. Cooperatives. Agrekon, 46(1): 40-52.

Rotan BL (2000). Net income declines in local cooperatives. Rural cooperatives. USDA rural development report, 25: 28.

Royer JS (2005). Economic nature of the cooperative association: A retrospective appraisal. J. Agric. Coop., 9: 82-89.

Thompson S (2002). Closing the gap. J. Rural Coop., 12: 16

Torgerson RE (1990). Human capital: Cooperatives build people, also farmers cooperatives, 57: 2.

Trechter D (1996). Impact of diversification on agricultural cooperatives in Wisconsin. Agribusiness, 12: 385-394.

United States Department of Agriculture (USDA) (2001). Food and agricultural policy: Taking stock for the new century.

United States Department of Agriculture (2005). Agricultural statistics. National Agricultural Statistics Service. USDA. Washington DC.

Walzer N, Merrett CD (2002). Collaboration: New generation cooperatives and local development. J. Community Dev. Soc., 33: 112-135.

3

Forest duiker (*Cephalophus spp.*) abundance and hunting activities in the Kakum conservation area, Ghana

Edward D. Wiafe* and Richard Amfo-Otu

Department of Environmental and Natural Resources Management, Presbyterian University College, P. O. Box 393, Akropong-Akuapem, Ghana.

The abundance of forest duikers (*Cephalophus* spp.) was compared to the incidence of hunting activities in the Kakum conservation area, Ghana. Transect surveys indicated that four duiker species were present: Maxwell's duiker (*Cephalophus maxwellii*), Bay duiker (*Cephalophus dorsalis*), Black duiker (*Cephalophus niger*), and Yellow-backed duiker (*Cephalophus silvicultor*). The indicators of hunting activities included the presence of empty cartridges, snares, gunshots, carbide powder, poacher's camps and arrest of poachers. Season's fluctuations and hunting activities appear to have effect on duiker abundance; the study could have strong evidence to establish it. A long term duiker population monitoring program is required in the area in order to come out with strong factors affecting duiker populations and their implications on conservation of wildlife resources.

Key words: Forest duikers, hunting activities, abundance, conservation, Kakum conservation area.

INTRODUCTION

Forest duikers (*Cephalophus* spp) are bovids of the sub family Cephalophinae, found only in Africa and primarily inhabiting forested areas (Grubb and Groves, 2001). The genus *Cephalophus* comprises of seventeen species, making it the most species-rich group of forest ungulates (Kingdon, 1997). Fruits and seeds constitute the bulk of the diet of duikers and they are thus potentially important seed dispersers (Eves et al., 2002). Many of the forest ungulates, including duikers, are intensively exploited as a source of game meat over most of the African forest zone (Gautier-Hion et al., 1980; Eves et al., 2002). Infield (1988) estimated that in the Korup National Park (Cameroon), duikers constituted 63.3% by weight of the total off-take by hunting, and that the number of animals killed by 15 households in six villages was estimated at 15,566 duikers from four species. Though conservation status of the duikers globally is least concern, apart from Ader's and Zebra duikers (*Cephalophus adersi* and

zebra), the populations keep decreasing (SSC-IUCN, 2009). In Ghana, wildlife populations are under constant threat from human activity, even in protected areas, and illegal activities have affected the populations of most wildlife species (Jachmann, 2008). The need for a better understanding of the population ecology of duikers is essential if these communities are to be managed and conserved. The objectives of this study were to provide baseline information on the abundance of different forest duiker species relative to human hunting activities and seasonal variations, in the Kakum conservation area in Ghana.

MATERIALS AND METHODS

The Kakum conservation area (KCA) is situated in the central region of Ghana. The KCA, established in 1992, is composed of the Kakum National Park, established in 1932, and the Assin-Attandanso resource Reserve, where selective logging took place from 1936 until 1989, when it was converted into a conservation area (Figure 1). The KCA covers an area of 360 km² of moist, semi-deciduous forest located between latitudes 05° 20'N and 05°40'N and longitude 001° 30'E and 001° 51'E. The rainfall pattern is bimodal, the major season occurring between April and July and

*Corresponding author. E-mail: edward.wiafe@presbyuniversity.edu.gh.

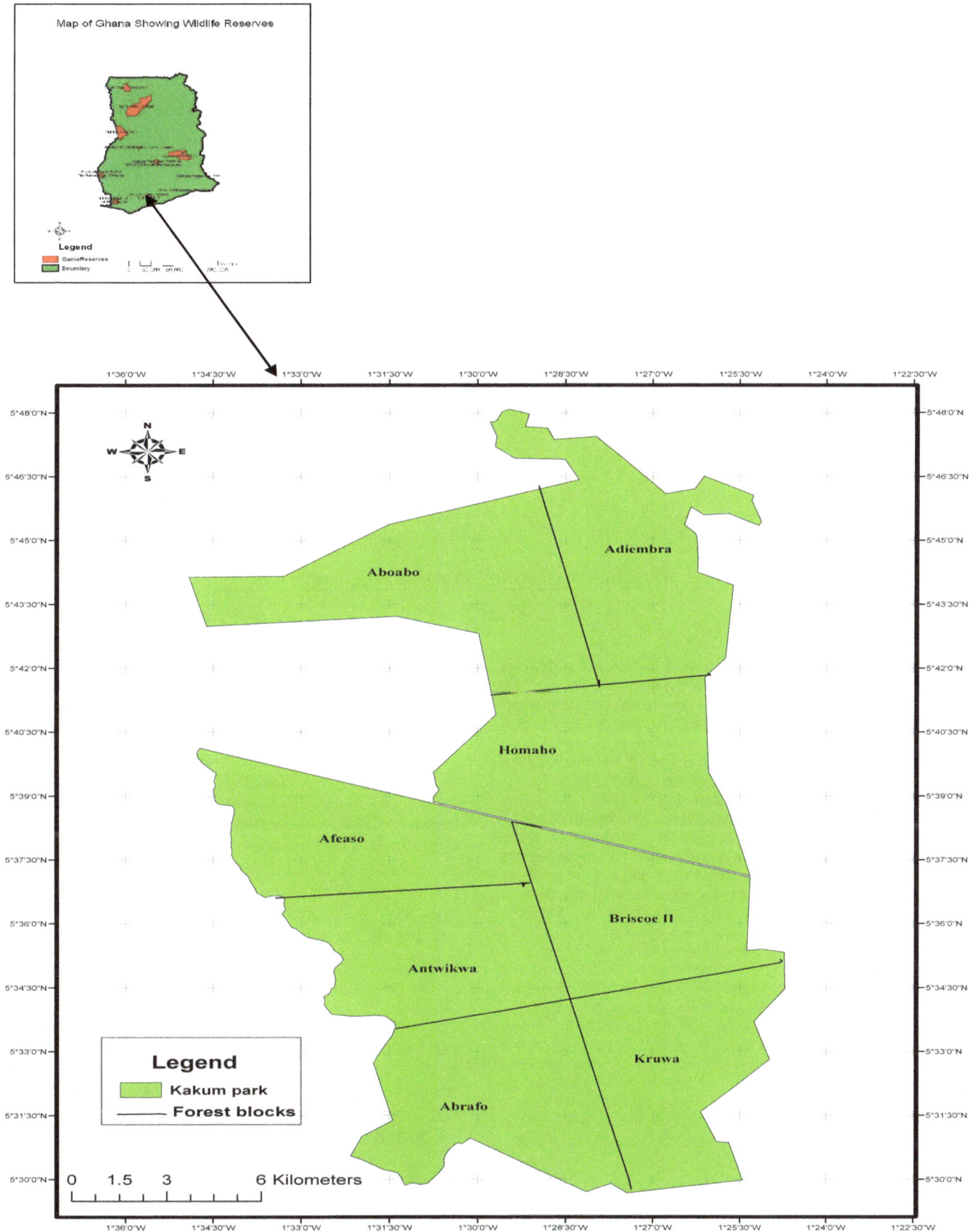

Figure 1. Map of Kakum conservation area showing the divisions into forest blocks. Inset: Map of Ghana indicating the position of Kakum conservation area and other protected areas in Ghana.

Table 1. Kilometric indices of abundance of the duiker species encountered in the wet season and dry season at the various forest blocks in the Kakum conservation area, Ghana. Numbers in parentheses indicate standard deviations.

Forest block	Bay duiker	Maxwell duiker	Black duiker	Yellow- backed duiker
		Wet season		
Antwikwa	0.63	1.25	1.13	0.00
Abrafo	0.63	1.38	0.50	0.00
Afeaso	0.00	2.88	0.13	0.38
Kruwa	1.50	1.75	0.50	0.00
Briscoell	0.75	1.13	0.88	0.00
Aboabo	0.75	1.75	0.50	0.13
Homaho	1.00	1.75	0.25	0.00
Adiembra	0.88	2.25	1.38	0.13
Mean	0.80 (0.39)	1.80 (0.54)	0.70 (0.40)	0.10 (0.12)
		Dry season		
Antwikwa	0.63	1.00	0.75	0.00
Abrafo	0.38	0.88	0.00	0.00
Afeaso	0.00	1.25	0.00	0.25
Kruwa	1.13	3.38	0.75	0.00
Briscoell	0.50	2.13	0.00	0.00
Aboabo	0.75	1.50	0.38	0.00
Homaho	0.38	1.25	0.00	0.13
Adiembra	1.00	2.75	1.50	0.00
Mean	0.60 (0.36)	1.80 (0.90)	0.40 (0.55)	0.05 (0.09)

the minor season between September and November (Ghana Wildlife Department, 1996). To equalize the sampling intensity, the KCA was divided into eight blocks of approximately equal sizes (Figure 1). In each of the blocks, four 2 km transects were randomly distributed perpendicular to the main drainage lines of the area. Transects followed compass lines, measured with a Geographical Positioning System (GPS) and censuses were conducted on foot by a team of up to three people from 6:00 to 16:00 GMT.

The team walked slowly (1 km/h), recording the presence of forest duiker species (by animal sighting) and signs of illegal hunting activities. A total of 64 km was surveyed during the dry season (December 2009 to January 2010) and the wet season (June to July 2010). Each transect was surveyed once per season. The software package DISTANCE (Thomas et al., 2005) is commonly used to analyze data from line transects (Gatti, 2010). However, the survey failed to meet the assumptions of the analysis (Buckland et al., 2001). The data was thus analyzed by calculating the Kilometric Indices of Abundance (KIA). This method consists in recording the number of animals and indicators of hunting observed per kilometre of transect (Groupe, 1991; Gatti, 2010). Mann-Whitney U-test was used to compare the abundance of duiker species in the dry and wet seasons. A Kruskal-Wallis test was used to evaluate the difference in medians of indices of hunting activities. The indices for duiker abundance and of all categories of hunting were summed up into a single index. Spearman's rank correlation was conducted to evaluate the relationship between these indices.

RESULTS AND DISCUSSION

Four species of duikers belonging to the genus *Cephalophus* were encountered during the survey and

were categorized into small, medium and large duikers in accordance with Estes (1991) and Hart (2001). They were Maxwell's duiker (*Cephalophus maxwellii*), Bay duiker (*Cephalophus dorsalis*), Black duiker (*Cephalophus niger*), and Yellow-backed duiker (*Cephalophus silvicultor*) (Table 1). Total abundance across all species and forest blocks did not differ between the two seasons ($U = 435$, $p>0.05$), rejecting the hypothesis that seasonal differences influence duiker abundance. This may be due to the fact that as the duikers do not migrate to or from the protected area, they may have evolved to adjust to the local conditions which are the fluctuations between the dry and wet seasons. Another explanation may be that the seasonal conditions were not extreme enough to bring about significant changes in the species abundance. The mean KIAs for the following illegal hunting activities enumerated in the wet season and dry seasons included the presence of empty cartridges, snares, gunshots heard, carbide powder location of poachers' camps and arrest of poachers (Table 2).

The incidence of hunting activities did not differ ($H = 7.93$, $p>0.05$) between categories during the wet season, however, there was a significant difference ($H = 15.99$, $p<0.05$) between the categories of hunting activities in the dry season (Table 2). A weak negative relationship ($r_s= -0.25$, $p>0.05$) was found between the duiker indices and the incidence of hunting activities in the conservation

Table 2. Kilometric indices of abundance of indicators of hunting activities encountered in the wet season and dry season in the forest blocks in the Kakum conservation area. Numbers in parentheses indicate standard deviations.

Forest block	Empty cartridges	Gunshot heard	Snares found	Carbide powder	Poacher's camps	Poacher's arrested
Wet season						
Antwikwa	1.50	0.63	1.25	0.00	0.00	0.13
Abrafo	0.13	0.25	0.00	0.00	0.00	0.00
Afeaso	0.25	0.50	0.00	0.50	0.13	0.00
Kruwa	2.50	1.25	6.13	0.25	0.13	0.00
Briscoell	4.13	0.88	5.25	0.00	0.25	0.38
Aboabo	0.63	0.13	0.00	0.13	0.13	0.13
Homaho	1.88	0.63	0.63	0.13	0.00	0.00
Adiembra	0.75	0.88	0.50	0.13	0.00	0.38
Mean	1.47(1.35)	0.64(0.36)	1.72(2.5)	0.14(0.17)	0.10(0.09)	0.13(0.16)
Dry season						
Antwikwa	1.13	0.25	0.00	0.00	0.13	0.13
Abrafo	0.63	0.25	0.38	0.00	0.00	0.00
Afeaso	0.00	1.13	0.00	0.38	0.38	0.00
Kruwa	1.88	0.75	0.88	0.38	0.00	0.00
Briscoell	2.88	1.25	1.00	0.00	1.25	0.00
Aboabo	0.00	0.00	0.00	0.00	0.00	0.13
Homaho	0.75	0.50	4.25	0.38	0.13	0.00
Adiembra	2.00	1.25	0.50	0.38	0.00	0.00
Mean	1.16(1.02)	0.67(0.50)	0.88(1.42)	0.19 (0.20)	0.23(0.43)	0.03(0.06)

area in the wet season (Table 2). However, during the dry season, this relationship was positive ($r_s = 0.45$, $p>0.05$) (Table 2). During the wet season, the negative relationship suggested a higher density of duikers in areas with low incidence of hunting activities. However, during the dry season, duiker abundance was higher when the incidence of hunting activities was greater. The relationships were found to be very weak and this might suggest that hunting alone does not influence the abundance of duikers in forest environment. This study could not establish any strong relationship between hunting activities and duiker population, probably due to the short period of the study, however, this is strongly suspected to affect the duiker population at one time or another. A long term study for monitoring duiker population and hunting activities is therefore, recommended. Further research investigating the role of other limiting factors, such as the availability of fruiting trees and water, are recommended.

ACKNOWLEDGEMENTS

We are indebted to the Presbyterian University College, Ghana (PUCG), for sponsoring this study. We are also grateful to the Ghana Wildlife Division for granting the permission to study in the area. We thank Dr. Frank Arku (PUCG, Akropong) for his support and encouragement, and the Manager and staff of the Kakum conservation area for their assistance during data collection.

REFERENCES

Buckland ST, Plumptre A J, Thomas L, Rexstad EA (2001). Design and analysis of line transect surveys for primates. Int. J. Primatol. 31:833-847.

Estes RD (1991). The behavior guide to African Mammals including Hoofed Mammals, Carnivores, primates. Berkeley, Los Angeles, London. The University of California press. p. 611.

Eves HE, Stein JT, BCTF (2002). BCTF Fact Sheet: Duikers and the African Bushmeat Trade. Bushmeat Crisis Task Force. Washington, DC. p. 2.

Gatti S (2010). Community Forest Biodiversity Project: Status of primate populations in Protected Areas targeted under CFBP.WAPCA and WD/FC, Accra. p. 42.

Gautier-Hion A, Emmons LH, Dubost G (1980). A comparison of the diet of three major groups of primary consumers of Gabon (primates, squirrels and ruminants) Oecologia (Berlin), 45: 182-189.

Ghana Wildlife Department (1996). Management plan for Kakum national park and Assin Attandanso resource reserve. Accra. Ghana Wildlife Department.

Groupe C (1991). Méthodes de suivi des populations de chevreuils en forêt de plaine: Exemple: L'indice kilométrique (I.K.). Bulletin Mensuel ONC, Supplément 157, Fiche. Office National de la Chasse, Paris. 70: 4.

Grubb P, Groves CP (2001). Revision and Classification of the Cephalophinae. In: V. J. Wilson (ed). Duikers of Africa: Masters of the African Floor. Chipangali Wildlife Trust, Bulawayo, Zimbabwe. pp. 703-728.

Hart TB (2001). Forest dynamics in the Ituri Basin (D. R. Congo): Dominance, Diversity and Conservation. In White JTL, Vedder A and Naughton-Treves L (eds). African Rainforest Ecology and Conservation. New Haven, London. Yale University Press. pp. 183-206.

Infield M (1988). Hunting, trapping, and fishing in villages within and on the periphery of the Korup national park. Paper no.6. of the Korup National Park socioeconomic survey. Publication 3206/a9.6, prepared by the World Wide Fund for Nature. Gland. Switzerland.

Jachmann H (2008). Illegal wildlife use and protected area management in Ghana. Biol. Conserv., 141: 1906-1918.

Kingdon J (1997). The Kingdon Field Guide to African Mammals. Great Britain. AP Natural World Academic press, p. 464.

SSC-IUCN (2009). Press Release - 06 February 2009. http://cmsdata.iucn.org/downloads/antelope_report.pdf. Assessed on 1/14/12.

Thomas L, Laake JL, Strindberg S, Marques FFC, Buckland ST, Borchers DL, Anderson DR, Burnham KP, Hedley SL, Pollard JH, Bishop JRB, Marques TA (2005). Distance 5.0. Release "x": Research Unit for Wildlife Population Assessment, University of St. Andrews, UK.

4

Microbial corrosion of steel coupons in a freshwater habitat in the Niger Delta

Odokuma L. O.* and Ugboma C. J.

University of Port Harcourt, Rivers State, Nigeria.

Microbial corrosion of stainless, mild and carbon steel coupons in a semi static freshwater habitat was carried out monthly for a period of one year. Corrosion rate was determined by weight loss. Carbon steel had corrosion rate of 44 g/year, mild steel 27 g/year, stainless steel 0.64 g/year. Ecological quality parameters of the river water for rainy season decreased from April to October, respectively in the following trend; total dissolved solids (23 to 2.31) mg/l, total suspended solids (40 to 12) mg/l, total organic carbon (78.24 to 4.33) mg/l, conductivity (43 to 4.73) μs/cm, salinity (31 to 3.10) mg/l, oil and grease (34.23 to 1.67) mg/l, Chloride (52.11 to 1.05) mg/l, Sulphate (33.14 to 1.15) mg/l, Sulphite (8.11 to 10.14) mg/l, pH(6.4 to 6.0), temperature (27 to 25°C), biochemical oxygen demand (64.13 to 1.34) mg/l. That of dry season (November to March) respectively showed this trend. Total dissolved solids (3.14 to 52.11) mg/l, total suspended solids (9 to 40) mg/l, total organic carbon (5.10 to 10.24) mg/l, conductivity (6.29 to 104) μs/cm, salinity (5.11 to 87.33) mg/l, oil and grease (2.14 to 1.82) mg/l, Chloride (1.02 to 2.11) mg/l, Sulphate (0.17 to 4.95) mg/l, Sulphite (11.10 to 16.23) mg/l, pH (6.1 to 6.0), Temperature (25 to 26°C), biochemical oxygen demand (1.53 to 7.31) mg/l. Weight loss of stainless, mild and carbon steel for rainy season (April to October) respectively, and showed the following trend (0 to 0.030), (0.695 to 1.568), and (1.000 to 2.316) g, while the dry season (November to March) respectively showed the following trend (0.044 to 0.111), (1.586 to 5.771) and (2.325 to 9.131) g. Corrosion rate of stainless, mild and carbon steel for rainy season (April to October) respectively showed the following trend (0 to 0.004), (0.695 to 0.244) and (1.000 to 0.331) g/month while the dry season (November to March) showed the following trend respectively (0.006 to 0.009), (0.198 to 0.481), and (0.291 to 0.761) g/month. The use of stainless steel, if not already in use should be encouraged as a material for pipelines, flowlines and bulklines in the Nigerian Petroleum industry. Integrity checks on these surface facilities should be increased in the dry season.

Key words: Freshwater, ecological quality parameters, corrosion rate, weight loss.

INTRODUCTION

Biocorrosion of steel materials is gaining tremendous importance in the petroleum sector of Nigeria's economy. It has been associated with the rupture/failure of pipelines used in transporting petroleum and its products from oil and gas fields, flow stations tank farms, depots and terminals in Nigeria. Unfortunately most of this corrosion has been reported as just corrosion or physicochemically induced corrosion. This is because of the limited knowledge in the Nigerian Petroleum industry that

corrosion is mainly physicochemically induced. There is however a dearth of information on biocorrosion in the Petroleum Industry in Nigeria. Biocorrosion is microbially mediated. Bacteria such as sulphate reducing bacteria, iron oxidizing bacteria and sulphur oxidizing bacteria have been implicated in anaerobic and aerobic bio-corrosion respectively. Internationally, biocorrosion causes billions of dollars in damage every year (Bitton and John, 2007). According to Gerald and Stams (2008), microbial corrosion of steel results in huge financial losses that amount to more than 100 million US dollar per annum in the USA alone.

According to ICEM (2011), it was told that Pakistan has

*Corresponding author. E-mail: luckyodokuma@yahoo.co.in.

been incurring an annual loss of RS 250 to 300 billion (about 3 billion US dollars) on account of infrastructure, industries and house hold corrosion, which does not include the damage to environment and loss of production due to unscheduled breakdown. Corrosion effects on different equipment or materials in different aspects of food processing and packaging are common (Ashassi-Sorkhabi, 2009). Smith and Hashemi (2006) reported that the materials used for most equipment in the manufacturing sector are mild steel because of its strength, ductility and weld-ability but it is prone to corrosion. In the freshwater habitats, microorganisms such as sulphate reducing bacteria, Iron bacteria and sulphur bacteria have been implicated in corrosion (Pitonzo et al., 2007; Lee and Newman, 2005).

In the freshwater habitat, organic pollutants such as biochemical oxygen demand (BOD), oil and grease, inorganic anionic pollutants including chloride, salinity, sulphate, sulphite and associated factors like temperature, pH, conductivity, total organic carbon, total dissolved solids and total suspended solids all affect microbial corrosion of steel products. Due to its economic importance, microbial corrosion of metals in freshwater has been studied for a long time (Dinh et al., 2004).

Badmos and Ajimotokan, (2009) demonstrated that low pH encouraged corrosion but higher pH reduced corrosion rates. The work of Birnin-Yauri and Garba (2006), demonstrated that low concentration of chloride ions caused corrosion of steel used for reinforcement. According to Jaganathan et al. 13% Cr is required for stable passivity of a Fe-Cr alloy in acidic and neutral solutions not containing inhibitors. Hamilton (2003) has demonstrated microbial influenced corrosion as a model system to study metal microbe interaction.

In this study, the river water from Ndoni River which is a fresh water habitat located in the Niger Delta was used for the test. The study was aimed at determining the corrosion rate of the different test metal types (stainless steel, mild steel and carbon steel) buried in the freshwater environment. In addition the study was to determine appropriate steel type that will be resistant to microbial corrosion in freshwater habitats which will be recommended for use as pipeline materials in the Petroleum Industry in Nigeria. Finally, the study was aimed at determining the effect of seasonal variations on corrosion rates. This was geared towards determining which season should witness more integrity checks as a mitigative action for pipeline failure/rupture.

MATERIALS AND METHODS

Area of study

The Ndoni River (Figure 15) is situated in the coastal environment of the Niger Delta in Rivers State, Nigeria. It is a fresh water site and influenced by River Niger influx. The river has a long rainy season from March to October. The dry season is very short, beginning in November and ending in March of the following year

referred to as the harmattan. The edges of the river are covered with typical rainforest vegetation and the soil is loamy in nature.

Industrial activities along the river

The main industrial activity in this area is dredging and sand wining (coded AML). Sand wining is a process where sand and water mixtures flow via dredger pipes as dredging goes on and the sand is preserved for other uses. The repair of marine boats also occurs here. Other minor industrial activities along the river include the lumbering, fish farming and a market situated by the river.

Water sample analyses

Surface water samples were collected on the 15th of every month between 10.00 and 12.00 a.m from April 2010 to March 2011. Water samples were collected with 100 ml sterile bottle. Samples for biochemical oxygen demand (BOD) determination were collected with 250 ml (BOD) bottles. For other chemical analysis, samples were collected with 500 ml glass bottles that had been sterilized at 121°C and 15 psi for 15 min. All samples were analyzed immediately on reaching the laboratory.

Chemical reagents

All chemical reagents employed in this study were products of Aldrich chemical co, Milwauke, USA, BDG Chemicals, Poole, England and Sigma chemical company, St Louis Missouri, USA.

Metal coupons

The steel materials used for this study namely stainless steel, mild steel and carbon steel were obtained from Nigeria Agip Oil Company (NAOC) and TOTAL FINA ELF Nigeria limited.

Ecological quality parameters

The following parameters were determined namely biochemical oxygen demand, oil and grease, pH, temperature, salinity, conductivity, total dissolved solids, total suspended solids, total organic carbon, chloride, sulphate, sulphite and heavy metals: Ni, V, Cr, Fe, Na, Ca, K, Zn, Cu. The heavy metals were determined using model AA320 atomic absorption spectrophotometer (Shangai analytical instrument Co). Oil and grease content was estimated using the method of Odu et al. (1988). A 10-ml portion of the water sample was shaken with 10-ml of toluene in a separator funnel. The hydrocarbon grease and oil content was then determined by the absorbance of the extract at 420 nm in a spectrophotometer. A standard curve of the absorbance of different known concentration of equal amount of crude oil and grease in the extractant was first drawn after taking reading from the spectrophotometer. Oil and grease concentrations in the water were then calculated after reading the optical density of the extract from the spectrophotometer.

The standard curve was used to estimate the oil and grease concentration after multiplying by an appropriate dilution factor. All other parameters such as sulphate, sulphite, total organic carbon, total suspended solids, total dissolved solids, salinity, chloride, and biochemical oxygen demand were determined employing methods from APHA (2000). Total dissolved solids by gravimetric method, sulphite by Iodometric method, sulphate by turbidimetric method, chloride and salinity by Argentometric method, total organic carbon by rapid oxidation method, biochemical oxygen demand by modified Winkler method, and total suspended solids by gravimetric method (APHA, 2000).

Microbiological analyses

Postgate broth (1985) was used to enumerate the sulphate reducing bacteria using the conventional five tube most probable number method (MPN). Water samples (10, 1 and 0.1 ml) were placed in a series of five tubes with nutrients. After sterilization they were inoculated with dilutions of scrapings from the metal coupons and enumerated after 7 days of incubations in an anaerobic gas jar at room temperature. Winograsky medium composition was used to enumerate iron bacterial, distilled water 1000 ml, NH_4NO_3 0.5 g/l, $NaNO_3$ 0.5 g/l, K_2HPO_4 0.5 g/l, $MgSO_4.7H_2O$ 0.5 g/l, $CaCl_2.6H_2O$ 0.2 g/l, Ferric ammonium citrate 10 g/l, Agar 15 g/l.

The sterile medium was poured in sterile Petri dish in duplicate and inoculated with dilution of scrapings from the metal coupons using spread plate method and incubated for 48 h at room temperature. Different selective media was used to enumerate the different sulphur bacteria. Rodina medium for *Thiosulphate* bacteria, distilled water 1000 ml, K_2HPO_4 3.0 g/l, $MgCl_2.6H_2O$ 0.5 g/l, $CaCl_2.6H_2O$ 0.2 g/l, pH 9.0, Agar 20 g/l, Na_2SO_4 10 g. Larsen's medium for colourless sulphur bacteria, distilled water 1000 ml, K_2HPO_4 0.2 g/l, NH_4Cl 0.1 g/l, $MgCl_2.6H_2O$ 0.1 g/l, KH_2PO_4 3.0 g/l, Na_2S 0.06 g/l, Agar 20 g/l. Vanniel medium for purple sulphur bacteria, distilled water 1000 ml, NH_4Cl 1.0 g/l, K_2HPO_4 0.5 g/l, $MgCl_2$ 0.2 g/l, Agar 20 g/l, $NaHCO_3$ 0.5 g/l. All the sterile media were poured into sterile Petri dishes in triplicate and inoculated with dilutions from metal coupon scrapings using spread plate method and incubated for 48 h at room temperature.

Statistical analyses

Correlation analyses and student t-test from Microsoft Excel 2010 was employed where value of P< 5% was considered to be significant and P> 5% was considered not significant for t-test. +1 (perfect correlation) through 0 (no correlation) to -1 (perfect negative correlation) for correlation analyses (Finney, 1978).

RESULTS

Results in Table 1 shows the bacterial count of microorganisms isolated in the freshwater habitat from April 2010 to March 2011. The sulphate reducing bacteria increased from April to June (7 to 14) MPN index per 100 ml before decreasing from July to November (7 to <2) MPN index per 100 ml and increased again from December to March (2 to 21) MPN index per 100 ml. The iron bacteria increased from April to May (261 to 278) 10^3 cfu/ml and decreased from June to October (270 to 65) 10^3 cfu/ml before increasing again from November to March (71 to 321) 10^3 cfu/ml. Sulphur bacteria decreased from April to May then increased by June and decreased from July to October (121 to 19) 10^3 cfu/ml and increased again from November to March (22 to 201) 10^3 cfu/ml.

Table 2 shows the chemical composition of the various metal coupons namely (stainless, mild and carbon steel). Figure 1 is the graphical representation of weight loss against time of various metal coupons used in this study. There was a significant weight loss between stainless

steel versus mild steel, stainless steel versus carbon steel, mild steel versus carbon steel. P = (8.839 E-06, 1.641E-05, 0.041031) respectively. The corrosion rate as shown graphically in Figure 14 shows a significance between stainless steel versus mild steel, stainless steel versus carbon steel, mild steel versus carbon steel, P= (1.959E-07, 3.228E-07, 0.016223) respectively. Graphically, Figures 2, 3, 4, and 5 is the result of oil and grease, pH, salinity and conductivity. Salinity showed an increase between April to May and decreased from November to March. Oil and grease showed an increase from April to May and decreased from June to October, increased in November and decreased again from December to March. The correlation values in weight loss and corrosion rate between ecological quality parameters and metal coupons namely stainless, mild and carbon steel respectively are weight loss; oil and grease r=(-0.5504,-0.4446,-0.4224), pH r=(0.2853, 0.7797, 0.7666), Salinity r=(0.6938, 0.8409, 0.8459), Conductivity r=(0.6986, 0.8309, 0.8494); Corrosion rate; oil and grease r= (0.2165, 0.7089, 0.7096), pH r=(0.2853, 0.7797, 0.7666), salinity r=(0.2583, 0.3889, 0.4553), conductivity r=(0.2365, 0.3864, 0.4528).

The ecological quality parameters represented graphically in Figures 6, 7, 8, and 9 are the results of chloride, total suspended solids, total organic carbon respectively. The chloride decreased from April to January and increased a little from February to March. Total dissolved solids decreased from April to October and increased from November to March. Total suspended solids increased from April to June, decreased from July to November and increased again from December to March. Total organic carbon decreased from April to October and increased from November to March.

The correlation values in weight loss and corrosion rate between ecological quality parameters and metal coupons namely (stainless, mild and carbon steel) respectively are weight loss; chloride r= (-0.5857, -0.4696, -0.4383), total dissolved solids r= (-0.06800, 0.8138, 0.8331), total suspended solids r= (-0.5159, -0.4603, -0.4625), total organic carbon r= (-0.5774, -0.4250, -0.4141). corrosion rate; chloride r= (-0.1032, 0.7571, 0.6621) total dissolved solids r= (0.4070, 0.8610, 0.8193), total organic carbon r= (0.0654, 0.8217, 0.7616). The results of temperature, sulphate, sulphite and biochemical oxygen demand are represented graphically in Figures 10, 11, 12, and 13 respectively.

The temperature decreased a little in the rainy season from April to October but remained constant from December to March. Sulphate decreased from May to November and increased a little from December to March. The sulphite level increased from April to May and decreased from five to increase again from August to March. The biochemical oxygen demand decreased from April to October and increased slightly from November to March. The correlation values in weight loss and

Table 1. Population of microorganisms isolated in the freshwater habitat.

Month	Sulphate reducing bacteria MPN index per 100 ml	Iron bacteria 10^3 cfu/ml	Sulphur bacteria 10^3 cfu/ml
April	7	261	186
May	11	278	180
June	14	270	245
July	7	184	121
August	2	113	87
September	<2	94	38
October	<2	65	19
November	<2	71	22
December	2	92	48
January	6	117	67
February	11	147	92
March	21	231	201

Table 2. Chemical composition of metal coupons.

Chemical element composition (%)	Carbon Steel	Mild steel	Stainless steel
Carbon	0.008-0.20	0.15-0.25	0.03 maximum
Manganese	0.45-0.65	0.45-0.65	2.0 maximum
Phosphorus	-	-	0.04 maximum
Sulphur	-	-	0.03 maximum
Chromium	-	-	16-18
Nickel	-	-	10-14
Silicon	0.25-0.60	0.25-0.60	1.0 maximum
Molybdenium	-	-	2-3
Copper	0.60	0.60	-

Lynch (1989).

corrosion rate between ecological quality parameters and metal coupons namely (stainless, mild and carbon steel) are weight loss; temperature r= (-0.04464, 0.0303, 0.1193), sulphate r= (-0.4228, -0.3398, -0.3206), sulphite r= (0.4485, 0.4560, 0.5262), biochemical oxygen demand r= (-0.5568, -0.3945, -0.3834). corrosion rate; temperature r= (0.1752, 0.5984, 0.7008), sulphate r= (0.4348, 0.9004, 0.3784), sulphite r= (0.4554, 0.1804, 0.3784), biochemical oxygen demand r= (0.0255, 0.8166, 0.7473).

DISCUSSION

Table 1 shows that organisms isolated were sulphur reducing bacteria, iron bacteria and sulphur bacteria. These organisms play a role in corrosion by inducing oxygen gradient which accelerates corrosion in two ways by acting as a depolarizer to form ferrous ions and oxidizing ferrous ions (Fe^{2+}) to ferric ions (Fe^{3+}). The latter reactions takes place in pH values higher than 4. A

species of sulphur reducing bacteria, *Desulfovibriode sulfuricans* isolated in this study directly removes corrosion product such as hydrogen formed at the cathode, this enhances biocorrosion because it causes depolarization thereby sustaining corrosion current (Battersby et al., 1985). *D. sulfuricans* produces hydrogen sulphide which causes hydrogen blistering and embrittlement in metals and structural fittings (Raloff, 1985). H_2SO_4 from *Thiobacillus ferroxidans* causes dissolution of metals from ores and alloys and also maintain pH levels favourable for Iron bacteria to corrode metals. Similar observations were made by Beech (2004) and Battersby et al. (1985).

The composition of metal coupons in Table 2 shows that stainless steel has alloying elements such as Chromium (16 to 18%), Nickel (10 to 14%) gave stainless steel lots of protection against corrosion because these elements are less reactive and occur in the lower part of the electrochemical series unlike the mild steel and carbon steel which does not have the alloying elements. The composition of mild steel and carbon steel namely

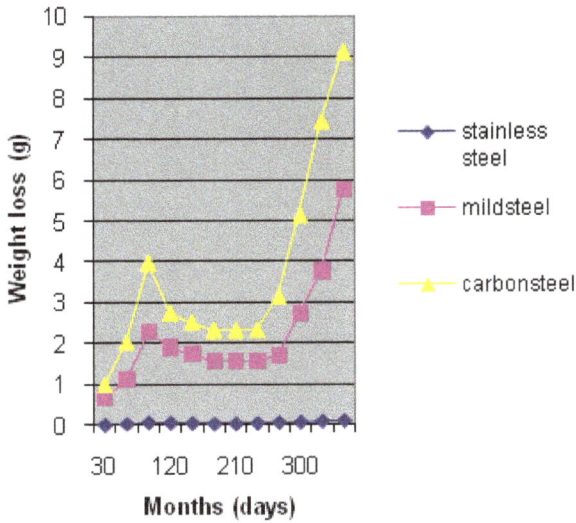

Figure 1. Weight loss against time.

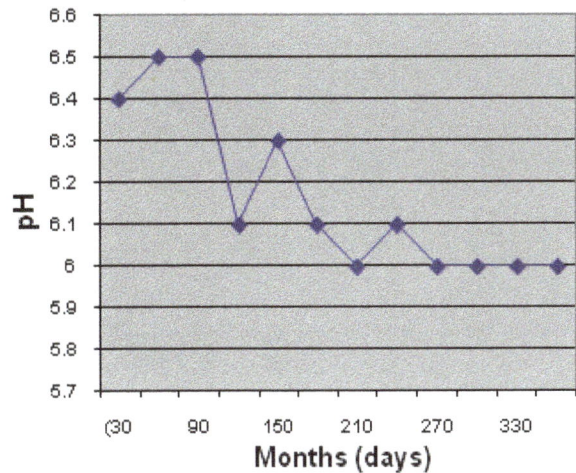

Figure 2. Oil and grease against time.

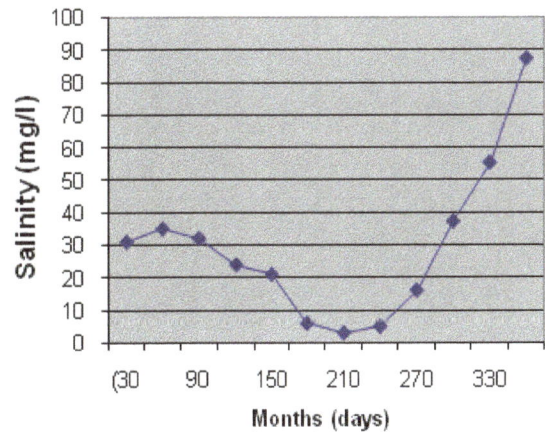

Figure 3. pH against time.

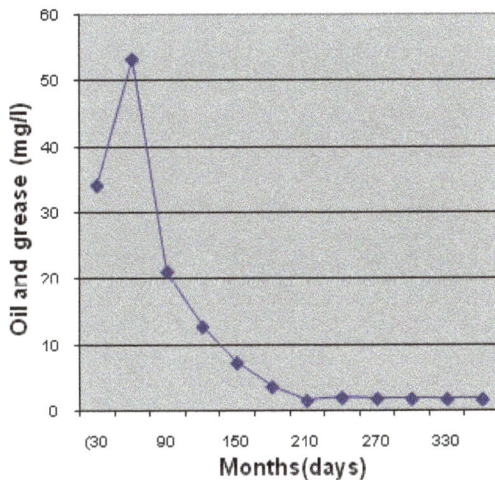

Figure 4. Salinity against time.

Manganese, Phosphorous, Copper, Carbon, Silicon are more reactive and hence easily corroded with a lot of weight loss. Similar observation has been made by Jaganathan et al. (2011). They showed that classically 13% Chromium is required for stable passivity of a Fe-Cr alloy in acidic and neutral solutions not containing inhibitors. Statistically, the corrosion rate and weight loss of stainless steel versus mild steel, stainless steel versus carbon steel, carbon steel versus mild steel respectively gave a significant difference because their P values were less than 5% even though the level of significance were higher in stainless steel versus mild steel, stainless steel versus carbon steel than in carbon steel versus mild steel. The P value percentages are given as follows respectively; Weight loss (0, 0, 4%), Corrosion rate (0, 0, 2%).

The ecological quality parameters shown in Figures 2 to 13 had positive correlation with corrosion rate in the three metal coupons used for this study, except chloride with negative correlation in stainless steel only. In weight loss, pH, salinity, conductivity, total dissolved solids, sulphite all had positive correlation with the metal coupons. Oil and grease, chloride, total suspended solids, total organic carbon, sulphate, biochemical oxygen demand all had negative correlation with the metal coupons. Temperature had negative correlation with stainless steel but mild steel and carbon steel had positive correlation with temperature. Positive correlation means that ecological quality parameters are directly proportional to weight loss and corrosion rate while negative correlation means that ecological quality parameters are inversely proportional to weight loss and corrosion rate.

The distribution of the ecological quality parameters shows that total dissolved solids, conductivity, salinity,

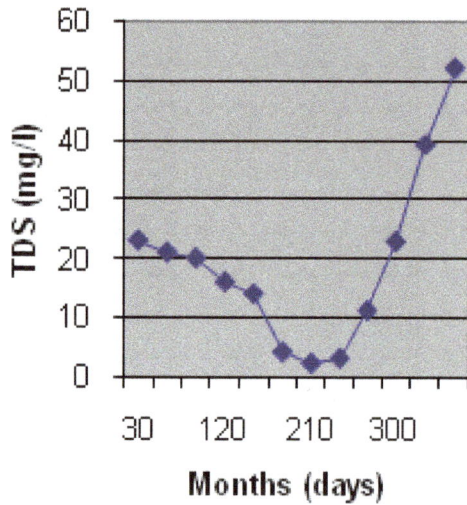

Figure 5. Conductivity against time.

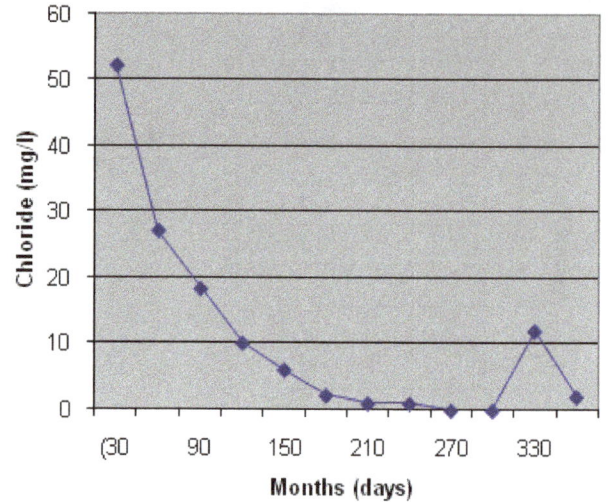

Figure 8. TSS against time.

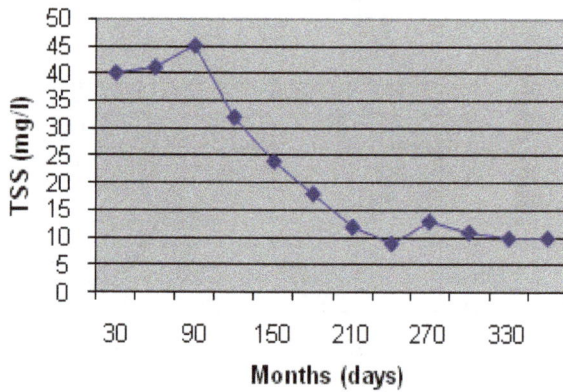

Figure 6. Chloride against time.

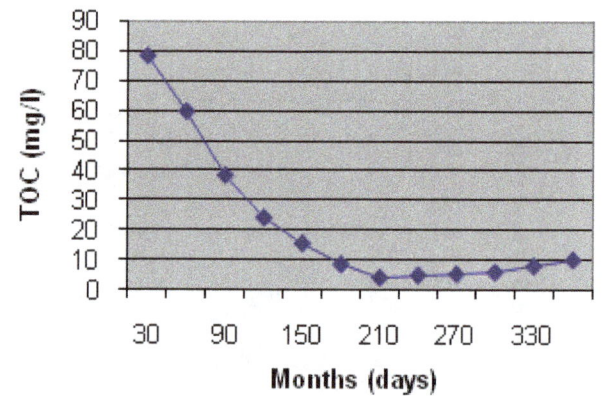

Figure 9. TOC against time.

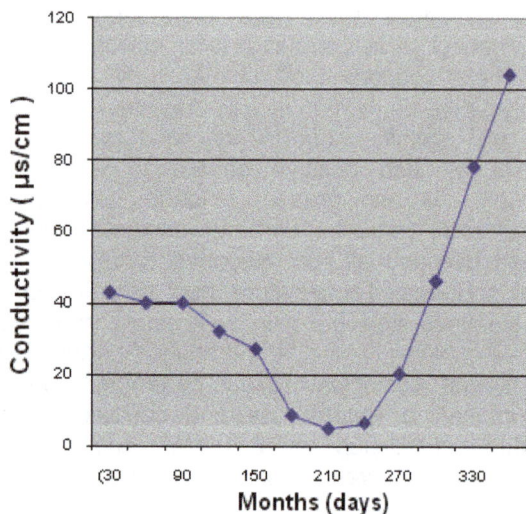

Figure 7. TDS against time.

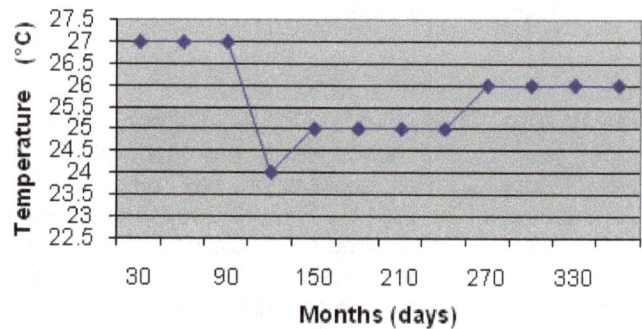

Figure 10. Temperature against time.

sulphite had higher values in the dry season than the rainy season. Total suspended solids, total organic carbon, oil and grease, chloride, sulphate, biochemical oxygen demand had higher values in rainy season than

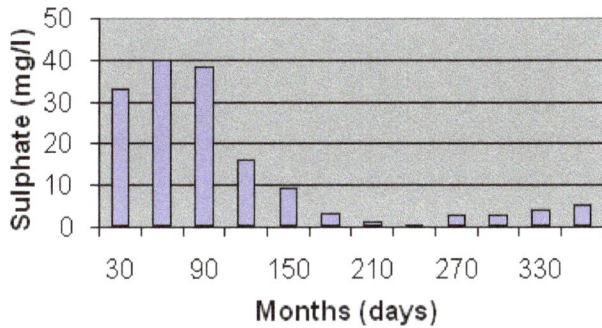

Figure 11. Sulphate against time.

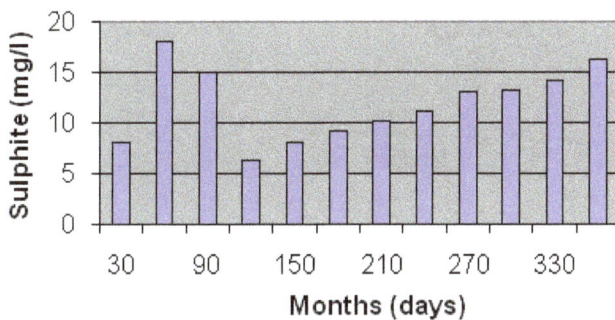

Figure 14. Corrosion rate against time.

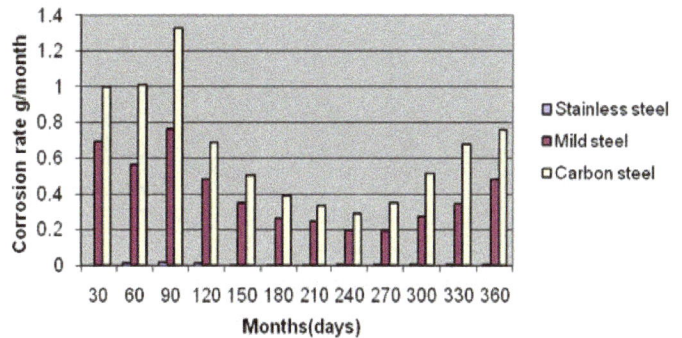

Figure 12. Sulphite against time.

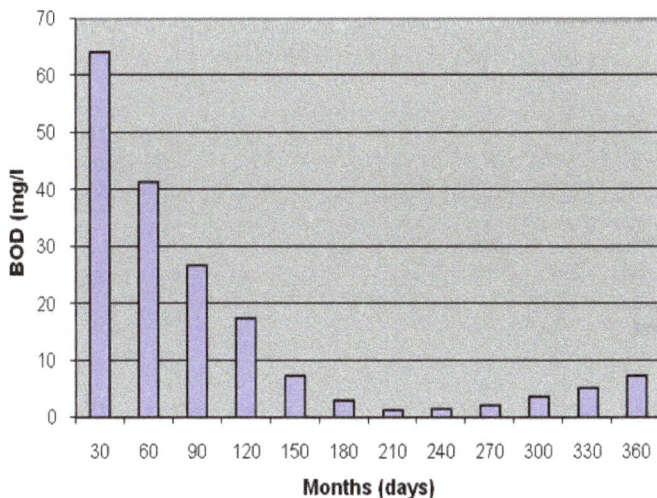

Figure 13. Biochemical oxygen demand against time.

dry season. All these could be attributed to commercial and domestic effluent discharges, run-offs and erosion in the freshwater habitats at different times thereby increasing or decreasing the ecological quality parameters at different seasonal variations. This agrees with the study of Grupa et al. (2008). They showed that

ecological quality parameters, nutrients and heavy metals are present in significant amount in all domestic wastes. Also Krishnan et al. (2007) stated that pH increase could be attributed to organic pollution, alkaline chemicals, soap and detergents produced due to commercial and residential activities.

Heavy metals monitored in this study showed that Ni, V, Cr was not detected throughout the study. Iron increased from April to May and decreased from June to October before increasing slightly from November to March respectively (8.13 to 9.06), (7.26 to 1.23) (1.32 to 1.55). Copper increased from April to May (0.61 to 0.81) and decreased from June to August (0.82 to 0.06) after which it was not detected again. Sodium and zinc increased from April to May and decreased from June to September in the manner respectively (1.13 to 1.19, 2.01 to 2.15), (1.16 to 0.09, 2.07 to 0.30). After which they were not detected again. Potassium decreased throughout from April to October (2.26 to 0.03) after which it was not detected again. Calcium decreased from April to May (0.84 to 0.83) and increased in June to (9.83) and decreased from July to September (0.71 to 0.22) after which it was not detected again.

Conclusion

The study showed that carbon steel had a higher corrosion rate than mild steel which also had a higher corrosion rate than stainless steel in the freshwater habitat. The corrosion rates were higher in the dry season than in the rainy season. The results show that the use of stainless steel for pipelines in freshwater environments should be promoted in the Nigerian Petroleum Industry or at least the composition of stainless steel should be incorporated in the current composition of steel used in the Petroleum Industry.

This will serve as a mitigative action against microbial mediated corrosion in brackish water environments. The frequency of integrity checks should be increased especially during the dry season, since rates of microbial induced corrosion are higher during the dry season than

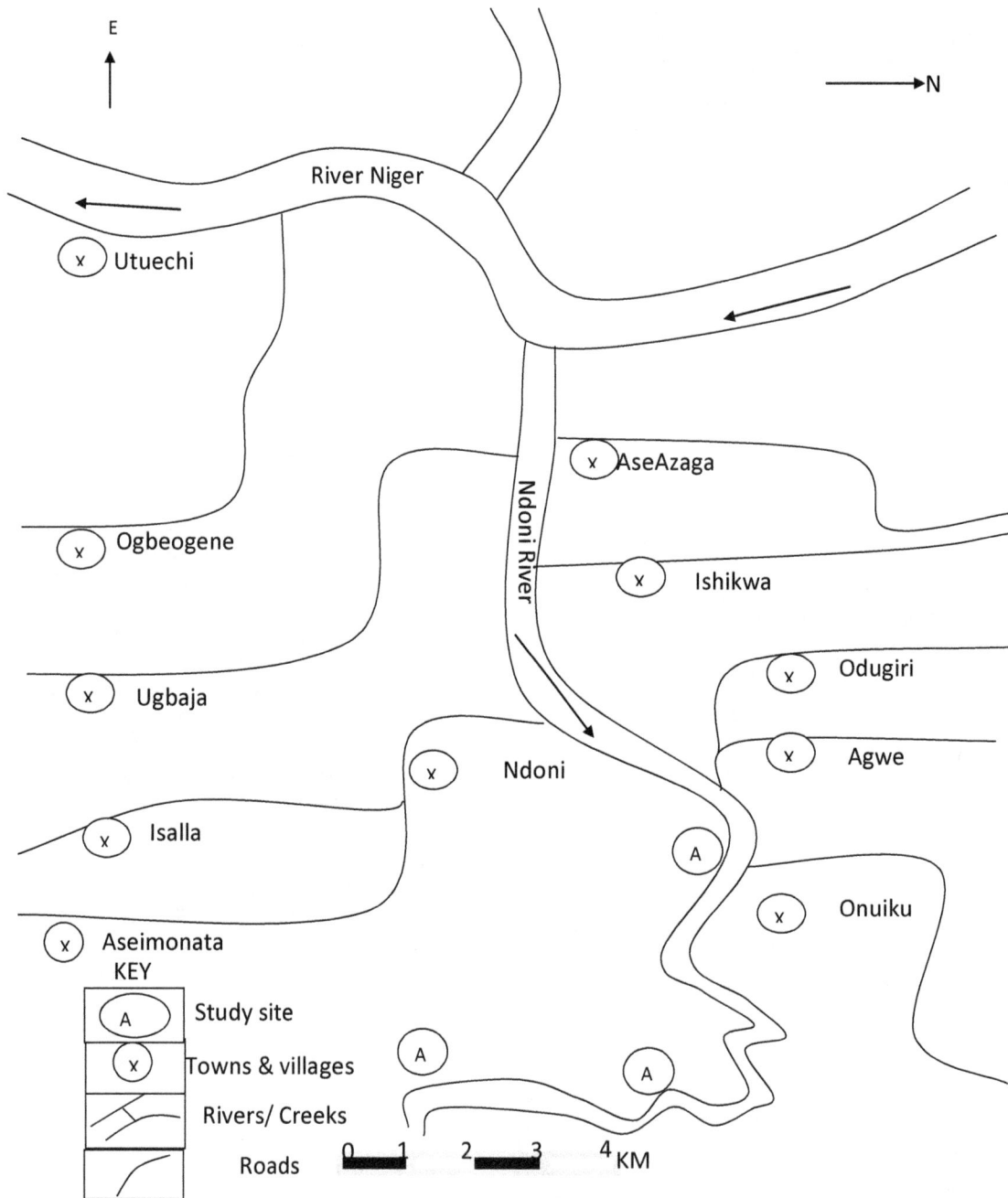

Figure 15. A map showing the sampling site the Ndoni River.

rainy season. These results also indicate that emphasis should not just be on physicochemical corrosion. Microbial induced corrosion is a potential cause of pipeline rupture/failure in the petroleum industry in Nigeria

REFERENCES

APHA (2000). Standard Methods for the Examination of Water and Waste Water. American Public Health Associated. Washington, D.C.

Ashassi SH (2009). Corrosion Inhibition of Mild Steel in Acidic Media by (BM/m) Br. Ionic Liquid. Mat. Chem. Phys., 114 (1) (15): 267-271, pp. 35-48.

Badmos AY, Ajimotokan HA (2009). The Corrosion of Mild Steel in Orange Juice Environment. www. Unilorin edu.ng/Publication.

Battersby NS, Stewart DJ, Sharma AP (1985). Microbiological Problems in the Offshore Oil and Gas Industries. J. Appl. Bacteriol. Symp. Suppl., pp. 227s-235s, Aberdeen, U.K.

Beech IB (2004). Corrosion of Technical Materials in the Presence of Biofilms Current Understanding and Staking-the-art- Method of Study. Int. Biodeterior. Biodegradation, 53: 117-183.

Birnin-Yauri UA, Garba S (2006). The Effects and Mechanism of Chloride Ion Attack on Portland cement Concrete and the

Structural Steel Reinforcement. Ife J. Sci. UN, 8: 2.

Bitton G, John W (2007). Biocorrosion: role of Sulphate reducing bacteria. Encyclopedia of Environ. Microbiol., pp. 465-475.

Dinh HT, Kuever J, Hassel AW (2004). Iron corrosion by novel anaerobic microorganisms. Nature, 427: 29-332.

Finney DJ (1978). Statistical Methods in Biological Assay 3rd edition, Charles Griffin London, pp. 1-252

Gerald M, Stams JM (2008). Nature review. Microbiol., 6: 441-458.

Hamilton WA (2003). Microbial influenced corrosion as a model System for the study of metal microbe interaction. Befouling, 19: 65-76.

International Conference and exhibition on corrosion management (ICEM-2011). Parkistani press international.

Jaganathan U, Nick S, Roges N (2011).improvement of passivity of Fe–xCr Alloys (x < 10%) by cycling through the reactivation potential, J. Appl. Electrochem., 41(7): 873-879.

Krishnan RR, Dharmaraj K, Ranjitha BD (2007). A Comparative Study on the Physicochemical and bacteria analysis of drinking bore well and sewage water in three different places of Surakasi. J. Environ. Biol., 28(1): 105-108.

Lee AK, Newman DK (2005). Microbial Iron respiration; impacts on corrosion processes, Appl. Microb. Biotechnol., 62: 134-139.

Lynch TC (1989). Practical Handbook of Material Science. 2nd edition, p. 2016.

Odu CTI, Esuruso OF, Nwoboshi LC, Ogu JA (1988). Environmental study of the Nigerian Agip Oil Company operational areas: Soils and Freshwater vegetation, Nigerian Agip Oil Company, Port Harcourt.

Pitonzo BJP, Amy PS, Castrol (2004). Microbiologically influenced corrosion. Corrosion, 60: 64-74.

Postgate JG (1985). Introduction to corrosion Preservation and Control. Rev. ed. Delft University press, Delft.

Raloff J (1985). The bugs of rust. Sci. News, 128: 42-44.

Smith WF, Hashemi J (2006). Foundation of material Science and Engineering. 4th edition, MC Graw- Hill, New York.

5

Distribution and abundance of small mammals in different habitat types in the Owabi Wildlife Sanctuary, Ghana

Reuben A. Garshong[1], Daniel K. Attuquayefio[1], Lars H. Holbech[1] and James K. Adomako[2]

[1]Department of Animal Biology and Conservation Science, University of Ghana, P. O. Box LG67, Legon-Accra, Ghana.
[2]Department of Botany, University of Ghana, P. O. Box LG55, Legon-Accra, Ghana.

Information on the small mammal communities of the Owabi Wildlife Sanctuary is virtually non-existent despite their role in forest ecosystems. A total of 1,500 trap-nights yielded 121 individuals of rodents and shrews, comprising five species: *Praomys tullbergi, Lophuromys sikapusi, Hybomys trivirgatus, Malacomys edwardsi* and *Crocidura buettikoferi,* captured in Sherman traps using 20 × 20 m grids. *P. tullbergi* was the most common small mammal species in all the four habitat types surveyed, comprising 63.6% of the total number of individual small mammals captured. The *Cassia-Triplochiton* forest had 61.2% of the entire small mammal individuals captured, and was the only habitat type that harboured higher abundances of the rare small mammal species in the sanctuary (*H. trivirgatus* and *M. edwardsi*). It also showed dissimilarity in small mammal species richness and abundance by recording a Sørenson's similarity index of less than half in comparison with the other three habitat types. Management strategies for the sanctuary should therefore be structured to have minimal impact in terms of development and encroachment on the *Cassia-Triplochiton* forest area in order to conserve the rare species and biodiversity of the Owabi Wildlife Sanctuary.

Key words: Small mammals, conservation, Ghana, Owabi Wildlife Sanctuary.

INTRODUCTION

The Owabi Wildlife Sanctuary (OWS) supports about 193 vascular plant species (Schmidtt and Adu-Nsiah, 1993), 191 bird species and some other key mammals such as Mona monkeys (*Cercopithecus mona*), Pottos (*Perodicticus potto*), Royal antelopes (*Neotragus pygmaeus*) and Cusimanses (*Crossarchus obscurus*) (Wilson and Kpelle, 1992). More emphasis has been placed on the above-mentioned vertebrate species probably because they serve as tourist attractants to the site, which generates some revenue. Information on small mammals like the rodents and shrews of OWS is virtually non-existent despite the significant role small mammals play in supporting vertebrate predators in forest ecosystems. Rodents and shrews, characterized by high productivity rates, serve as vital food sources for a large number of medium-sized predators such as mongooses (*Herpestes* spp.) and civets (*Nandinia* spp.), raptors like owls (*Strix* spp.) and goshawks (*Accipiter* spp.), and some reptiles like snakes (e.g. *Python regius*) (Laudenslayer and Fargo, 2002), even at relatively low densities. For example, Rabinowitz and Walker (1991) reported that dry tropical forest murid rodents accounted for 33% of prey items in scats of small carnivores in Thailand. Predation on seeds and seedlings by murids in forest ecosystems influences which tree species grow to maturity as well as plant regeneration rate (Davies and

Howell, 2004). Above all, small mammals are good bio-indicators of environmental condition due to their (i) rapid turnover rate (Happold, 1979), (ii) high biotic potential, (iii) ability to invade reclaimed areas and (iv) sensitivity to environmental disturbance (Malcom and Ray, 2000). Hence, this study will contribute to improve conservation strategies for the OWS.

The objectives of this research were to determine: (i) the variations in small mammal compositions in different floristic habitats at the OWS based on the hypothesis that floristic composition influences food availability and, hence, the distribution of small mammal species in an area (Ahmad et al., 2002) and (ii) the similarity of the small mammal communities occupying the different floristic habitats to help know which habitat type was distinct.

METHODS

Study area

The Owabi Wildlife Sanctuary (6°45'N, 1°43'W), an inland Ramsar site, is located about 10 km north-west of Kumasi, Ghana's second largest city, in the Ashanti Region. The area was designated a Ramsar site in 1988 and is now managed by the Ramsar Focal Point Section of the Forestry Commission of Ghana. The mean annual rainfall was 1,402 mm for the period 1961 to 1991 and the average monthly temperature varies little (24.6 to 27.8°C), while the diurnal range is up to 9.1°C. The general vegetation is a moist semi-deciduous Forest (north-west subtype) (Hall and Swaine, 1976). The study was conducted in the 13 km^2 inner sanctuary of the reserve where park guards mount their regular surveillance. For the purpose of this study, four vegetation subtypes were identified—monodominant forests, mixed forest and old farms, and labeled as follows:

Cassia siamea forest (CF)

This zone, located on the north of the reserve, was dominated by dense and uniform stands of the exotic tree *Cassia siamea*, which formed about 50% of the plant species in the area. Other plant species recorded at the site were *Triplochiton scleroxylon*, *Piptadeniastrum africanum*, *Funtumia elastica*, *Elaeis guineensis*, *Ficus exasparata*, *Trichilia monadelpha*, *Antiaris africana*, *Baphia nitida*, *Culcasia* sp., *Cola gigantea*, *Acacia* sp., *Combretum* sp., *Bambusa vulgaris* and *Cnestis ferruginea*. The herb layer was dominated by *Culcasia* sp. Many small woody plants were also present.

Cassia-Triplochiton forest (CTF)

This forest area, located westward, was dominated by *C. siamea* and *T. scleroxylon* (although they were abundant, none constituted about 50%). Other fairly abundant tree species recorded were *P. africanum*, *F. elastica*, *E. guineensis*, *Cleistopholis* sp., *Albizia zygia*, *Cola gigantea*, *Lonchocarpus sericeus*, *Terminalia superba*, *Terminalia ivorensis*, *Acacia* sp., and *Ficus exasperata*. The main shrubs of the site included *T. monadelpha*, *Bafia nitida* and *Macrodesmis puberula*, and *Culcasia* sp. was the dominant herb. Climbing plants such as *Pollinia pinnata*, *Griffonia simplicifolia*,

Combretum racemosum and *Piper guineensis* were also present.

Abandoned farmlands (AF)

These are lands reclaimed by the government in 1972, consisting of secondary regrowth of the former Moist Semi-deciduous Forest. Some areas were dominated by banana (*Musa* sp.), while others have been invaded by 'Acheampong' weed (*Chromolaena odorata*). Some other areas also had a few of the indigenous tree species: *Panicum maximum*, *G. simplicifolia*, *F. exasperata*, *Myrianthus arboreus*, *Ceiba pentandra*, *Triplochiton scleroxylon*, *Celtis zenkeri*, *Acacia* sp., *Antiaris toxicaria*, *Cola gigantea*, *Culcasia* sp., *Blighia sapida*, *Cnestis ferruginea* and *C. siamea*.

Bamboo cathedral (BC)

This area had dense clumps of the exotic bamboo (*B. vulgaris*) which were up to about 10 m high with a patchy shrub and herb layer dominated by broad-leaved species (*Culcasia* sp.). The following sparsely distributed floral species were also recorded: *T. ivorensis*, *Baphia pubescens*, *Piptadeniastrum africanum*, *Milicia excelsa* and *Trichilia* sp.

Small mammal live-trapping

Small mammals were live-trapped, five consecutive nights in a month, from October, 2009 to February and then April, 2010 using Sherman collapsible traps (23 × 9 × 7.5 cm). In each of the first three months, six 20 × 20 m grids were established to cover the *C. siamea* and *Cassia-Triplochiton* forests (three for each), while the Abandoned Farmlands and Bamboo Cathedral vegetation subtypes had one each due to relative differences in habitat size and trap availability. An additional grid of the same area was established for each vegetation zone in the subsequent months. Each square grid contained five traps— one at the centre and one at each vertex.

The grid system was used in order to obtain a fair idea of the biomass of the small mammal species. The traps were baited with a mixture of groundnut paste and dried grated cassava (gari). Traps were set at 1600 h GMT, and inspected daily from 0630 h GMT. Captured animals were shaken gently out of the trap into a mesh bag, anaesthetized with chloroform (to kill unidentified species humanely and daze aggressive species for ease of measurements), examined for reproductive condition (abdominal or scrotal testes in males and enlarged nipples, perforate vaginas and pregnancy in females), identified on the spot (when possible) and released after toe-clipping. One individual of every new specimen both identified and unidentified species were preserved in formalin to be transferred to the Animal Biology and Conservation Science Department museum of the University of Ghana for use as voucher specimens.

Recaptured individuals had their initial marks recorded. The following standard morphometric measurements (body, tail, ear, and hind limb lengths) were taken: (i) TOTL (total body and tail length; from nose-tip to end of tail), (ii) TL (tail length; from base of tail at right angle to body to the tip of tail), (iii) HBL (head and body length, TOTL − TL), (iv) HFL (hind foot length; from heel to the tip of the longest toe), (v) EL (ear length; from basal notch to the distal tip of pinna) and (vi) WT (weight in grams). The sex (using ano-genital distance which is longer in males) and age class (assigned into three broad age-classes: juvenile, sub-adult and adult) of the captured small mammal species were also determined.

The study by Rosevear (1969) and Kingdon (1997) were used as key references for rodent taxonomy and identification while another study by Hutterer and Happold (1983) was used for shrew identification. Small mammal field handling techniques were followed as described by Davies and Howell (2004).

Table 1. Small mammal capture data in the different habitat types in the Owabi Wildlife Sanctuary.

Species	Habitat type				Total (%)
	CF (%)	CTF (%)	AF (%)	BC (%)	
Insectivora					
Crocidura buettikoferi	0 (0.0)	1 (1.4)	1 (11.1)	0.0	2 (1.7)
Rodentia					
Praomys tullbergi	14 (53.8)	49 (66.2)	4 (44.4)	10 (83.3)	77 (63.6)
Lophuromys sikapusi	9 (34.6)	4 (5.4)	2 (22.2)	1 (8.3)	16 (13.2)
Hybomys trivirgatus	1 (3.8)	12 (16.2)	0 (0.0)	0 (0.0)	13 (10.7)
Malacomys edwardsi	2 (7.7)	8 (10.8)	2 (22.2)	1 (8.3)	13 (10.7)
Number of captures	26	74	9	12	121
Number of species	4	5	4	3	5
Number of trap-nights (TN)	525	525	225	225	1500
Trapping success/100 TN	4.95	14.10	4.00	5.33	8.07
Biomass (g/ha)	982.143	2796.429*	858.333	1283.333	1757.958
DIVERSITY SCORE					
exp (H^l)	1.132	1.509	1.439	1.300	
Pielou's Index (J^l)	0.092	0.413	0.309	0.200	

Values in brackets indicate capture percentage (%) of that species relative to total number of captures for that site; *indicate significant difference; CF = *Cassia siamea* forest; CTF = *Cassia-Triplochiton* forest; AF = Abandoned farmlands; BC = Bamboo cathedral).

Analyses of data

1. Capture percentage (CP) = (N_i / N_t) × 100.................................. [1]

Where, N_i = number of individuals of each species in each habitat; N_t = total number of individuals caught during the entire study. The CP gives an indication of the abundance of each species relative to the habitat type with reference to the total capture.

2. Trapping success (TS) = (N_i/ Tn) × 100................................ [2]

Where, Tn = total number of trap-nights (one trap set for one night). The TS tells how many of the traps set at a site were able to capture the target species.

3. Biomass = average adult fresh weight of live-trapped small mammals within a habitat × estimated mean density (no. of individuals/ha) (Fleming, 2009). The area of each grid and the corresponding individuals therein was used to calculate the number of individuals per hectare. Biomass relates to the productivity of the environment since it is weight oriented, and weight gives an indication of food availability.

Diversity indices

1. Shannon-Wiener index = expH^1, where H^1 = -$\sum$ p$_i$ Inp$_i$.............. [3]

The 'p$_i$' refers to the proportion of species 'i' in the sample (the relative abundance of that species [N_i/N_{tot}]). This index is species richness weighted. Shannon-Wiener measures the amount of uncertainty in predicting what species an individual chosen at random from a sample would belong to. It also measures the effective number of species. The index is highly correlated to the evenness index (Jost, 2006).

2. Evenness, J^1 = H^1/H^1$_{max}$...[4]

Where, H^1$_{max}$ = In S (S is the total number of different species in the sample). It focuses on how evenly the species are distributed in the community, that is, how evenly resources or niches are divided among the species. It is also called the Pielou's index.

3. Sørenson's similarity index, C$_N$ = 2$_{jN}$/(aN + bN) ....[5]

Where, $_{jN}$ = sum of the lower of the two abundances recorded for the species found in both sites; aN = total number of individuals in site A and; bN = total number of individuals in site B. It measures how different a range of habitats are in terms of the variety and abundance of the species found in them by comparing the species shared by the different communities. The lower the C$_N$ value, the less similar the communities under comparison (Magurran, 1991).

Statistical analysis

Inferential statistics involved the use of nonparametric test, Kruskal-Wallis (H) and Fisher (LSD) test (Ashcroft and Pereira, 2003) to help establish significant differences where necessary.

RESULTS AND DISCUSSION

A total of 121 individuals of small mammals belonging to five species and two mammalian orders (Rodentia and Soricomorpha) were recorded in 1,500 trap-nights, with a trapping success of 7.5% (Table 1). There were 77 Tullberg's soft-furred rats (*Praomys tullbergi*), 16 Rusty-bellied rats (*Lophuromys sikapusi*), 13 Temminck's back-striped or hump-nosed mice (*Hybomys trivirgatus*), 13 Edward's long-footed, big-eared or swamp rats (*Malacomys edwardsi*), and two Buettikofer's shrews

Table 2. Coefficient of similarity in small mammal communities in the Owabi Wildlife.

Habitat type	Similarity coefficient (C_N)
CF + CTF	0.420
CF + BC	0.632
CF + AF	0.457
CTF + BC	0.279
CTF + AF	0.217
AF + BC	0.571

Values in brackets indicate capture percentage (%) of that species relative to total number of captures for that site; *indicate significant difference; CF = *Cassia siamea* forest; CTF = *Cassia-Triplochiton* forest; AF = Abandoned farmlands; BC = Bamboo cathedral).

(*Crocidura buettikoferi*) (Table 1). Sampling was concentrated in the core area because it contained the rare and true forest species of the site, which may serve as the main determinants of habitat disturbance. Rodents constituted 98.3% of the total captures. This may mean that the shrew density in the OWS was probably low during the study period.

P. tullbergi was the most dominant small mammal species, with a capture percentage of 63.6% (Table 1). Similarly, most studies of terrestrial small mammals in West African forests found *P. tullbergi* to be the dominant species. For example, in western Ghana, Cole (1975) and Jeffrey (1977) recorded *P. tullbergi* as the most dominant rodent in a lowland evergreen forest and primary forest, new farms and cocoa plantations. The same was reported for some Nigerian forests (Iyawe, 1989; Oguge, 1995). *P. tullbergi* was referred to as "West African forest mouse" by Happold (1975, 1978), probably due to its dominance in West African forests. The occurrence and dominance of *P. tullbergi* in all four habitat types as compared to the other species captured may be attributed to their generalized habitat and dietary requirements as reported by Iyawe (1989). *P. tullbergi* are also polyoestrous (Happold, 1978). The dominant small mammal species contribute to forest richness the most, by positively influencing diversity and species composition of other species and life forms due to their linkage in complex ways with other biotic and abiotic components of the ecosystem (Sieg, 1987). Changes in the relative abundance of *P. tullbergi* in the OWS should therefore provide an indication of habitat disturbance, which may lead to biodiversity loss.

Cassia-Triplochiton forest (CTF) recorded the highest number of individuals of each small mammal species captured at the OWS except for *L. sikapusi*, which is regarded as an atypical forest species, inhabiting open and drier areas of forests (Rosevear, 1969; Okia, 1992) (Table 1). CTF also recorded the highest diversity (exp H') and evenness of 1.509 and 0.413, respectively (Table 1). It was also the only habitat type that recorded the two

rarest rodent species known to be restricted to primary forests (*H. trivirgatus* and *M. edwardsi*) in higher numbers (Jeffrey, 1977), suggesting that the CTF habitat type is a remnant of the original forest of the OWS. Habitats such as the *Cassia-Triplochiton* forest that exhibit high floral diversity are reported to provide variable feeding options and microhabitats that probably offer cover and nesting sites to different species of small mammals (Kasangaki et al., 2003). This was supported by the significant difference (N = 4; H = 10.915; Chi-critical $_{(3, 0.05)}$ = 7.815; p < 0.05) in its biomass (2796.429 g/ha) as compared to the other habitat types (Table 1).

The *Cassia-Triplochiton* forest (CTF) differed from the remaining three habitats by recording the lowest similarity coefficients when compared with the other habitat types, which showed much similarity in small mammal communities (Table 2). This may mean that the floristic composition of CT, AF and BC are similar, resulting in similar small mammal composition since separated habitats exhibiting similar floral features could have similar faunal communities too (Reed and Clockie, 2000). This makes the CTF habitat type a distinct habitat, supporting the rare small mammal species (*H. trivirgatus* and *M. edwardsi*) and 40.5% of the most abundant species (*P. tullbergi*) of the sanctuary. In case of *L. sikapusi*, CF is the only habitat type where it has higher abundance as compared to the other habitat types. Hence, for the conservation of *L. sikapusi,* CF is important too.

We therefore suggested that management strategies adopted to conserve the site should be streamlined to put little or no development on the *Cassia-Triplochiton* forest habitat type in order to conserve the biodiversity of the site.

REFERENCES

Ahmad MY, Batin Z, Shukor N (2002). Influence of elevational habitat changes on non- volant small mammal species: distribution and diversity on mountain Nuang, Hulu langat, Selangor Malaysia. Pakist. J. Biol. Sci. 5:819-824.

Ashcroft S, Pereira C (2003). Practical Statistics for the Biological Sciences: simple pathways to statistical analyses. Palgrave Macmillan, New York.

Cole LR (1975). Foods and foraging places of rats (Rodentia: Muridae) in the lowland evergreen forest of Ghana. J. Zool. (Lond.). 175:453-471.

Davies G, Howell K (2004). Small Mammals: Bats, Rodents and Insectivores. In: African Forest Biodiversity, (G Davies, K Howell, Eds.) Earthwatch Institute, UK.

Fleming TH (2009). The role of small mammals in tropical ecosystems. In: Small Mammals: their productivity and population dynamics, (FB Golley, K Petrusewicz, R Ryszkowski, Ed.). Cambridge University Press, UK.

Hall J, Swaine MD (1976). Classification and ecology of closed-canopy forest in Ghana. J. Ecol. 64:913-951.

Happold DCD (1975). The effects of climate and vegetation on the distribution of rodents in western Nigeria. Z. Säugetierkd., 40:221-242.

Happold DCD (1978). Reproduction, growth and development of a West African Forest Mouse, *Praomys tullbergi* (Thomas). *Mammalia*, 42(1):74-95.

Happold DCD (1979). Age structure of a population of *Praomys tullbergi* (Muridae, Rodentia) in Nigerian Rainforests. Rev. Ecol.-Terre Vie, 33:

253-274.

Hutterer R, Happold DCD (1983). *The Shrews of Nigeria (Mammalia: Soricidae)*. Bonne. Zool. Monogr., Nr. 18, Germany.

Iyawe JG (1989). The ecology of small mammals in Ogba Forest Reserve, Nigeria. J. Trop. Ecol. 5:51-64.

Jeffrey SM (1977). Rodent ecology and land use in Western Ghana. J. Appl. Ecol. 14:741-755.

Jost L (2006). Entropy and diversity. *Oikos*, 113(2):363-375.

Kasangaki A, Kerbis J, Kityo R (2003). Diversity of rodents and shrews along an elevational gradient in Bwindi Impenetrable National Park, Southwestern Uganda. Afr. J. Ecol. 41:115-123.

Kingdon J (1997). The Kingdon Field Giude to African Mammals. Academic Press/Harcourt Brace, San Diego, U. S. A.

Laudenslayer WF, Fargo RJ (2002). Small Mammal Populations and Ecology in the Kings River Sustainable Forest Ecosystems Project Area. USDA Forest Service Gen. Tech. Rep. PSW-GTR-183. pp. 133-142.

Magurran AE (1991). Diversity and Its Measurements. Chapman and Hall, London. pp. 90-95.

Malcom JR, Ray JC (2000). The influence of timber extraction routes on central African small-mammal communities, Forest Structure and Tree Diversity. Conserv. Biol. 14:1623-1638.

Oguge NO (1995). Diet, seasonal abundance and microhabitats of *Praomys (Mastomys) natalensis* (Rodentia: Muridae) and other small rodents in Kenyan sub-humid grassland community. Afr. J. Ecol. 33:211-225.

Okia NO (1992). Aspects of rodent ecology in Lunya Forest, Uganda. J. Trop. Ecol. 8:153-167.

Rabinowitz AR, Walker S (1991). The carnivore community in a dry tropical forest mosaic in Huai Khakhaeng Wildlife Sanctuary, Thailand. J. Trop. Ecol. 7:37-47.

Reed MS, Clokie MRJ (2000). Effects of grazing and cultivation on Afromontane plant communities on Mount Elgon, Uganda. Afr. J. Ecol. 38(2):154-162.

Rosevear DR (1969). Rodents of West Africa. British Museum (Nat. Hist.), London. pp 604.

Schmidtt K, Adu-Nsiah M (1993). The vegetation of Owabi Wildlife Sanctuary. Forest Resource Management Project, Ghana Wildlife Division/International Union for Conservation of Nature Project 9786.

Sieg CH (1987). Small Mammals: Pests or Vital Components of the Ecosystem. Wildlife Damage Management, Internet Center for Great Plains Wildlife Damage Control Workshop Proceedings, University of Nebraska – Lincoln. Retrieved from http://digitalcommons.unl.edu/gpwdcwp/97. (Accessed: 30/09/10)

Wilson VJ, Kpelle DG (1992). Report on zoological survey of Owabi Wildlife Sanctuary, Kumasi-Ghana. Forest Resource Management Project, Ghana Wildlife Division/International Union for Conservation of Nature Project 9786.

Indigenous knowledge on fuel wood (charcoal and/or firewood) plant species used by the local people in and around the semi-arid Awash National Park, Ethiopia

Tinsae Bahru[1]*, Zemede Asfaw[2] and Sebsebe Demissew[2]

[1]Ethiopian Institute of Agricultural Research (EIAR), Forestry Research Center (FRC),
Non-Timber Forest Products (NTFPs) Case Team, P. O. Box 30708, Addis Ababa, Ethiopia.
[2]The National Herbarium, Department of Biology, Faculty of Science, Addis Ababa University, P. O. Box 3434,
Addis Ababa, Ethiopia.

Fuel wood (charcoal and/or firewood) species used by the Afar and Oromo (Kereyu and Ittu) Nations in and around the semi-arid Awash National Park (ANP), Ethiopia was conducted ethnobotanically. The study aimed to investigate and document various aspects of indigenous knowledge (IK) on fuel wood species and their associated threats. A total of 96 informants between the ages of 20 and 80 were selected using prior information. Data were collected using semi-structured interview, guided field walk, discussions and field observation. Preference ranking, paired comparison, Jaccard's coefficient of similarity and priority ranking were applied for data analysis. A total of 100 species belonging to 71 genera and 38 families were collected within the study area. Of these, 10 species were reported by the Afar Nation, 11 by the Oromo Nation and the rest by both of them. Family Fabaceae was represented by the highest number of fuel wood species, which accounted for 20%. From 27 species used for charcoal and firewood production, 11 species (40.7%) belonged to the genus Acacia. Preference ranking and paired comparison showed that Acacia tortilis is the most selected Acacia species as perceived by key informants within the park for charcoal production. Overgrazing, followed by deforestation were the major threats in the study area, which scored 21.7% and 19.9%, respectively.

Key words: Acacia species, Awash National Park (ANP), charcoal, Ethiopia, indigenous knowledge (IK).

INTRODUCTION

The most important sources of fuel, which are the necessities for human kind, are fuel wood (charcoal and firewood), petroleum and peat. Of these, wood makes an outstanding fuel as it is 99% flammable if completely dry (Hill, 1952; Kochhar, 1998). It is the cheapest, the most suitable and accessible energy source in many rural areas (Abbiw, 1990; Cotton, 1996). For example, Abbiw (1990) reported that 90% of the wood cut is used for fuel wood. However, an inefficient and wasteful method of traditional open fire cooking accounts mainly for the

consumption of relatively a higher proportion of fuel wood. So, to combat the problem of deforestation designing fuel-saving stoves is one of the practical solutions in many developing countries. Although the majority of wild woody plant species can be used as a source of fuel for indigenous peoples, many species are recognized for particular burning qualities (Cotton, 1996).

Charcoal is a valuable and a chief domestic fuel in most tropical countries (Hill, 1952). Due to this reason, it is a common source of fuel wood in urban centers. In the absence of fossil fuel, charcoal is more advantageous and much preferred fuel wood than firewood due to being of lighter weight, less bulky and more compact, thereby easier to store indefinitely and cheaper to transport

*Corresponding author. E-mail: batinsae@gmail.com.

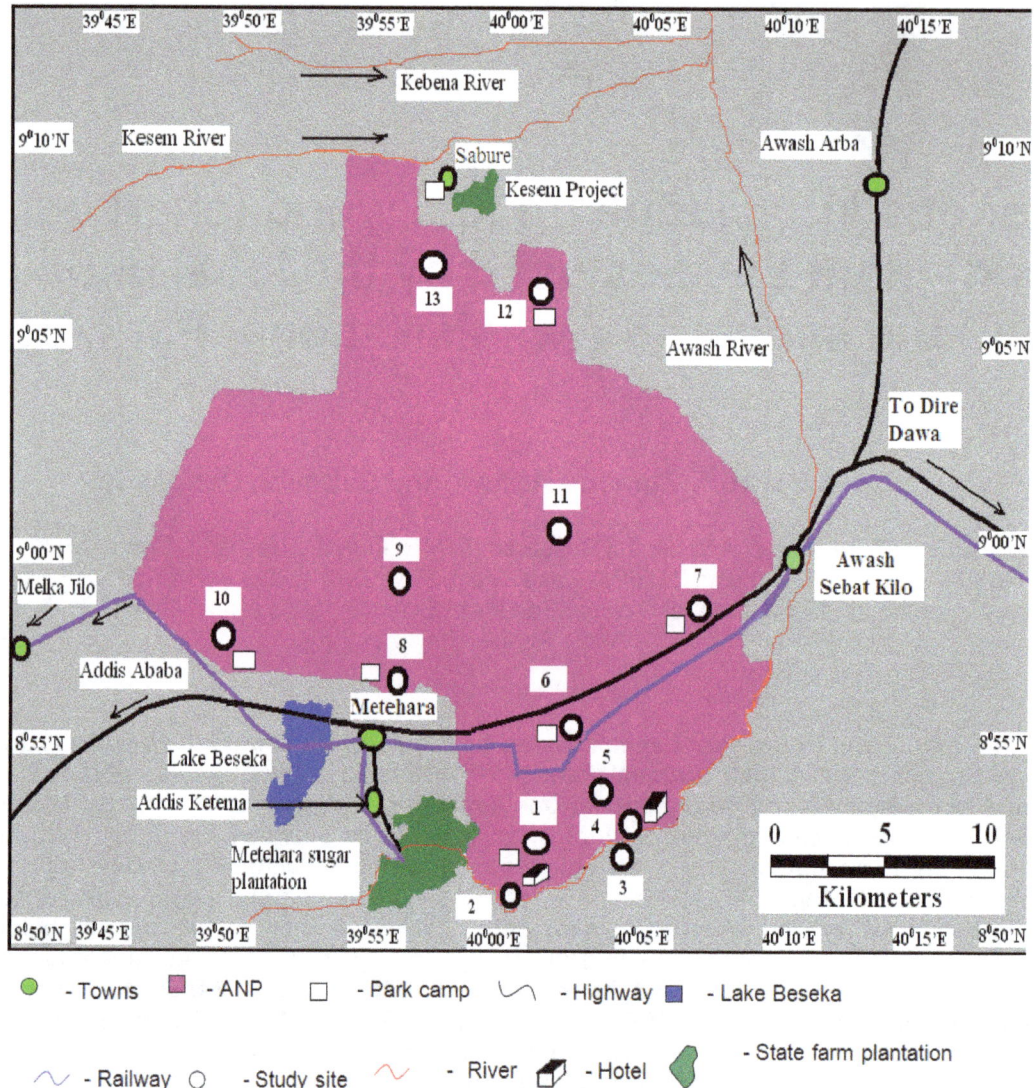

Figure 1. Map of ANP modified from EMA (1992), Jacobs and Schloeder (1993) and Berihun and Solomon (2005).

(Abbiw, 1990). It is more efficient and produces a steady heat with little or no smoke or soot (Hill, 1952; Abbiw, 1990; Cotton, 1996; Kochhar, 1998). On the contrary, the long distance transportation makes for its high cost as compared to firewood. During charcoal preparation, about half of the wood's energy is wastefully burned away (Abbiw, 1990). Consequently, extensive woodland has to be cleared to meet the high charcoal demand. Moreover, charcoal making causes many accidental fires on forests. Thus, both charcoal harvesting and accidental fire contribute to global warming, deforestation and land degradation (Silayo et al., 2008). Therefore, the present study aimed to conduct an ethnobotanical study of fuel wood species used by indigenous peoples of the Afar and the Oromo Nations in and around ANP and record, compile and document the associated IK to assist in the proper utilization, management and conservation of

useful plants and the settings of the Park as a whole.

The study area

Geographical location

The study was conducted in ANP, Ethiopia, which is 225 km away from Addis Ababa and situated between latitudes 8°50' and 9°10' north and longitudes 39°45' and 40°10' east (EMA, 1992) (Figure 1). The Park covers approximately 756 km^2 and is bordered by the Sabober plain to the west, the Awash River to the south and southeast and Kesem River and Filwuha Hot Springs to the north (Jacobs and Schloeder, 1993). ANP is characterized by semi-arid climate or Qolla Zone and bimodal rainfall with the annual rainfall ranging between

400 and 700 mm (Jacobs and Schloeder, 1993). Out of the nine vegetation types of Ethiopia, the vegetation type of ANP is classified under *Acacia-Commiphora* woodland (Sebsebe and Friis, 2009) in the Somali-Masai Regional Center of endemism (White, 1983). Thirteen data collection sites in ANP were: 1. Gotu, 2. Awash River, 3. Awash Gorge, 4. Karreyu Lodge, 5. Ilala Sala plain, 6. Hamareti, 7. Geda, 8. Sogido, 9. Mt. Fentale, 10. Sabober, 11. Dunkuku (Kudu Valley), 12. Filwuha, and 13. Sabure (Figure 1).

MATERIALS AND METHODS

Ethnobotanical data collection

After a reconnaissance survey from August 15 to 30th, 2008, 13 study sites (Figure 1) were selected and established as data collection sites. Following this, ethnobotanical data were collected between September, 2008 and March, 2009, on three field trips that were carried out in each study site, following the methods by Martin (1995), Cotton (1996) and Cunningham (2001). Semi-structured interview, guided field walk, discussions and observation, with informants and key informants were applied based on a checklist of questions using the Afar and Oromo languages with the help of translators to obtain IK of the local people on fuel wood species in and around the ANP. Voucher specimens were collected, identified and kept at National Herbarium, Addis Ababa University. Consequently, informants were selected from the Afar and/or the Oromo Nations based on the vicinity of their Kebeles to the Park. Four Kebeles from the Afar Nation (Awash, Doho, Dudub and Sabure Kebeles), whereas five Kebeles from the Oromo Nation (Benti, Fate Leidy, Gelcha, Ilala and Kobo Kebeles) were taken. Of these, 96 informants 7 or 8 individuals for each study site (76 men and 20 women) between the ages of 20 and 80 were selected using prior information. Out of these, 36 key informants (32 men and 4 women) were selected. Basic information on fuel wood species including their local names, part (s) used, species used for charcoal and/or firewood and other additional uses was/were collected from informants.

Ethnobotanical data analysis

The data were analyzed and summarized using simple preference ranking, paired comparison and direct matrix ranking, following Martin (1995) and Cotton (1996). The Jaccard's Coefficient of Similarity (JCS) was also calculated and the similarity in fuel wood composition between the Afar and the Oromo Nations were compared as it was described in Kent and Coker (1992). Accordingly, JCS was calculated between paired habitat types (A and B) as follows:

$$JCS = \frac{c}{c + b + a}$$

Where   a - is the number of species found only in habitat A,
           b - is the number of species found only in habitat B and
           c - is the number of common species found in habitat A and B.

Finally, JCS was multiplied by 100 in order to obtain the percentage similarity in species composition between the Afar and the Oromo Nations as applied by Kent and Coker (1992).

RESULTS AND DISCUSSION

Diversity and distribution of fuel wood species

Most of the local communities around the ANP directly or indirectly rely on the Park's resources for energy source. In this field study, a total of 100 fuel wood (charcoal and/or firewood) species were recorded, being distributed in 71 genera and 38 families. Of these, 10 species were reported by the Afar Nation, 11 by the Oromo Nation and the rest by both of them. About 80% of the species were reported with their vernacular (local) names, where 63% were reported by the Afar Nation and 80% by the Oromo Nation (Appendix 1). Firewood was the major source of energy which accounted for 73% and an income generating activity in the livelihoods of many rural dwellers. The majority of the local communities use fuel wood to cook their food and heat and light up their houses. The source of fuel wood was found to be from woodland reserve, bushland, shrubland and market areas. The urban dwellers in Awash Sebat Kilo, Metehara, Addis Ketema, Sabure and Melka Jilo towns buy sticks or splits/bundles of firewood or sacks of charcoal from the local markets. On the other hand, the rural peoples harvest firewood and/or charcoal from the study area. According to Zerihun and Mesfin (1990), the Rift Valley vegetation is an important source of charcoal making for the nearby towns and Addis Ababa. Informants during individual or group discussions revealed that selection of firewood mainly relied on availability, burning quality, little/no smoke/soot production and moisture content.

Accordingly, the most widely identified plant species for the source of firewood comprise *Acacia*, *Ziziphus* and *Grewia* species, *Balanites aegyptiaca*, *Olea europaea* subsp. *cuspidata*, *Tamarindus indica*, *Terminalia brownii*, *Manilkara butugi*, *Berchemia discolor*, *Trilepisium madagascariense*, *Dobera glabra* and so forth. In the same way, *Acacia* species such as *Acacia dolichocephala* and *Acacia brevispica* are the most preferred firewood species based on burning qualities in their respective orders (Hussien, 2004). Similarly, species like *Acacia nilotica*, *Balanites aegyptiaca* and *Acacia tortilis* are more preferable for both firewood and charcoal making. According to informants responses' *Lantana camara* and *Prosopis juliflora* are used as sources of firewood and firewood/charcoal making respectively. For instance, informants explained that the wood (stem) of *Prosopis juliflora* serve as an excellent source of fuel wood (firewood and charcoal making), house construction (timber, house posts) and dry fencing. This is a good practice in order to control and manage these noxious invasive alien plant species.

This was followed by both charcoal and firewood that come up with 27%. The informants stated that if the wood of a plant is used for charcoal making most of the time it also serves for firewood. Of these results, from 27 plant species used for charcoal and firewood production, 11

Table 1. Preference ranking for seven most preferred *Acacia* species used for charcoal production as perceived by key informants in the study area.

Major plants used for charcoal production	Key informants															Total score	Rank
	R_1	R_2	R_3	R_4	R_5	R_6	R_7	R_8	R_9	R_{10}	R_{11}	R_{12}	R_{13}	R_{14}	R_{15}		
Acacia mellifera	4	1	2	4	1	1	5	7	1	2	4	1	2	7	3	45	5th
Acacia nilotica	3	6	7	6	5	7	4	4	4	6	7	6	4	5	4	78	2nd
Acacia oerfota	1	3	4	1	2	2	1	1	2	1	1	2	5	2	1	29	7th
Acacia prasinata	7	7	3	5	7	6	2	3	5	3	3	4	3	6	6	70	3rd
Acacia senegal	2	2	1	2	3	3	6	2	3	5	2	3	1	1	5	41	6th
Acacia seyal	5	4	6	3	4	4	3	5	6	4	5	5	6	3	2	65	4th
Acacia tortilis	6	5	5	7	6	5	7	6	7	7	6	7	7	4	7	92	1st

Table 2. Paired comparison of five major important *Acacia* species based on degree of tree cutting for charcoal production as perceived by key informants in the study area.

Acacia species used for charcoal production	Key informants										Total score	Rank
	R_1	R_2	R_3	R_4	R_5	R_6	R_7	R_8	R_9	R_{10}		
Acacia mellifera	1	1	0	0	2	1	2	3	0	3	13	5th
Acacia nilotica	4	5	3	4	3	4	5	3	5	4	40	2nd
Acacia prasinata	4	0	2	4	4	3	5	2	3	4	31	3rd
Acacia seyal	3	2	1	2	0	4	4	5	1	3	25	4th
Acacia tortilis	5	4	5	5	5	5	3	4	5	4	45	1st

5= Series; 4=Very high; 3= High; 2= Medium; 1= Least.

species (40.7%) belonged to the genus *Acacia*. This indicated that charcoal production was one of the major factors responsible for cutting of trees within the Park. The family with the highest number of fuel wood was represented by Fabaceae with 20 species (20%), followed by Capparidaceae 7 (7%), Tiliaceae 6 (6%) and 21 families had only 1 species each. The most frequently reported plant parts used for firewood and both firewood and charcoal production were shrubs 47 species (47%), followed by trees 33 (33%) and Herbs 11 (11%).

Both preference ranking and paired comparison showed that *Acacia tortilis* is the most selected *Acacia* species within the park for charcoal

production (Tables 1 and 2). This might be due to its high abundance in the study area as well as preference for high quality charcoal in the local towns or other distant towns like Walenchit, Adama and Addis Ababa. As a result, charcoal producers more selectively use this species. According to informants, Park's administrators and scouts, *Acacia nilotica, Acacia prasinata, Acacia seyal* and *Acacia mellifera* may be compared in terms of the production of high quality charcoal with that of *Acacia tortilis*. For example, in the Rift Valley area, *Acacia seyal* is extensively used for charcoal production due to its softer wood (Zerihun and Mesfin, 1990). Again, Hussien (2004) revealed that *Acacia nilotica* stood

the 2nd rank for production of high quality charcoal in the study area. But, due to their scarcity in the study area charcoal producers are more inclined towards the harvesting of *Acacia tortilis*. Similar observation was also reported by Kebu et al. (2004) that *Acacia tortilis* was more commonly used for firewood and charcoal making instead of medicinal uses. Similarly, Makenya (2006) reported that *Acacia tortilis* was the best plant species for charcoal making in Tanzania. On the other hand, the paired comparisons showed that there was a deviation of R_1, R_3, R_4, R_6 and R_9 in their evaluation from the remaining key informants as they were older peoples, illiterate and charcoal producers and sellers. While, R_2, R_5, R_7, R_8 and

Table 3. The species similarity between the Afar and the Oromo Nations for fuel wood and the JCS in the study area

Total number of species	Total number of species reported			Jaccard's coefficient of similarity	Percentage similarity
	The Afar Nation	The Oromo Nation	Both Nations		
100	10	11	79	0.79	79

R_{10} were less experienced and less knowledgeable as compared to their counterparts due to literacy, younger ages and even older ages but different living professions (Table 2). This revealed that there was a difference in IK among key informants in evaluating each species for charcoal production due to their age, literacy, experience and profession. Such deviation could be attributed to the fact that older peoples, illiterate and charcoal producers and sellers are more experienced and knowledgeable as compared to younger ages, literate and even older ages, but different living professions. This is because most young informants particularly those who go to school were not interested to know and learn about plant resources from their parents. In general, socio-cultural factors such as age, whether literate or not and occupation affect the distribution of IK among individuals (Cotton, 1996).

In the present study, information from informants as well as field observation indicated that the commercial sale of firewood and charcoal from *Acacia* species and many other woody species is a common practice in the area. As a result, charcoal producers most frequently targeted by cutting the trees of *Acacia* species, which accounted the highest proportion and the least on shrubs. Hussien (2004) reported that trees are the most widely used life forms that accounted for about 69%. Widespread sale of firewood and charcoal in the area, in turn leads to threatening and gradual extinction of plant resources (Kebu et al., 2004). Widespread sale of

firewood and charcoal also have environmental problems in the Rift Valley area (Zerihun and Mesfin, 1990; Ensermu et al., 1992). Furthermore, the cutting of *Acacia* trees and other plant resources for charcoal production leads to not only the destruction of wildlife habitats but also the complete disappearance of species from the Park (Andeberhan, 1982). Again, charcoal producers are major cause of accidental and deliberate forest fire in the ANP. This is because as they set the fire for making charcoal, the fire suddenly escapes and damages a large area of woodland resources.

Fuel wood species use diversity

Some of the surveyed fuel wood species in the study area were found to have multi-purpose values in various ways such as forage/fodder species, medicine, food, material culture and miscellaneous uses. Out of the total recorded fuel wood species, about 33% of the species were found to have 4 distinct uses, 14% with 5 uses and 13% with 6 uses to the local people (Appendix 1).

Variation of indigenous knowledge between the Afar and the Oromo Nations

Research findings during data collection showed that 10 fuel wood species were reported by the Afar Nation, 11 species by the Oromo Nation and

79 species were common to both Nations (Table 3). The percentage similarity (about 79%) for the species, in turn, indicated that since the two groups situated almost in close geographical settings, there is a cultural diffusion and sharing of experiences and knowledge between them. Thus, they commonly utilize the same species.

Threats to fuel wood species and associated indigenous knowledge

Since the local peoples have an intimate relationship towards their natural environment, they are familiar with the threats for fuel wood species. Therefore, during both group and individual discussions, key informants identified seven major threats by priority ranking in the ANP. Consequently, overgrazing/over browsing, followed by deforestation scored 21.7 and 19.9%, respectively. This was followed by deforestation for different purposes (for example, firewood and charcoal production, building and construction, household furniture and farm tools, fencing materials and others), human settlement and agricultural expansion as well as forest fire in their respective orders.

CONCLUSION AND RECOMMENDATIONS

The indigenous people mainly depend on fuel wood species for their house hold consumptions and income generation. As a result, high diversity of species is recorded even if human-induced and natural factors influence the species. Overgrazing and deforestation are the major threats within the Park. Planting of fuel wood species around homesteads and farmlands for household supply and sale; conservation of threatened species such as *Acacia tortilis*, *Acacia nilotica*, *Acacia prasinata* and *Acacia negrii* to alleviate from threat; practical application of cooking with wood saving stoves and improved charcoal stove as well as enhancing the utilization of invasive alien plant species for various purposes to control further spread are recommended.

ACKNOWLEDGEMENTS

The main author would like to acknowledge the Horn of Africa Regional Environment Center and Network (HoA-REC/N), members of the Afar and the Oromo Nations, Awash-Fentale Wereda and Fentale Wereda Offices, all the staff members of National Herbarium and ANP, the Department of Biology and others which directly or indirectly offered their various supports.

REFERENCES

Abbiw DK (1990). Useful plants of Ghana: West African uses of wild and cultivated plants. Intermediate technology publications, London and the Royal Botanic Gardens, UK, p. 337.

Andeberhan K (1982). Wildlife management problems in Ethiopia. Walia: J. Ethiop. Wild. Nat. Hist. Soc., 8: 3-7.

Berihun G, Solomon Y (2005). A study on abundance, group size and composition of Soemmering's Gazelle (*Gazella soemmerringii*) in Awash National Park and Alledeghi wildlife reserve, Ethiopia. *SINET: Ethiop. J. Sci.*, 28(2): 161-170.

Cotton CM (1996). Ethnobotany: Principles and Applications. John Willey and Sons Ltd. Chichester, p. 424.

Cunningham AB (2001). Applied Ethnobotany: People, Wild plant Use and Conservation. People and Plants Conservation Manuals. Earthscan publications Ltd., London and Sterling, VA, p. 300.

EMA (1992). Map of Awash National Park. Addis Ababa, Ethiopia. Ethiopian Mapping Authority. Four Sheets of Paper 1:50,000.

Ensermu K, Sebsebe D, Zerihun W, Edwards S (1992). Some threatened endemic plants of Ethiopia. In: The Status of Some Plant Resources in Parts of Tropical Africa, pp. 35-55, (Edwards, S. and Zemede Asfaw, eds). Published by NAPRECA, Addis Ababa University, Addis Ababa. Botany 2000: East and Central Africa. NAPRECA Monograph Series p. 2.

Hill AF (1952). Economic Botany: A Textbook of Useful Plants and Plant Products. 2nd ed. McGraw-Hill Book Company, Inc., New York, p. 560.

Hussien A (2004). Traditional use, management and conservation of useful plants in dryland parts of North Shoa Zone of the Amhara National Region: An Ethnobotanical Approach. M. Sc. Thesis, AAU, p. 174.

Jacobs MJ, Schloeder CA (1993). The Awash National Park Management Plan, 1993-1997. Nyzs-The Wildlife Conservation Society International and The Ethiopian Wildlife Conservation Organization. Ministry of Natural Resources Development and Environmental Protection, Addis Ababa, Ethiopia. NYZS-The Wildlife Conservation Society, New York, USA and the Ethiopian Wildlife Conservation Organization, Addis Ababa, Ethiopia, p. 285.

Kebu B, Ensermu K, Zemede A (2004). Indigenous medicinal plant utilization, management and threats in Fentalle area, Eastern Shewa, Ethiopia. Biological Society of Ethiopia. Ethiop. J. Biol. Sci., 3(1): 37-58.

Kent M, Coker P (1992). Vegetation Description and Analysis: A Practical Approach. CRC press, Boca Raton Ann Arbor and Belhaven press, London. p. 354.

Kochhar SL (1998). Economic Botany in the Tropics. 2nd ed. Macmillan India Limited, New Delhi, p. 604.

Makenya CA (2006). Wild plant use by local communities within the 'Kwakuchinja' wildlife corridor in Tarangire-Manyara Ecosystem, Tanzania. In: Drylands Ecosystems: Challenges and Opportunities for Sustainable Natural Resources Management. Proceedings of the Regional Workshop held at Hotel Impala, Arusha, Tanzania, June 7-9, 2006, (Nikundiwe, A. M. and Kabigumila, J. D. L., eds). pp. 85-94.

Martin GJ (1995). Ethnobotany: A Method of Manual. Chapman & Hall, London, p. 268.

Sebsebe D, Friis I (2009). Natural vegetation of the Flora area. In: Flora of Ethiopia and Eritrea.Vol. 8. General part and Index to (Hedberg, I., Friis, I. and Persson, E., eds). National Herbarium, Biology Department, Science Faculty, Addis Ababa University, Addis Ababa and Department of systematic Botany, Uppsala University, Uppsala, Sweden. 1-7: 27-32,

Silayo DA, Katani JZ, Maliondo SMS, Tarimo MCT (2008). Forest plantation for biofuels to serve natural forest resources. Research and development for sustainable management of semiarid Miombo woodlands in East Africa. Working Papers of the Finnish Forest Research Institute, 98: 115 -124.

White F (1983). The vegetation of Africa. A Descriptive memoir to accompany the UNESCO/AFTAT/UNSO. Vegetation Map of Africa, Paris, UNESCO, Natural Resource Research, p. 356.

Zerihun W, Mesfin T (1990). The status of the vegetation in the Lakes Region of the Rift Valley of Ethiopia and the possibilities of its recovery. SINET: Ethiop. J. Sci., 13(2): 97-120.

Appendix 1. List of plant species used for fuel wood (firewood and/or charcoal) in the study area (ANP).

Scientific name	Family name	Ha	Vernacular name	U	MUC	CN
Acacia brevispica Harms	Fabaceae	S	HAMARESA (Or)	B	F, Fu, Fo, Mc	TB204
Acacia dolichocephala Harms	Fabaceae	T	-	B	F, Fu, Mc, Mi	TB058
Acacia mellifera (Vahl) Benth.	Fabaceae	S	MAKA'ARTO/MA'EGHERTO (Af); SEPENE GURO (Or)	B	F, Fu, Mc	TB011
Acacia negrii Pic.- Serm.	Fabaceae	S	KESEL-E (-TO) (Af); KESELE (Or)	B	F, Fu, Mc	TB051
Acacia nilotica (L.) Willd. ex Del.	Fabaceae	T	KESEL-E (-TO) (Af); BURKUKE (Or)	B	F, Fu, M, Fo, Mc, Mi	TB003
Acacia oerfota (Forssk.) Schweinf.	Fabaceae	S	GOMERTO (Af); AJO (Or)	B	F, Fu, M, Fo, Mc, Mi	TB045
Acacia prasinata Hunde	Fabaceae	T	SEKEKTO (Af); DODOTI (Or)	B	F, Fu, Mc	TB201
Acacia robusta Burch.	Fabaceae	T	GERE'INITO (Af); WANIGAYO (Or)	B	F, Fu, Mc	TB180
Acacia senegal (L.) Willd.	Fabaceae	S	ADADO (Af); SEPENSA DIMA/SEPESA (Or)	B	F, Fu, M, Fo, Mc, Mi	TB001
Acacia seyal Del.	Fabaceae	T	ADIGENTO/MAKANI (Af); WACHU (Or)	B	F, Fu, Fo, Mc	TB190
Acacia tortilis (Forssk.) Hayne	Fabaceae	T	E'IBITO/BEHBEY (Af); DEDECHA (Or)	B	F, Fu, M, Fo, Mc, Mi	TB026
Acalypha fruticosa Forssk. *	Euphorbiaceae	S	CHIRI (Or)	B	F, Fu, Mc, Mi	TB090
Azadirachta indica A. Juss.	Meliaceae	T	MIMI HARA (Af); KININI (Or)	F	F, Fu, M, Mi	TB207
Balanites aegyptiaca (L.) Del.	Balanitaceae	T	UDAYITO/ALA'ITO (Af); BEDENO (Or)	B	F, Fu, M, Fo, Mc, Mi	TB004
Barleria acanthoides Vahl **	Acanthaceae	S	BALIWERANITI (Or)	F	F, Fu	TB123
Berchemia discolor (Klotzsch) Hemsl.	Rhamnaceae	T	YEYEBITO (Af); JEJEBA (Or)	B	F, Fu, Fo, Mc	TB191
Bidens biternata (Lour.) Merr. & Sherff	Asteraceae	H	CHOGOGE (Or)	F	Fu, M	TB143
Boscia salicifolia Oliv.	Capparidaceae	S	-	F	F, Fu, Fo, Mc	TB107
Boswellia papyrifera (Del.) Hochst.	Burseraceae	T	LUBATEN (Af); MUKE ITANA (Or)	F	F, Fu, Fo, Mc	TB099
Cadaba farinosa Forssk.	Capparidaceae	S	FURA (-YITO)/NUMHELE (Af); KELIKNATIONHA (Or)	F	F, Fu, M, Fo, Mc, Mi	TB031
Cadaba rotundifolia Forssk.	Capparidaceae	S	ANAGALI/ADENGELITA (Af); ARANGILLE (Or)	F	F, Fu, Mi	TB052
Calotropis procera (Ait.) Ait.f.	Asclepiadaceae	S	GELE'ATO/GHULA'ENTO (Af); FELFELA ADAL (Or)	F	Fu, M, Mc, Mi	TB012
Capparis cartilaginea Decne.	Capparidaceae	S	DELENSISA (Or)	F	Fu, M, Fo	TB117
Capparis tomentosa Lam.	Capparidaceae	S	HARENIGEMA (Or)	F	F, Fu, M, Fo, Mc, Mi	TB084
Caucanthus auriculatus (Radlk.) Niedenzu	Malpighiaceae	C	GALE (Or)	F	F, Fu	TB005
Ceiba pentandra (L.) Gaertn. *	Bombacaceae	T	FERENJI TUTI (Af)	B	F, Fu, Fo, Mc, Mi	TB083
Celtis toka (Forssk.) Hepper & Wood	Ulmaceae	T	GUDIBI'ATO (Af); METEKOMA (Or)	F	F, Fu, Fo, Mc	TB192
Cleome brachycarpa Vahl ex DC. *	Capparidaceae	H	-	F	F, Fu	TB013
Combretum molle R. Br. ex G. Don	Combretaceae	T	WE'IBA'ITO (Af); RUKESA (Or)	B	F, Fu, Mc, Mi	TB197
Commiphora erythraea (Ehrenb.) Engl.	Burseraceae	T	YEYEBITO (Af); CHELANKA (Or)	F	F, Fu, Mc	TB187
Commiphora habessinica (Berg) Engl.	Burseraceae	S	HEDAYITO (Af); HAMESA (Or)	F	F, Fu, Fo, Mc, Mi	TB086
Cordia monoica Roxb.	Boraginaceae	S	MINE GURE/SUBULA (Af); MEDERO (Or)	F	F, Fu, Fo, Mc	TB025
Crotalaria incana L.	Fabaceae	H	IJISISE (Or)	F	Fu, M, Mc, Mi	TB101
Cryptostegia grandiflora Roxb. ex R. Br.	Asclepiadaceae	S	HALI MERO (Af); HAKONKOL (Or)	F	Fu, Mc, Mi	TB018
Cynanchum gerrardii (Harv.) Liede **	Asclepiadaceae	C	HIDA KELA/MUKA JINI (Or)	F	Fu, M	TB188
Cynanchum hastifolium N.E.Br. **	Asclepiadaceae	C	SARA KORPO (Or)	F	Fu	TB106

Appendix 1. Continues.

Species	Family	Habit	Local name		Uses	Code
Dalbergia lactea Vatke	Fabaceae	S	Dilo Lelafa (Or)	F	F, Fu, Mc	TB198
Dichrostachys cinerea (L.) Wight & Arn.	Fabaceae	S	Jirme (Or)	B	F, Fu, Mc, Mi	TB009
Dicoma tomentosa Cass. *	Asteraceae	H	-	F	F, Fu	TB131
Dobera glabra (Forssk.) Poir. **	Salvadoraceae	T	Ghersa (Af); Ade (Or)	F	F, Fu, Fo, Mc, Mi	TB195
Echinops pappii Chiov. **	Asteraceae	S	Bilingi (Or)	F	Fu	TB006
Ehretia cymosa Thonn.	Boraginaceae	S	Mine Gure (Af); Ulaga (Or)	F	F, Fu, M, Fo, Mc	TB097
Eucalyptus globulus Labill.	Myrtaceae	T	Bahir Zafi (Af & Or)	F	Fu, M, Mc	TB210
Euclea racemosa Murr. subsp. *schimperi* (A. DC.) White **	Ebenaceae	S	Miessa (Or)	F	F, Fu, Fo, Mc	TB200
Euphorbia polyacantha Boiss.	Euphorbiaceae	S	-	F	Fu, Mi	TB142
Euphorbia tirucalli L.	Euphorbiaceae	T	Lihaso (Af); Ano (Or)	F	F, Fu, M, Mi	TB046
Ficus sycomorus L.	Moraceae	T	Subula (Af); Oda (Or)	F	Fu, M, Fo, Mc, Mi	TB043
Ficus vasta Forssk.	Moraceae	T	Mara'ito (Af); Kiltu (Or)	F	Fu, M, Fo, Mc, Mi	TB047
Flacourtia indica (Burm.f.) Merr. **	Flacourtiaceae	S	-	F	F, Fu, Fo, Mc	TB014
Forsskaolea viridis Webb. **	Urticaceae	H		F	F, Fu	TB030
Grewia bicolor Juss.	Tiliaceae	S	Adibi'ato (Af); Haroresa (Or)	B	F, Fu, Fo, Mc	TB185
Grewia ferruginea Hochst. ex A. Rich.	Tiliaceae	S	Adibi'ato/Fo (Af); Haroresa (Or)	F	F, Fu, Fo, Mc	TB186
Grewia schweinfurthii Burret	Tiliaceae	S	Adibi'ato (Af); Mudhe Gure (Or)	B	F, Fu, Fo, Mc	TB181
Grewia tenax (Forssk.) Fiori	Tiliaceae	S	Hedayito/Huda/Mine Gure (Af); Deka Tuntuna (Or)	F	F, Fu, Fo, Mc	TB038
Grewia velutina (Forssk.) Vahl	Tiliaceae	S	Adibi'ato (Af); Haroresa (Or)	F	F, Fu, Fo, Mc	TB054
Grewia villosa Willd.	Tiliaceae	S	Garwa (Af); Ogomdi (Or)	F	F, Fu, M, Fo, Mc, Mi	TB024
Hagenia abyssinica (Bruce) J.F. Gmel.	Rosaceae	T	Begala (Af); Heto (Or)	F	Fu, M, Mc	TB209
Hibiscus micranthus L. f.	Malvaceae	H	Akilehena (Af)	F	F, Fu, Fo, Mc	TB145
Hippocratea africana (Willd.) Loes.	Celastraceae	C	Misi (Af); Tero (Or)	F	F, Fu, Mc	TB196
Hyphaene thebaica (L.) Mart. *	Arecaceae	T	Unga/Gara'ito (Af); Meti (Or)	F	F, Fu, Fo, Mc	TB128
Indigofera arrecta Hochst. ex A. Rich.	Fabaceae	H	Herchumen (Or)	F	Fu, M, Fo, Mc	TB008
Indigofera coerulea Roxb.	Fabaceae	H	Adulala (Or)	F	Fu, M, Fo	TB120
Ipomoea carnea Jacq. *	Convolvulaceae	S	Biroli (Af)	F	Fu	TB015
Jatropha curcas L.	Euphorbiaceae	S	Abete Bulk (Or)	F	Fu, Mi	TB102
Kleinia odora (Forssk.) DC.	Asteraceae	S	Luko (Or)	F	F, Fu, Mc, Mi	TB206
Lantana camara L.	Verbenaceae	S	Baduwa Hara (Af); Midan Dubra (Or)	F	F, Fu, Fo, Mc, Mi	TB050
Maerua angolensis DC. *	Capparidaceae	S	Dunibayito/Sekileli'a (Af)	F	F, Fu, Fo, Mc	TB136
Manilkara butugi Chiov.	Sapotaceae	T	Butuye (Af); Butuui (Or)	F	F, Fu, Fo, Mc, Mi	TB194
Melhania ovata (Cav.) Spreng.	Sterculiaceae	S	Hambokito (Af)	F	F	TB033
Moringa stenopetala (Bak.f.) Cuf.	Moringaceae	T	-	F	Fu, Fo, Mc	TB096

Appendix 1. Continues.

Scientific name	Family	Habit	Local names	MUC	Uses	Coll. No.
Morus mesozygia Stapf	Moraceae	S	-	F	F, Fu, Mc	TB105
Ocimum spicatum Deflers	Lamiaceae	S	MISE (Af); KORCHA MICHI (Or)	F	Fu, M, Fo	TB139
Ocimum stirbeyi Schweinf. & Volk. *	Lamiaceae	S	BIRITELI (Af)	F	Fu, Mc	TB021
Olea europaea L. subsp. *cuspidata* (Wall. ex G.Don) Cif.	Oleaceae	T	WEYIBO (Af); EJERSA (Or)	B	F, Fu, M, Mc, Mi	TB132
Oncocalyx schimperi (A. Rich.) M. Gilbert	Loranthaceae	P	HATOTE (Af); DERTU HARORESA (Or)	F	Fu, M, Mc	TB028
Otostegia integrifolia Benth.	Lamiaceae	S	TUNGIT (Af); TINJITI (Or)	F	Fu, M, Mi	TB215
Parkinsonia aculeata L.	Fabaceae	S	-	B	F, Fu, Fo, Mc	TB057
Plicosepalus sagittifolius (Engl.) Danser	Loranthaceae	P	HATOTE (Af); DERTU DEDACHA (Or)	F	Fu, M, Mc	TB087
Premna resinosa (Hochst.) Schauer	Lamiaceae	S	BOBA'O (Af); URGESA (Or)	F	F, Fu, Fo, Mi	TB035
Prosopis juliflora (Sw.) DC. *	Fabaceae	S	WEYANE (Af & Or)	B	F, Fu, M, Fo, Mc, Mi	TB020
Rhus vulgaris Meikle **	Anacardiaceae	S	DEBOBESA (Or)	B	F, Fu, M, Fo, Mc	TB103
Ricinus communis L.	Euphorbiaceae		SHERBETI (Af); KOBO (Or)	F	Fu, M, Fo, Mc	TB048
Salvadora persica L.	Salvadoraceae	S	HADAYITO/DADAHO (Af); ADE (Or)	F	F, Fu, M, Fo, Mc	TB039
Schinus molle L.	Anacardiaceae	T	KUNDO BERBERE (Or)	B	Fu, M, Mc, Mi	TB114
Sesbania sesban (L.) Merr.	Fabaceae	S	ENCHINI/HARCHA (Or)	F	F, Fu, Mc	TB135
Sida schimperiana Hochst. ex A. Rich.	Malvaceae	S	WELAYINEBA (Af); KORCHA IJOLE (Or)	F	F, Fu, M, Mc, Mi	TB094
Solanum coagulans Forssk.	Solanaceae	H	-	F	F, Fu	TB104
Solanum hastifolium Hochst. ex Dunal	Solanaceae	S	BURI BOLO (Or)	F	F, Fu	TB088
Solanum incanum L.	Solanaceae	S	AMBOKO ASO (Af); HIDI LONI (Or)	F	F, Fu, Mc, Mi	TB016
Steganotaenia araliacea Hochst. ex A. Rich. **	Apiaceae	T	-	F	Fu	TB055
Sterculia africana (Lour.) Fiori	Sterculiaceae	T	KERERI (Or)	F	F, Fu, Fo, Mc, Mi	TB022
Tamarindus indica L. *	Fabaceae	T	SEGENTU (Af); ROKA (Or)	B	F, Fu, M, Fo, Mc, Mi	TB126
Tamarix nilotica (Ehrenb.) Bunge	Tamaricaceae	S	SEGE'ITO (Af)	F	F, Fu	TB202
Terminalia brownii Fresen.	Combretaceae	T	WE'IBA'ITO (Af); BIR'ENSA (Or)	B	F, Fu, M, Mc, Mi	TB098
Trilepisium madagascariense DC. **	Moraceae	T	SELAWETA (Or)	F	Fu, Mc	TB189
Vernonia cinerascens Sch. Bip.	Asteraceae	S	FILE NEME'A (Af); KERTATUME (Or)	F	F, Fu, Mc	TB049
Vernonia uncinata Oliv. & Hiern	Asteraceae	H	FILE NEME'A (Af)	F	Fu	TB110
Ximenia americana L.	Olacaceae	T	HUDHA (Or)	B	F, Fu, Fo, Mc	TB199
Ziziphus mucronata Willd.	Rhamnaceae	T	KUSIR-A (-TO) (Af); KURKURA HADO (Or)	B	F, Fu, M, Fo, Mc, Mi	TB056
Ziziphus spina-christi (L.) Desf.	Rhamnaceae	T	KUSIR-A (-TO) (Af); KURKURA (Or)	B	F, Fu, M, Fo, Mc, Mi	TB041

Major use category (MUC) [F = Forage/fodder; Fu = Fuel wood; M = Medicine; Fo = Food; Mc = Material culture; Mi = Miscellaneous uses]; Uses (U) [F - Species used for firewood; B - Species used for both firewood and charcoal]; Part (s) used in all species of fuel wood = Stem and branches]; Habit (Ha) [C-Climber; H-Herb; P - Semi-parasitic; S-Shrub; T-Tree]; [* Species used for fuel wood without asterisks are reported by the Oromo Nation; Species used for fuel wood reported by the Afar Nation; ** Species used for fuel wood reported by both Nations]; [Collection No. (CN)].

7

Gill net selectivity in the White Nile fisheries, Khartoum State, Sudan

Mohammed M. O.[1]* and Ali M. E.[2]

[1]Environment and Natural Resources Research Institute, the National Centre for Research, Khartoum, Sudan.
[2]Fisheries Research Centre, Agricultural Research Corporation, Khartoum, Sudan.

Gill-net selectivity is a size and type of fish that caught depending upon specific mesh size of a used gill-net. In this study, four different mesh sizes (4, 6, 8 and 15 cm) were used in al-Kalakla Fishery (KF) and Jabel Awlia Dam Fishery (JADF) in the White Nile for studying their selectivity. The results of a catch per unit effort (CPUE) study showed that meshes 4 and 6 cm had higher productivity in fishing in KF than that in JADF; whereas, meshes 8 and 15 cm caught more fishes in JADF than in KF for the entire year. In KF, autumn season was the best for fishing with meshes 6 and 8 cm. In JADF, summer and autumn were the best for fishing by meshes 4 and 6 cm. The gill-nets of the study were highly selective for fish species according to their body size. Five fishes of small sizes (body depth at around 4 cm) as Tilapias (*Oreochromis niloticus*: Linnaeus, 1758; *Tilapia zilli*: Gervais, 1848; *Sarotherodon galilaeus*: Linnaeus, 1758), Kas (*Hydrocynus forskali*: Cuvier, 1819; *Hydrocynus vittatus*: Castelnau, 1861; *Hydrocynus brevis:* Günther, 1864), Nile Perch (*Lates niloticus:* Linnaeus, 1758), Dabis (*Labeo vulgaris*: Heckel, 1847) and Bayad (*Bagrus bayad:* Forsskål, 1775) were caught intensively by mesh of 4 cm during all seasons in both KF and JADF; whereas, meshes of 8 and 15 cm caught occasionally bigger fishes ranged between 8 to15 cm as Nile Perch and Bayad. The selectivity of these nets reflected over-fishing caused in the White Nile.

Key words: Fisheries, freshwater fish, River Nile, fishing gear, fishing nets.

INTRODUCTION

Gill-net selectivity affect fish stocks through (a) decrease in yield due to intensive fishing, (b) direction of catching effort toward specific species and (c) over-fishing due to long-term practice (Potter and Pawson, 1991). Nowadays, there is a considerable interest in improving the selectivity of fishing gear to reduce the capture and discarding of unwanted sizes and species of both freshwater and marine fishes (Steward, 2001).

Twine thickness was correlated with mesh size to establish a comparison (Ali, 1975; Ali and Abu-Gideiri, 1984). The physical properties of synthetic nets may change gradually with use. The abrasion of the nets removes the transparency from the material so; the accumulation of knots and roughened surface of the yarn makes the netting more visible in water. Monofilament nets are more effective than multifilament nets, because they are less visible (Potter and Pawson, 1991).

In the Sudan, with its large bodies of water, there are no regulations governing fishing gear (Ali and Abu-Gideiri, 1984). As a result of the intensive fishing with a wide range of mesh sizes, especially the smaller ones which are used irregularly have led to over-fishing of the White Nile fishery (Kawai, 1994; Mohammed, 2004). The commonest types of fishing nets used in Khartoum State fisheries are fixed nets, drift nets, trammel nets and beach nets. Most of them are made of monofilament; while, few are made of multi-filament (Mohammed, 2004; 2008).

This study was done to determine which mesh size of gill-net causes the greatest depletion of fish stocks and to study the effect of water level and its transparency on gill-net efficiency.

*Corresponding author. E-mail: moh.9808@gmail.com.

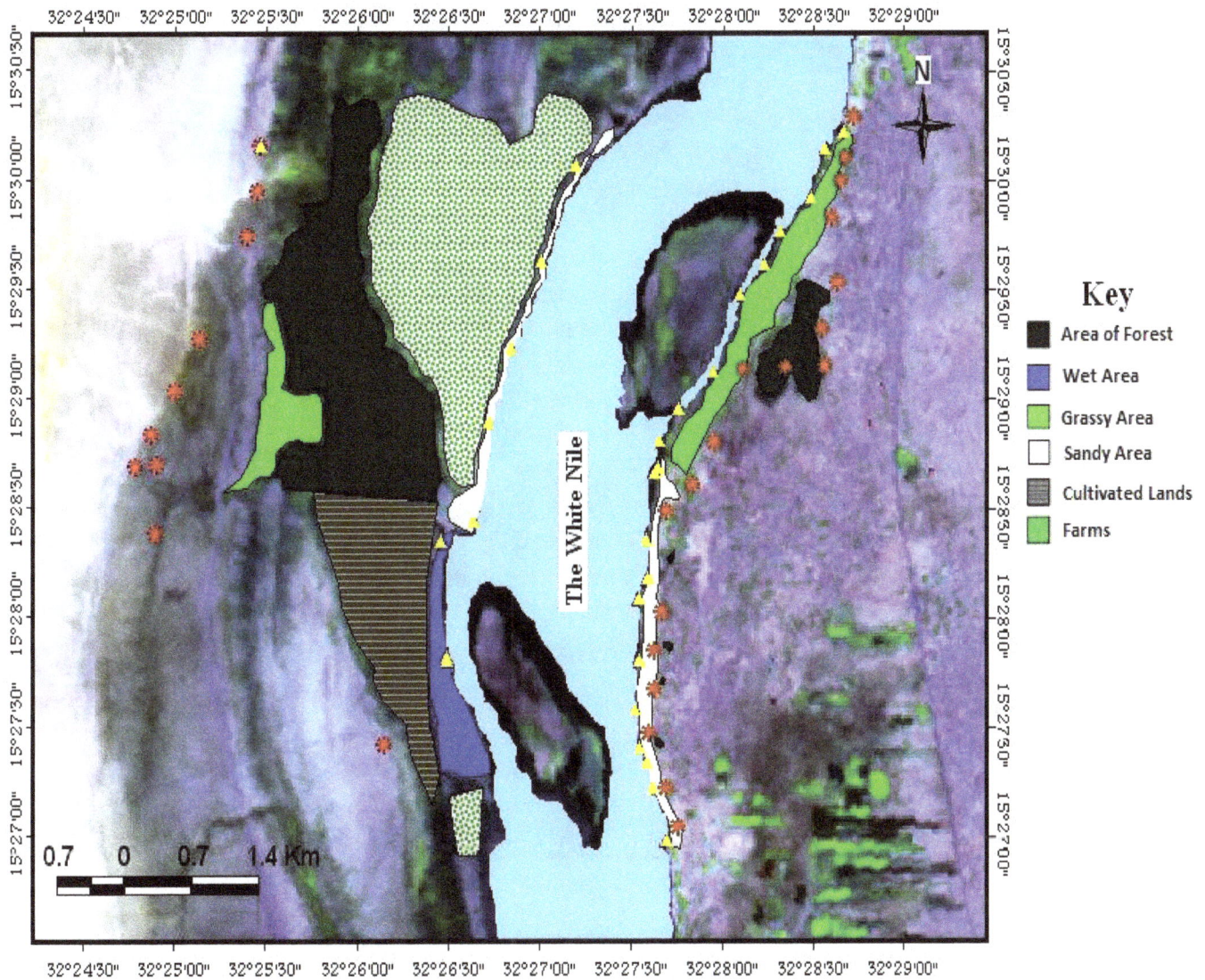

Figure 1. A satellite map of Al-Kalakla Fishery (KF) in the White Nile, Khartoum State (cited by Mohammed 2006).

MATERIALS AND METHODS

This study was conducted in JADF (dam reservoir and five km downstream from the dam's barrier) and KF in the White Nile (Figures 1 and 2, respectively) described by Mohammed (2006: 34 to 35, Figure 6a and b). Four different mesh sizes of multi-filament gill-nets were used. Their mesh sizes were 4, 6, 8, and 15 cm made of appropriate twines of numbers 2/210, 3/210, 6/210, and 12/210 respectively. Each net measured 50 m in length and 1.5 m and half in depth.

The gill-nets were set for three days every mid-month for the entire year of the study (October, 2005/2006). They were checked to record the catches twice a day at sunrise and sunset as described by Mohammed (2006).

The caught fishes were identified according to Sandon (1950). They were measured using a dissecting board and weighed using a Salter Balance (25 kg × 100 g) for large specimens and a One-Pan Balance for small specimens. A Secchi Disk was used to determine transparency of water and a Bulb thermometer of 50 × 1°C for water temperature. Catch per unit effort (CPUE) was computed

using a following formula:

$$CPUE = \frac{\text{Gross weight of catch of fish (g)}}{\text{Surface area of the net (m}^2\text{) long of time of fishing (h)}}$$

RESULTS

Catch per unit effort (CPUE)

Results of CPUE showed that mesh 4 cm netted with twine 6/210 was the most effective particularly during the night in both JADF and KF. JADF was the best fishery to use mesh 15 cm netted with twine 12/210. Generally, KF appeared to be a more productive fishery than JADF (Table 1).

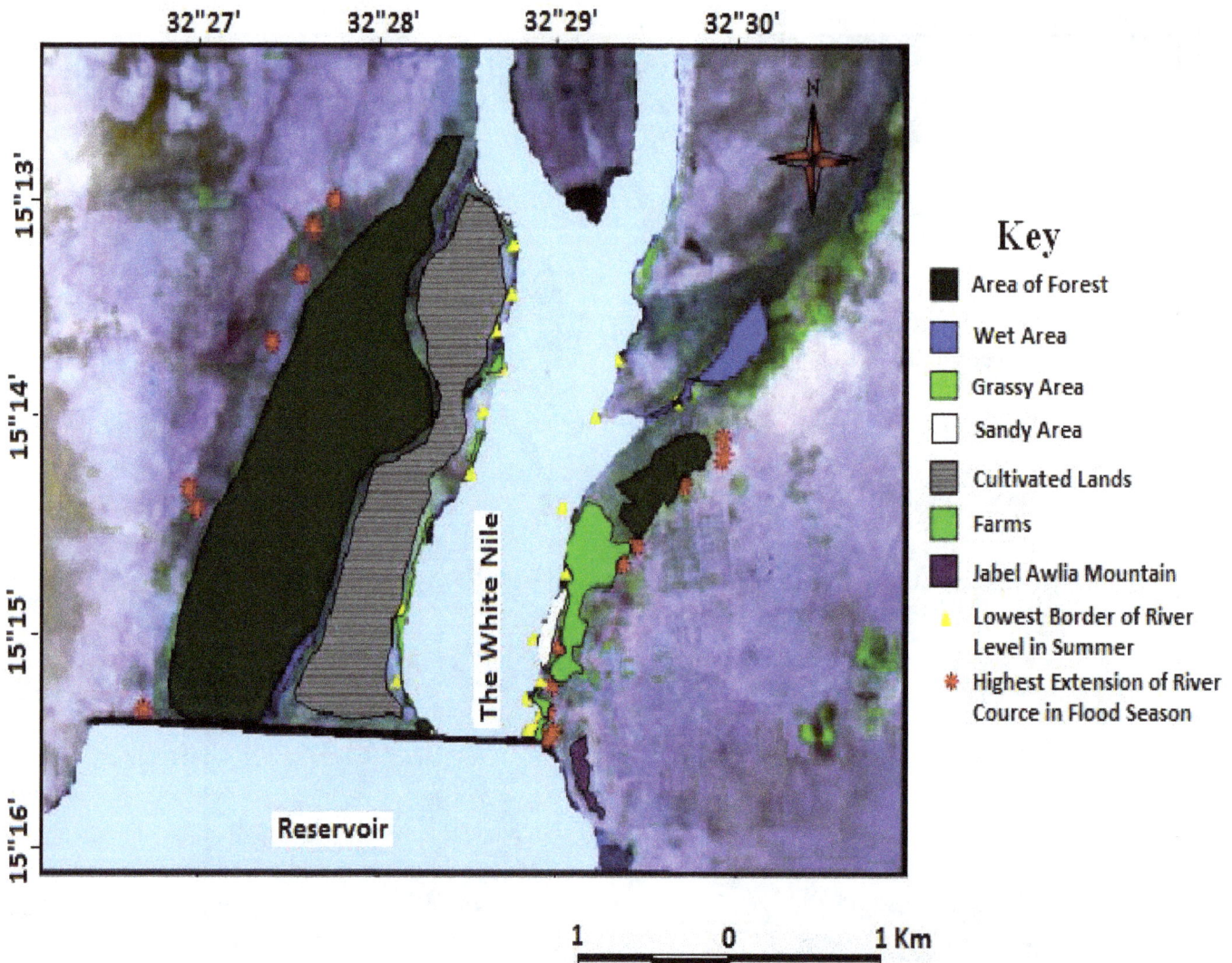

Figure 2. A satellite map of Jabel Awlia Dam Fisheries (JADF) in the White Nile, Khartoum State (cited by Mohammed 2006).

CPUE per day and night

In JADF, mesh 4 cm was most productive in fishing at night for the entire year, but daytime productivity in December, February, and June was also high. Mesh 8 cm showed high productivity at night for the entire year, but mesh 6 and 15 cm were ineffective. In KF, mesh 4 cm showed high productivity when used nightly for the entire year, but mesh 8 and mesh 15 cm were ineffective in fishing both night and day (Table 2).

Selectivity of each mesh/water characteristics

Mesh 4 cm showed the highest selectivity towards five small Nile fishes (body depth at around 4 cm) in JADF and KF during all seasons of year, followed by mesh 6 cm. Those fishes were Tilapias (O. niloticus, T. zilli, and

S. galilaeus), Kas (H. forskalii, H. vittatus & H. brevis), Nile Perch (L. niloticus), Dabis (L. vulgaris), and Bayad (B. bayad). Mesh 15cm showed selectivity toward bigger fishes (body depth at around 15 cm) and also reflected over-fishing of the White Nile (Table 3).

KF recorded lower temperatures during the entire year than the JADF. On the other hand, JADF showed highest transparency during the entire year (Table 4).

DISCUSSION AND RECOMMENDATION

The results of CPUE showed that KF was more productive than JADF, probably, due to three reasons: one, KF is a breeding region for herbivorous fishes such as Tilapias, so that, all members of chain food are available, Two, there are fewer fishermen in KF and three, the JADF is rocky and consequently prohibits

Table 1. Catch per unit effort (CPUE) of multi-filament gill-nets used in Jabel Awlia Dam Fishery (JADF) and Al-Kalakla Fishery (KF) in the White Nile, Khartoum State (2005/2006).

CPUE	Mesh size							
	Mesh 4		Mesh 6		Mesh 8		Mesh 15	
	Day	Night	Day	Night	Day	Night	Day	Night
KF	4.19	5.61	0.05	1.44	0.52	0.80	0.0	0.48
JADF	0.92	2.24	0.09	1.06	0.61	1.36	0.0	1.28

Table 2. Seasonal fluctuations of catch per unit effort (CPUE) of the different mesh sizes in Jabel Awlia Dam Fishery (JADF) and Al-Kalakla Fishery (KF) in the White Nile, Khartoum State (2005/2006).

Month mesh	Fishery	October, 2005	December, 2005	February, 2006	April, 2006	June, 2006	August, 2006
Mesh 4	KF	$D^{\dagger}$ 5.87	1.14	0.15	0.66	0.42	0.08
		N^{*} 6.06	2.40	0.08	2.04	0.51	0.06
	JADF	$D^{\dagger}$ 0.27	0.30	0.49	0.16	0.38	0.23
		N^{*} 1.28	0.38	0.38	1.15	0.59	0.69
Mesh 6	KF	$D^{\dagger}$ 0.0	0.0	0.0	0.0	0.09	0.0
		N^{*} 2.88	0.0	0.0	0.0	0.0	0.0
	JADF	$D^{\dagger}$ 0.0	0.76	0.11	0.0	0.0	0.0
		N^{*} 1.63	0.0	0.0	0.48	0.0	0.0
Mesh 8	KF	$D^{\dagger}$ 0.76	0.0	0.0	0.0	0.28	0.0
		N^{*} 0.32	0.0	0.0	0.0	0.64	0.32
	DSF	$D^{\dagger}$ 0.0	0.0	0.45	0.38	0.19	0.19
		N^{*} 0.80	0.80	0.16	0.32	0.32	0.32
Mesh 15	KF	$D^{\dagger}$ 0.0	0.0	0.0	0.0	0.0	0.0
		N^{*} 0.96	0.0	0.0	0.0	0.0	0.0
	DSF	$D^{\dagger}$ 0.0	0.0	0.0	0.0	0.0	0.0
		N^{*} 2.56	0.0	0.0	0.0	0.0	0.0

† = day, *=night.

Table 3. Seasonal selectivity of the different mesh sizes in Jabel Awlia Dam Fishery (JADF) and Al-Kalakla Fishery (KF) in the White Nile, Khartoum State (2005/2006).

Species	Mesh 4cm			Mesh 6cm			Mesh 8cm			Mesh 15		
	Aut† (%)	Win* (%)	Sum¥ (%)	Aut. (%)	Win. (%)	Sum. (%)	Aut. (%)	Win. (%)	Sum. (%)	Aut. (%)	Win. (%)	Sum. (%)
Tilapias	0.32	4.81	0.32	-	-	-	-	-	-	-	-	-
Hydrocynus sp.	35.58	3.21	1.60	0.32	0.32	-	-	0.32	-	-	-	-
Lates niloticus	2.56	2.88	1.60	0.96	-	-	064	-	-	-	-	-
Labeo vulgaris	2.88	6.73	4.81	-	-	-	0.32	-	-	-	-	-
Bagrus Bayad	2.56	1.28	0.32	-	-	-	1.28	-	-	-	-	-
Others	8.01	1.92	9.94	0.9	-	-	0.32	-	-	0.32	-	-

†= Autumn, *=Night, ¥= Summer.

Table 4. Mean temperature (°C) and transparency (m) readings in Jabel Awlia Dam Fishery (JADF) and Al-Kalakla Fishery (KF) in the White Nile, Khartoum State (2005/2006).

Fishery		Max. value (night)			Min. value (night)			Max. value (day)			Min. Value (day)		
		Aut†	Win*	Sum¥	Aut.	Win.	Sum.	Aut.	Win.	Sum.	Aut.	Win.	Sum.
°C	KF	26	20	24	25	17	23	26	19	24	25	16	22
	JADF	28	23	24	25	18	23	26	23	26	25	17	24
m.	KF	32	42	31	32	39	30	33	41	32	30	39	30
	JADF	28	46	45	24	35	30	23	43	44	47	36	38

†= Autumn *= Winter ¥= Summer

fishing by seine nets or even drifting nets. This observation confirms the findings of Mohammed (2006).

Mesh 4cm and mesh 6 cm showed higher CPUE than other meshes used in KF, particularly during autumn and winter. This may be for many reasons: flooded water is considered one of the most effective factors, because rates of migrated fishes from both the upper parts of the White Nile and Blue Nile enrich KF. This fishery is rich with breeding ground areas, so fishing by finer meshes was more productive.

At the end of flood season, the high level of the river begins to decline and fishes accumulate from the flooded banks to the main river. This observation also confirms the findings of Mohammed (2006).

Mesh 4 cm appeared effective in JADF due to the diverse environment which is suitable for many kinds of fishes through the year. Fishermen used both fixed gill-nets and drift gill-nets. Winter and summer seasons had low levels of water and high transparency, so that, fishing by mesh 8 cm was the best because it could easily be

determined which place contained the largest number of fish. This agrees with the findings of Ali and Abu-Gideiri (1984) and Mohammed (2006). The CPUE of mesh 4 cm in both fisheries indicated that this mesh size select towards small fish sizes; whereas mesh 15 cm selects toward big fishes. The small sizes of most fishes and fewer numbers of large fishes indicate an over-fishing phenomenon. This agrees with results of Kawai (1994) and Mohammed (2006).

As discussed above, the results of this study suggests a prohibition of mesh sizes 4 and 6 cm in both fisheries during all the year except in flooded water due to the migration of fishes from Blue Nile and upper parts of the White Nile.

REFERENCES

Ali ME (1975). Gill net selectivity and fish population in Jebel Aulia Reservoir. M.Sc. (Qualifying) dissertation, University of Khartoum, p. 155.

Ali ME, Abu-Gideiri YB (1984). Gill net selectivity in Lake Nubia fisheries. J. Hydrobiologia, 110: 315-317.

Kawai MK (1994). Fishing effort and fishing intensity in the northern part of Jebel Aulia reservoir. M. Sc. dissertation, University of Khartoum, p. 120.

Mohammed MO (2008). A study on fishing gear and methods used in the White Nile, Khartoum state. Sudanese J. Standards Metrology, 2: 56- 62.

Mohammed MO (2006). Effects of gill-nets and fishers on fisheries of Al-Kalakla and Jabel Awlia Dam. M. Sc. Thesis, Sudan Academy of Sciences, Khartoum, p. 130.

Mohammed MO (2004). Studies on fishing gear, fish compositions & fishermen sector in the fishery of Khartoum State. B.Sc. (Hons.) dissertation, University of Khartoum, p. 45.

Potter ECE, Pawson MG (1991). Gill netting. Lab. Leafl., MAFF Direct. Fish. Res., Lowestoft, 69: 34.

Sandon H (1950). An illustrated guide to the freshwater fishes of the Sudan. 1st edn (17- 61) Sudan Notes and Records, McCorguodale Co. London.

Steward PAM (2001). A review of studies of fishing gear selectivity in the Mediterranean. Aberdeen, Scotland.

8

Wildlife use of Bharandabhar forest corridor: Between Chitwan National Park and Mahabharat foothills, Central Tarai, Nepal

Ram Chandra Kandel

Department of National Parks and Wildlife Conservation, P. O. Box 860, Kathmandu, Nepal.

Barandabhar forest is a wildlife corridor connecting Chitwan National Park and Mahabharat foothills in Nepal's Inner Tarai. Chitwan harbors the largest population of the great one-horned rhinoceros (*Rhinoceros unicornis* L.) in Nepal. Barandabhar forest serves as a highly potential alternative habitat to enable wildlife to move up to Mahabharat foothills mainly during the rainy season. The whole forest area was divided into four blocks from south to north and sampled plots along 1.5 km length transects spaced at intervals of 250 m apart for wildlife and disturbance signs. Wildlife signs were higher near the National Park (ANOVA, P < 0.001; Tukey's HSD, P <0.05). Disturbance signs were lower near the National Park and the Mahabharat Foothill forests and highest in the central part of the corridor. Wildlife signs were also affected by the distance of a sample plot from the edge of the corridor (ANOVA, P = 0.032), while disturbance signs were similar irrespective of the distance of a sample plot from the edge of the corridor (ANOVA, P = 0.56). The results illustrated that the central portion of the corridor near the East-West Highway is the weakest point in the corridor being flanked by the township of Bharatpur Municipality and the relocated village of Padampur.

Key words: Barandabhar, corridor, Chitwan, Tarai, Bishazaari, Padampur, highway.

INTRODUCTION

The lowland Tarai region of Nepal covers about 15% of the total area wherein Chitwan, the central stretch has been famous since time immemorial for its rhinoceros (*Rhinoceros unicornis* Linn.) population that harbored about 1000 animals until 1950 (Gee, 1959). Significant declines in the rhinoceros population were noticed due to the accelerated rate of habitat loss during 1950s when malaria was eradicated and degradation of forest was increased dramatically (Gee, 1963). In addition, several factors contributed to decline in animal populations and deteriorating habitats such as encroachment for human habitation and cultivation, habitat shrinkage, extensive poaching and other human induced activities that led to the fragmentation. With the suppression of malaria, settlement of Chitwan valley was initiated by both spontaneous immigration and planned government

resettlement in early 1960s. Chitwan forests, other than those in the National Park had all been converted to human settlements and farmland, as had most of the low-lying forests and remnant patches of forests along the mountain chains (Panwar, 1986). Consequently, Barandabhar forest between Chitwan National Park (CNP) and Mahabharat Mountain range remained the only forest strip connecting two different ecological systems. It serves to function as a wildlife corridor for some animals and alternative or seasonal/and temporal habitat for others ((Litvaitis et al., 1996). The East-West National Highway passes across the Barandabhar forest mid way in the corridor has been highly disturbed spot and is under the most severe human pressure. It is also the weakest link in the corridor due to the township of Bharatpur. This forest has been frequently utilized as a wildlife corridor by mega-herbivores like rhinoceros (*Rhinoceros unicornis*), carnivores like tigers (*Panthera tigris*) and leopards (*Panthera pardus*), reptiles like mugger crocodiles (*Crocodylus palustris*), waterfowls and

*Corresponding author. E-mail: rckandel01@yahoo.com.

wintering birds. It also serves as a refuge during the monsoon floods (Kandel, 2003).

Periodic floods occurring in the Rapti flood plain force rhinoceros to move towards the highlands through? these corridor forests (Kandel, 2003). The Barandabhar forest, the north-south narrow strip has a minimum width of about 1.5 km, an average width of 4 km and length of 20 km. It narrows down towards Mahabharat foothills and this area was chosen for studying use by rhino and other wildlife (Picture 1). This strip has been exploited by a variety of wildlife species as a day to day movement path to reach the foothills. The forest patch has been flanked by human habitation along either side and is being degraded and compressed as the anthropogenic pressure increased.

Collections of grass, lopping fodder trees for the livestock, cutting for firewood, livestock grazing are the prominent disturbance factors (Heinen and Kandel, 2006). The only conservation measure taken in this forest until a decade ago was management of a few patches of Community Forest by local communities. Rhinoceros, one of the prominent immigrants frequently moved up to foothills and a small population of around 15 to 20 resident rhinoceros had been found at the uppermost portion of this forest corridor till 1990s (Kandel, 2003). With the relocation of Padampur village to Saguntole at the north-eastern border of Barandabhar forest from the National Park area in late 1990s, this resident rhino population has vanished. This forest corridor is still used seasonally by more than 30 rhinos and 10 different tigers according to recent field studies during the monsoon in 2002 (Bishnu Lama's pers. comm.). This area was previously declared as Mahendra Mrig Kunja (Deer Park) in 1959 (Spillett and Tamang, 1966) before the CNP was designated as the National Park.

There has not been any intensive study of wildlife use in Barandabhar forest to date. The aim of this study is to measure the habitat use by wildlife in this forest so that it would be possible to predict weak links within it which could have significant implications for wildlife management in the future. It is important that the animal presence, distribution and habitat use by animals from CNP in this forest strip are monitored regularly. This study will serve as a baseline for the habitat use by wild animals in the Barandabhar forest. Such short-term studies taken up at regular intervals will update and augment the information used for management.

MATERIALS AND METHODS

Rhinoceros and other animal's use of the Barandabhar forest corridor

To assess rhinoceros and other wildlife use, human and other

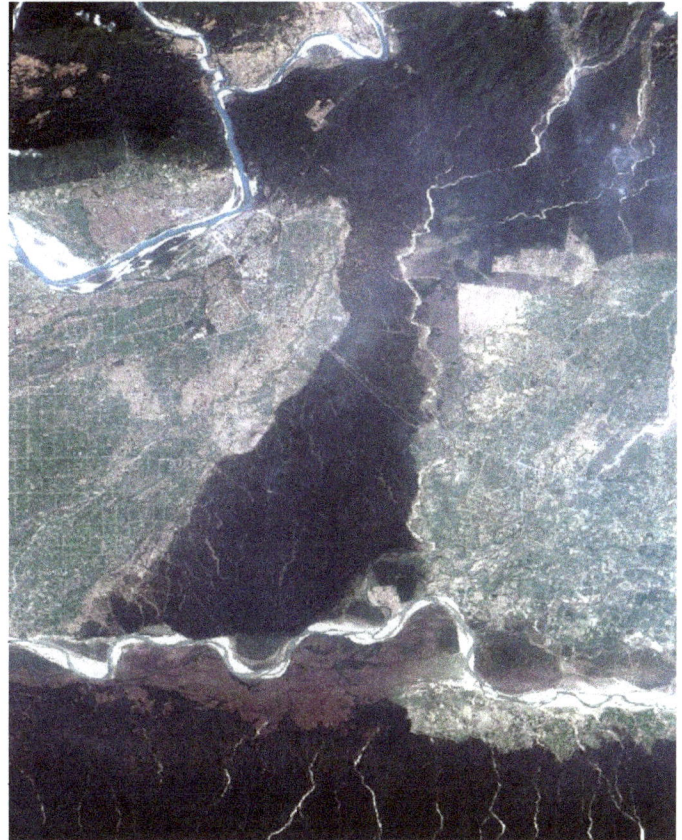

Picture 1. Aerial photograph of the study area.

pressures in the corridor, transects were spaced 1.5 km apart parallel to the corridor. Presence of animals and habitat usage by them was assessed with systematic sample plots of size 2 × 25 m at every 250 m interval on each transect. Indirect evidence like footprints, dung piles, feeding signs, scratch/markings and digging signs were recorded. The distance along the transects was measured with a laser range finder and standard plots were calibrated with nylon rope. A 'silva compass' was used to determine the bearing of the transects and a handheld GPS unit was used to determine the location of the sample plots in the corridor. The wild animal signs which were recorded from the sample plots were rhinoceros, tiger, leopard, sambar (*Cervus unicolor*), spotted deer (*Axixi axix*), barking deer (*Muntiacus muntajak*), hog deer (*A. porcinus*), wild boar (*Sus scrofa*), Indian hare (*Lepus nigricollis*), Small Indian civet (*Viverricula indica*), sloth bear (*Melursus ursinus*), yellow throated martin (*Martes flavigula*), Indian grey mongoose (*Herpestes edwardsi*) and jungle cat (*Felis chaus*). Each sign of wildlife presence and human disturbance was given a weight based on the rarity of the species and perceived magnitude of the disturbance (Tables 1 and 2; and Pictures 2 and 3).

All weighted values of animal presence and disturbances were summed up separately. Distances of each sample plot (inbetween the middle of the two sample plots) from the National Park border and nearest edge of the forest along the farm land were measured in a geographic information system (GIS) domain. Each plot was categorized into one of four groups based on its distance from the

Table 1. Weightage given to the signs of the animals recorded in the Barandabhar Forest.

Animals recorded	Weighted values
Rhino	10
Tiger	10
Leopard	8
Sambar	6
Spotted deer	7
Barking deer	5
Hog deer	3
Wild boar	5
Indian hare	3
Civet	3
Sloth beer	8
Martin	4
Mangoose	2
Jungle cat	5

Table 2. Weightage given to the signs of the disturbances recorded in the Barandabhar Forest.

Signs of disturbances	Weighted values
Presence of cattle	5
Grass cutting	5
Thatch collection	5
Lopping signs	7
Human trails	6
Wood cutting	8
Debarking	7
Cart trail	10

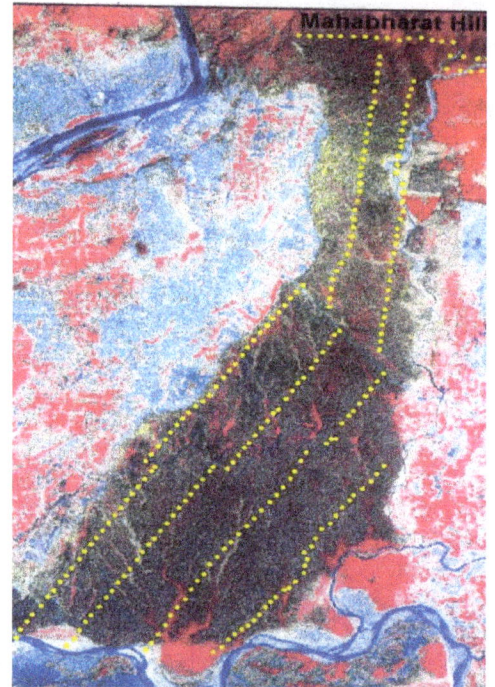

Picture 2. Sample plots along the transects in the study area.

Picture 3. Researcher showing the transect marking in the study area.

National Park (in 4-km distance groups). The categories were created based on the intensity of disturbances due considering the distance from the core habitat area for the animals and from the farmland. One-way ANOVA followed by Tukey's honestly significant difference and analysis of covarience (Sokal and Rolph, 1995) were applied to understand the effect of increasing distance from National Park on wildlife use of the corridor (fixed effects) and increasing distance from the edge of the corridor (covariate) using SPSS (1999) software. The Ancova repeats the results of the one-way Anova for the distance categories and is unnecessary (ok).

RESULTS

Use of Barandabhar forest corridor by rhinoceros and other animals

Signs of sloth bear, leopard, tiger, rhinoceros and five species of ungulates were recorded in the Barandabhar forest (Figure 1). Spotted deer and wild boar signs were seen most commonly in plots. Large carnivore signs were rare but were clearly observed while rhinoceros signs were fairly common. The most abundant disturbance signs were in the form of livestock grazing and grass cutting for thatch and fodder (Figure 2). The occurrence

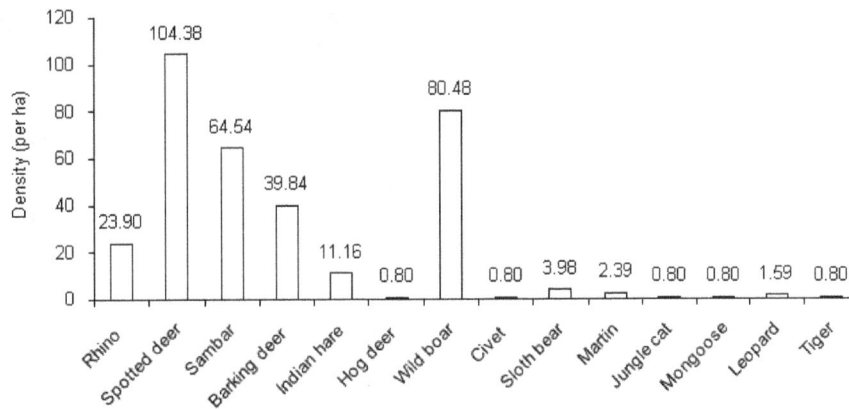

Figure 1. Density of wildlife signs in Barandabhar Forest corridor (per ha).

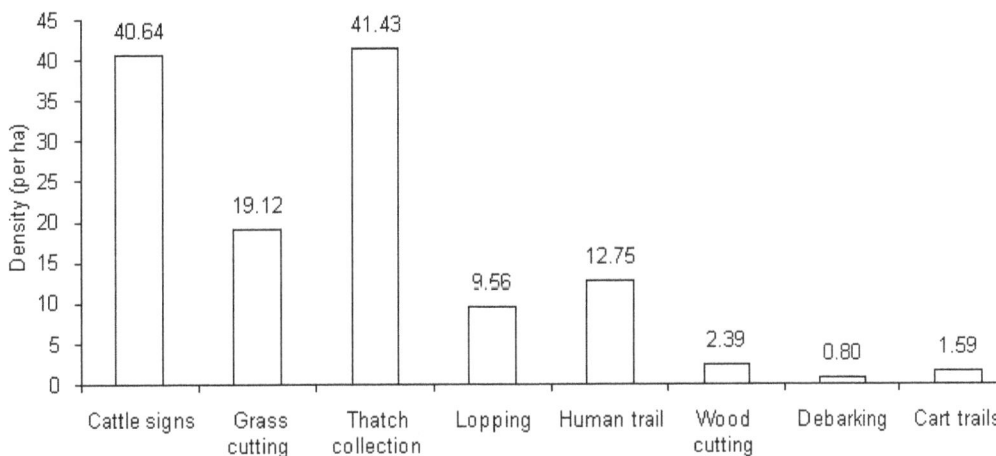

Figure 2. Density of disturbances in Barandabhar Forest corridor (per ha).

of wildlife signs were different between the four different distance categories (F = 30.277; df = 3, 247; P≤ 0.001). A post hoc Turkey's honestly significant test showed that wildlife signs from 0 to 12 km from the National Park boundary were similar, while wildlife signs between 13 to 16 km distances from National Park were lower than the first group (Tukey's test P≤ 0.001). The signs of disturbance were also different between different distance categories (F = 3.59; df = 3, 247; P≤ 0.014). The post hoc Turkey's test showed that the level of disturbance is significant in the area of third category (mid section of the corridor) at 12 km distance from the National Park boundary. The results of the analysis of covariance for the wildlife sign index and disturbance index as the dependent variables, distance from National Park in 4 categories as fixed effect and distance from the nearest edge of the corridor showed that wildlife signs decreased with distance from National Park, while disturbance was maximum in the middle part of the corridor (it was less near the National Park and near the Mahabharat foothill forests).

The covariate (distance from the nearest edge) was not significant for either wildlife signs or disturbance index suggesting that the corridor was too narrow to have a

Picture 4. Habitat used by wild and domestic animals in the Baradabhar Forest.

core of less disturbed area with more wildlife usage.

DISCUSSION

Function of Barandabhar forest as a wildlife corridor

At one time, the Tarai forests were connected to the Himalayan foothill forests and reported to be used by wildlife to move freely between these different ecological systems in this forest to exploit seasonal food resources and exchange genetic material (Patridge, 1978). The role of a corridor is vital to allow outward dispersal of individuals from a source population, reduce local concentrations and overcome environmental stochasticity (Noss, 1987). It also allows ecological separation and resource partitioning between different animal species. Most wildlife corridors have lost their effectiveness as a result of biotic pressures and developmental activities (Dendy, 1987). Corridors allow access to refuges and sources of recolonization in the event of floods, fire or diseases. Neglected corridors may detract from the value of wildlife use and increase the interface between animals and humans (Johnsingh et al., 1990). Corridors provide both temporary and permanent habitat for the animals. Moreover, the forest in these areas is essential to reduce habitat fragmentation (Dendy, 1987). The status of wildlife populations likely to be vulnerable to sporadic diseases, flood and resource crunch could be

ensured by migration to alternative habitats. In the case of Chitwan, the foothills of Mahabharat range linked by the Baradabhar forest corridor could provide an alternative habitat in contrast to the view of Yonjan (1996). Periodic floods occurring in the Rapti flood plain forces rhinoceros towards the highlands of the Baradabhar forest corridor (Kandel and Upadhyay, 2010, Picture 4).

Gee (1959) has emphasized the importance of the inclusion of present study area and at least southern portion of it in the current CNP area. But at the time of establishment of CNP, policy makers and other concerned people might not have been conscious of the importance of Barandabhar forest as a corridor for the wildlife, although it was previously designated as Mahendra Mrig Kunja in 1959, covering parts of Barandabhar forest areas. The idea of using habitat corridors for elephant movements was introduced first in Sri Lanka in 1959 (Anon, 1959). Ironically, developmental projects like east-west national highway, irrigation canal, and other east-west crossing roads have fragmented wildlife habitats, though the concept of corridor was being developed elsewhere during that time. Yes it is related to the case of Sri Lanka.

Though, there were plenty of wildlife signs recorded during this study, the use of the corridor was likely to be higher during the monsoon as also reported by Gee (1959, 1963)? (No, it was studied during winter). This fact was highlighted by interviews of local people living on the fringes who stressed that rhinos intensively use the corridor once more water becomes available and the flood plains of Rapti river are waterlogged (Dinerstein, 1991). This alternative habitat could be of critical use for rhinoceros (Kandel and Jhala, 2008) and one should never underestimate its importance by only considering the overall habitat preference throughout the year (Kandel and Jhala, 2008). Rhinoceros have different habitat preferences for different activities (Kandel and Upadhyay, 2010) and require a landscape that has a mosaic of different habitats to meet all their various requirements. The result of one-way ANOVA shows that the amount of wildlife signs is high up to 12 km from the National Park border and the level of disturbance is low. The habitat mid way in the corridor to Mahabharat Mountain beyond the East-West National Highway had the highest disturbance index and is under the most severe human pressure. It is the weakest link in the corridor due to the township of Bharatpur on one side and the recently rehabilitated village of Padampur (which had been previously enclaved in the CNP) on the other side.

If these pressures continue to increase unchecked then the impact would be severed and could no longer be used by wildlife. Floods occur almost every year compelling the rhinoceros and other animals to take refuge in the nearby Barandabhar forest. It is conceivable that heavy flooding in a single season or within a short span of years) could kill many animals by washing them away (Kandel, 2003). This corridor forest area offers a refuge for the species like rhinoceros and deer that depend on the grasslands. An average daily traffic of over 5082 vehicles has been recorded on the East-West National Highway (bus/truck: 2262, car/jeep/tractor: 551, motorbikes: 1040, micro/mini bus: 600, autorikshaw: 64, cycles: 535, rickshaw: 20, bullock cart: 10 per day) (Kandel, 2003). It is apparent that the absences of under-storey cover as well as ground cover mainly along the highway is a serious impediment for animal movements. Animals have been reported to be killed while crossing the highway. Likewise the chances of migrating animals being exposed to poachers are very high. If entire stretch of the corridor can not be preserved intact, the animals may soon be left with no refuge in the Barandabhar Forest. Forest above the East-west Highway should gradually be conserved to maintain a corridor (Poudel et al., 1998). It is critical to conserve the potential and alternative habitats of corridors for endangered animals like rhinoceros and tigers according to the higher protection status as of other areas. Implementation of the proposed management plan and regular co-ordination and communication between agencies such as local governments, conservation-based organizations and other close stakeholders could lead to better prospects for the larger mammals that depend on humans to ensure their survival. There needs to be political will and sustainable participation of the local communities (Sharma, 1989). This should be achieved through establishing alternatives for fuel wood and fodder to support the livelihoods of people inhabiting the area. Strict enforcement practices through community-based planning should also be implemented. A committee comprising social workers, local community representatives, local administration and political units, Forest Department and the park authority should be formed and given full responsibility for managing the forest corridor immediately. Such an approach has been extremely successful in managing the buffer zone of Chitwan which consists primarily of community owned forestlands (Sharma, 1989). I also suggest the implementation of strict control over vehicle movement on the highway and to construct an overhead bridge in the long term. Similarly, grassland sites and water holes at uniform distances should be maintained for use by rhinoceros during seasonal movements. Otherwise the meta-populations of endangered animals like rhinoceros and tiger would be threatened.

Management of short grasslands for productivity (Kandel and Jhala, 2010) and reducing livestock pressure (Kandel, 2003) in the Baradabhar forest is very important. Intermittent intensive studies in Barandabhar forest corridor could be very useful to understand the seasonal patterns? of tropical ecosystems (Kandel and Jhala, 2008) and to update and augment the baseline that would help to inform better management in the future.

REFERENCES

Anon (1959). Report of the Committee on Preservation of Wildlife. Session paper XIX, Government Press, Ceylon.

Dendy (1987). The value of corridor and design features of same and small patches of habitats. In Nature Conservation, the role of remnants of native vegetation (D.A. Saunders, G.W. Arnold, A.A. Burbidge and A.J.M. Hopkins. eds.) Survey Beatty, Sydney, Australia, pp. 357-9.

Dinerstein E (1991). Demography and habitat use by Greater One-horned Rhinoceros in Nepal. J. Wildlife Manage., 55(3): 401-411

Johnsingh AJT, Prasad SN, GoyaL SP (1990). Conservation Status of the Chilla-Motichur corridor for elephant movemnts in Rajaji-Corbett National Park area, India. Biological Conserv., 52:125-138.

Gee EP (1959). Report on a Survey of the Rhinoceros area of Nepal. Oryx, 5: 51-85.

Gee EP (1963). Report on brief Survey of the Wildlife Resources of Nepal, including the Rhinoceros, March 1963. Oryx 7(2,3):67-76

Heinen JT, Kandel RC (2006). Threats to a small population: a census and conservation recommendations for wild buffalo Bubalus arnee in Nepal. Oryx, July 2006. 40(3): 8.

Kandel RC (2003). Aspects of foraging, activity,habitat use and demography of Rhinoceros (Rhinoceros unicornis Linn.) in Chitwan National Park, Nepal. Masters Dissertation,Wildlife Institute of India, Dehradun. Kandel RC, Jhala YV (2008). Demographic structure, activity patterns, habitat use and food habits of Rhinoceros unicornis in Chitwan National Park, Nepal. J. Bombay Hist. Soc. January-April 2008. 105(1): 9.

Kandel RC, Upadhyay GP (2010). Food habits, habitat use and behavioral activities of Rhinoceros unicornis in Chitwan National

Park, Nepal. In Biodiversity Conservation Efforts in Nepal. Special issue published on the occasion of 15th Wildlife Week 2067. Department of National Parks and Wildlife Conservation, Babarmahal, Kathmandu, April 2010. pp.13.

Litvaitis JA, Titus K, Anderson EM (1996). Measuring Vertebrate Use of Terrestrial habitats. In Research and Management techniques for Wildlife and Habitats (Bookhout, T.A. ed.) The Wildlife Society, Bethesda, USA. pp. 254-274.

Noss (1987). Corridors in real landscape: a reply to Simberloff and Cox. Conserv. Biol., 1:159-164.

PANWAR HS (1986). Forest cover mapping for planning tiger corridors between Kanha and Bandhavgarh-a proposed project. In Proc. of the seminar-cum-workshop on wildlife habitat evaluation using Remote Sensing Techniques, Oct. 22-23. IIIRS and WII, Dehradun.

PATRIDGE L (1978). Habitat selection. In Behavioral ecology: an evolutionary approach. (Krebs, J.R. and N.B. Davies eds.). Sinaur Associates, Sunderland, USA. pp. 351-376

Poudel NP, GP Upadhyay, BB Thapa, RC Kandel, TB Khatrl, UR Regmi (1998). Proposed Master Plan for the Conservatiion and Management of Barandabhar Forest in Chitwan, Nepal. Submitted to DNPWC, Nepal. Unpub. p. 38.

Sharma UR, (1989). An overview of park-people interactions in Royal Chitwan National Park, Nepal. Landscape Urban Plann., 19:133-144.

Sokal RR, FJ Rohlf (1995). Biometry. W.H. Freeman and Co. New York, USA.

Spillett JJ, KM Tamang (1966). Wild Life Conservation in Nepal. JBNHS, 1966 December, 63(3).

SPSS Inc (1999). SPSS Statistical Package, Version 8.0. Chicago, USA.

Yonjan P (1996). Eastern Himalaya, corridors and climate change. Habitat Himalaya, 3(1).

9

Grazing management systems and their effects on savanna ecosystem dynamics: A review

Olaotswe Ernest Kgosikoma[1,2], Witness Mojeremane[3] and Barbra A. Harvie[1]

[1]University of Edinburgh, Crew Building, West Mains Road, EH9 3JN, Scotland.
[2]Department of Agricultural Research, Private Bag 0033, Gaborone, Botswana.
[3]Botswana College of Agriculture, Private Bag 0027, Gaborone, Botswana.

The savanna ecosystems support livestock production and livelihood of pastoral communities. The degradation of savanna ecosystems due to overgrazing had lead government commercializing communal grazing land to privately owned ranches. However, grazing policies in Southern Africa had recently been debated, and yet there are few studies comparing grazing management systems on ecological system. This article provides an overview of current knowledge on effect of grazing management systems on savanna ecosystems. Ranching and communal grazing do not necessarily affect soil, herbaceous and woody vegetation differently. Thus current management systems do not promote sustainability of savanna ecosystems and there is a need for further research and participation of local communities on addressing land degradation.

Key words: Fencing policy, rangeland degradation, sustainability, livestock production.

INTRODUCTION

Globally, managed grazing lands comprises the largest land use (Liebig et al., 2006) estimated to cover about 25% of Earth's land surface (Asner et al., 2004). The extensive area covered by rangelands makes them an essential resource for maintaining biodiversity (O' Connor, 2005) and a source of livelihood, especially for rural communities (Eriksen and Watson, 2009a; Muhumuza and Byarugaba, 2009). They are utilised for livestock production which has continually played a significant role in the economic development of rural communities world-wide. It is estimated that approximately 76% of Botswana's total land surface area is used for grazing by both domestic and wild animals (Asner et al., 2004). Thus, grazing land, especially communal rangelands, are being degraded due to overgrazing which threaten their sustainability (Vetter, 2005; Darkoh, 2009) and rural communities whose livelihoods depends on livestock rearing.

The savanna ecosystems are highly dynamic, characterized by erratic rainfall and high rate of vegetation dynamics (Herlocker, 1999; Dahdough-Guebas et al., 2002), soil nutrient levels, fire and herbivory (Sharpe, 1992). But, livestock management systems can exert a considerable change on the diversity, composition, structure, and development of native plant communities (Popolizio et al., 1994; Vavra et al., 2007) in rangelands. Most savannas are degraded and dominated by unpalatable and annual herbaceous plant species and encroached by woody plants (van Vegten, 1984; Abule et al., 2005). The changes in the composition of plant species in savanna ecosystems has a significant influence on the sustainability of livestock production (Sankaran et al., 2005). Proper understanding of effects of grazing management systems on savanna ecosystem dynamics is therefore essential in maintaining productivity and biodiversity (Sternberg et al., 2000;

Mohammed and Bekele, 2010). The objective of this paper is to synthesis existing knowledge on grazing management systems and their effects on savanna ecosystem.

GRAZING MANAGEMENT SYSTEMS AND RELATED POLICY

Grazing management systems refer to all production systems that are used to exploit the rangeland through grazing. Savanna systems have been used for many centuries extensively for grazing livestock, with communal grazing land and commercial ranching being the dominant land use management systems (Rohde et al., 2006; Masike and Urich, 2008; Terefa, 2011). In southern Africa, the communal grazing rangelands are located in tribal land and shared by all pastoral farmers (Rohde et al., 2006; Masike and Urich, 2008). The communal grazing system used to be mainly dominated by subsistence farmers, but has shifted towards a cash economy (Wigley et al., 2010) in recent years.

Livestock management in communal grazing land is mainly influenced by local ecological knowledge (Smet and Ward, 2005). Mixed livestock, dominated by cattle and goats, is the traditional practices in the communal grazing system (Wigley et al., 2010). Livestock herding used to be a key part of managing communal rangelands which was based on mobility, splitting and dispersing livestock over the landscape during the wet and dry seasons (Oba et al., 2000), to ensure limited dry concentrated continuous grazing (Kioko et al., 2012). However, with recent developments in some developing countries (for example, Botswana), herding is no longer a common livestock management practice (Reed et al., 2008), instead livestock is allowed to continuously and selectively graze without any control (Parsons et al., 1997) around water sources such as boreholes. Borehole rights provided by land authorities such as Land Boards in Botswana indirectly give the borehole owners control of grazing resources around the water points.

The communal rangelands are used for grazing throughout the year (Oba et al., 2001) by pastoral communities without paying any levy associated with livestock grazing (Weimer, 1977). Consequently, several studies have suggested that they are poorly managed and degraded (Ellis and Swift, 1988; Abel, 1997; Dougill et al., 1999; Hendricks et al., 2007). As illustrated by "Tragedy of Commons" (Hardin, 1968), it is argued that each pastoralist find it profitable to increase his herd size in communal rangeland. However, as pastoralists increases their herd size, the livestock density increases to exceed the rangeland's carrying capacity resulting in overgrazing and land degradation (van Vegten, 1984). Several countries have introduced policies and laws in response to land degradation due to poor management of communal rangelands. In Botswana, the government introduced commercial ranches through the Tribal

Grazing Land Policy (TGLP) (Botswana Government, 1975). The communal grazing land was demarcated into ranches owned by individuals or a group of farmers who paid a levy for exclusive use of fenced ranches (Dougill et al., 1999). According to the TGLP (1995) and Tsimako (1991), the TGLP served the following purposes:

a) To control overgrazing and reduce land degradation through improved rangeland management such as rotational grazing and optimal stocking rates in commercial ranches and shifting large herds of livestock from already overstocked communal lands.
b) To improve livestock productivity and farmers' income through better management practices (for example, controlled breeding and early weaning).
c) To secure interest of the poor (social equity) by reserving the communal grazing land for small scale farmers and have reserve land for future generations.

The management of commercial ranches is based on a range-succession model, whereby the goal of management aligns stocking rates to the carrying capacity of ranches (Mphinyane et al., 2008). Commercial ranches are characterised by rotational grazing which, consist of alternating periods of use and rest, to promote vegetation growth. To facilitate rotational grazing, several paddocks are demarcated within ranches to spread the livestock grazing intensity uniformly across the rangeland. The main focus in commercial ranches is on cattle rearing (Smet and Ward, 2005). Ranches have been promoted as a sustainable livestock management system by policies in Southern Africa (TGLP, 1975; Tsimako, 1991; Rohde et al., 2006), although they have high costs associated with fencing, drilling and water reticulation (Motlopi, 2006). Commercial livestock ranches are suitable for farmers with adequate financial resources. Their establishment has marginalised poor pastoral farmers because the total area for animal production in communal grazing land has declined dramatically in recent years (Eriksen and Watson, 2009b). This is further exacerbated by dual grazing rights which allow farmers allocated ranches to continue grazing their livestock in communal rangelands (Thomas and Sporton, 1997).

Several studies have criticised the conversion of communal rangelands to commercial livestock ranches (Abel, 1997; Rohde et al., 2006). Abel (1997) indicated that fencing policy is based on wrong assumptions and subsequently has failed to reduce rangeland degradation (Dahlberg, 2000). The productivity and sustainability of communal rangelands and their contribution to the livestock industry is also being underestimated (Abel, 1997). The communal grazing system is suitable for arid-land ecosystems because it is adapted to rainfall variability and spatial heterogeneity through opportunistic management such as mobility (Westoby et al., 1989). The TGLP had a provision protecting the interests of the poor pastoral farmers through reserve areas and removal of farmers with large herd from the communal rangeland

(Weimer, 1977). However, it was not clearly documented how this was to be achieved, given the fact that communal rangelands continue to shrink as more ranches are being demarcated.

The appropriateness of commercial livestock ranches in dry savannas has been debated in recent years (Ellis and Swift, 1988; Dahlberg, 2000; Rohde et al., 2006). Processes in savanna ecosystems are primarily influenced by rainfall variability and thus suited to traditional practices in communal land (Ellis and Swift, 1988; Westoby et al., 1989). Yet, few studies have been conducted to compare the effects of livestock grazing management systems of savanna ecosystems dynamics (Dahlberg, 2000; Smet and Ward, 2005; Tefera et al., 2010). Most of these studies are site specific (Asner et al., 2004), despite the high variability of the savanna ecosystem. A broad-scale multivariate analysis of relationships between diversity, environmental variables and management systems are required to understand savanna ecosystem dynamics (van der Heijden and Phillips, 2009). Analysis of interactions between natural factors (for example, rainfall and soil type) and anthropogenic drivers (Scholes and Archer, 1997) may improve our understanding of how a particular factor influences vegetation conditions (Groffman et al., 2007) in arid-land ecosystem.

EFFECT OF LIVESTOCK GRAZING MANAGEMENT SYSTEMS ON SOIL HEALTH

Soils with good physical and chemical characteristics are essential in maintaining productivity in terrestrial ecosystems and driving processes that maintain environmental quality (Moussa et al., 2008) and sustainability (Hopmans et al., 2005; Liebig et al., 2006). The biological, physical, and chemical properties of soil can be modified by livestock grazing. It has been demonstrated that intensive livestock grazing profoundly affects soils as it increases soil compaction, soil erosion and loss, decreases soil organic matter, affect nutrient cycling and reduces water infiltration (Kauffman and Kruger, 1984; Stephenson and Veigel, 1987; Fleischer, 1994; Belsky and Blumenthal, 1997; Ingram et al., 2008). Livestock grazing cause disturbances to surface soils and can influence savanna ecosystem productivity and fertility by altering the soil physical and chemical properties and thus cause land degradation (Neff et al., 2005; Liebig et al., 2006).

Livestock grazing compacts soil particularly under high grazing intensity (Fleischer, 1994; Kauffman and Krueger, 1984; Robertson, 1996; Asner et al., 2004; Fatunbi and Dube, 2008). Most studies have reported significant increases in bulk density in grazing land, especially in finer textured soils and in the soil surface layers (Warren et al., 1986; Abdel-Magid et al., 1987; Steffens et al., 2008; du Toit et al., 2009), caused by hoof traffic of livestock (Walker and Desanker, 2004). Compaction is

directly related to soil productivity (Liebig et al., 2006) because it reduces water and air movement into and through the soil, and therefore reduces water availability to plant roots, restricts and reduces soil microorganisms, reduces soil nutrient availability and increase soil surface runoff and soil erosion (Fleischer, 1994; Kauffman and Krueger, 1984; Robertson, 1996; Asner et al., 2004; Fatunbi and Dube, 2008).

The soil surface erosion has profound effects on soil productivity and the ecosystem function because microorganisms, organic matter, soil fauna and roots are all concentrated in the surface soil (Brandy and Weil, 2007). Research has shown that soil erosion increases with livestock grazing intensity (Bari et al., 1995; Belsky and Blumenthal, 1997). Studies conducted in grazing land and ungrazed enclosures have reported significantly higher sediments production rates in many plant communities under grazing land (Bohn and Buckhouse, 1985; Pluhar et al., 1987), and production was observed to be significantly related to grazing intensity (Beeskow et al., 1995; Bari et al., 1995; Warren et al., 1986).

Heavy grazing reduces vegetation cover and limit organic matter inputs into the soil, and subsequently affect soil structure stability, resistance to rainfall impact, infiltration rate and soil microbial activity (Roose and Barthes, 2001; Snyman and du Preez, 2005). Overgrazing caused by livestock grazing reduces plant biomass accumulation and cause a shift in plant species composition (Owen-Smith, 1999; Klumpp et al., 2009) by replacing highly palatable grass species with their unpalatable counterparts (Owen-Smith, 1999). The shift in species composition can affect soil fertility (Scholes, 1990) because of changes in root biomass (Klumpp et al., 2009) and quality of organic matter, and decrease soil's capacity to sequester carbon (Lal, 2002; Northup et al., 2005; Savadogo et al., 2007; Klumpp et al., 2009). Research has shown that soil nutrient depletion reduces the primary production of rangelands (Girmay et al., 2008) which in turn affect their carrying capacity.

Effects of livestock grazing management systems on soil quality are poorly understood. Kgosikoma (2011) showed that livestock grazing management systems do not affect soil properties in savanna ecosystems differently. The study observed that soil texture, bulk density and pH did not differ between management systems despite differences in grazing intensities, though significant differences may occur occasionally between and within the study sites. Another comparative study conducted by Tefera et al. (2007) found no significant effect of livestock grazing management on soil texture, bulk density and pH. Warren et al. (1986) found no relationship between soil bulk density and livestock stocking rates, which is in agreement with Kgosikoma (2011). A study conducted in South Africa also found no significant differences in soil chemical properties (for example, pH) between grazed communal land and ungrazed land (Moussa et al., 2008). The results of these studies could suggest that livestock management systems do not have significant effect on soil properties especially in sandy regions. In contrast, other studies have shown that livestock grazing management systems affect soil pH (Geissen et al., 2009). The differences between these studies could probably be attributed to differences in environmental conditions such as rainfall and soil or management practices.

EFFECTS OF LIVESTOCK GRAZING MANAGEMENT SYSTEMS ON HERBACEOUS VEGETATION

Savannas in Africa are largely exploited through livestock grazing (Scholes and Archer, 1997; Bilotta et al., 2007), and the grazing intensity (removal of plant biomass by livestock) influence their sustainability (Mphinyane et al., 2008). Most grasses in savanna ecosystems are fairly tolerant to grazing, however, prolonged intense grazing eventually lead to shift in species composition (Skarpe, 1992) and reduction in grass biomass especially when soil nutrients are depleted (van Auken, 2009). Overgrazing affect the botanical composition and species diversity by depressing the vigour and presence of dominant species, which then enables colonization by less competitive, but grazing tolerant plant species (Sternberg et al., 2000). Selective grazing of palatable herbaceous plants by livestock enhances the growth of annuals and unpalatable herbaceous plants as well as woody plants (Skarpe, 1992) resulting in the decline of palatable species (Fensham et al., 2010). Overgrazed rangelands are normally dominated by Increaser II species such as *Aristida congesta* (Trollope et al., 1989) which are indicators of poor rangeland conditions (Fatunbi and Dube, 2008), while Increaser I species dominate undergrazed rangelands or selectively utilised rangelands (Trollope, 1990; du Plessis et al., 1998).

Tefera et al., (2007) observed a higher density of palatable herbaceous plant in ranches than in communal grazed rangelands in Ethiopia. In contrast, Kgosikoma (2011) found no significant differences in palatable herbaceous plant cover between communal grazing land and ranches at two different sites in Botswana, which was supported by other studies in other African savannas (Parsons et al., 1997; Tefera et al., 2008b). This could suggest that rangeland vegetation does not always respond in a linear way to grazing intensity (Sasaki et al., 2010), partly because local environmental conditions such as high rainfall and soil fertility regulate the plants' ability to cope with grazing pressures. However, herbaceous biomass appears to be more responsive to differences in grazing intensities between communal grazing land and ranches. Ranches exhibited a higher herbaceous biomass than the communal grazing land at most sites, which could be due to the higher grazing intensities of the communal grazing lands compared with ranches. Kgosikoma (2011). Sternberg et al. (2000) and Mphinyane et al. (2008) have also demonstrated that the biomass of herbaceous plants is highly responsive to grazing pressures.

EFFECTS OF LIVESTOCK GRAZING MANAGEMENT SYSTEMS ON BUSH ENCROACHMENT

Woody plant encroachment has been widespread in grasslands and savanna ecosystems worldwide (Archer et al., 2001): Bush encroachment is an important indicator of land degradation (van Vegten, 1984; van Auken, 2009) which has become a global concern (Moleele and Perkins, 1998; Moleele et al., 2002; Sankaran, 2009). Woody plant encroachment into grassland-dominated savannas has contributed to a decrease in the productivity of rangelands (Wiegnand et al., 2005; Douglass et al., 2011), and jeopardizes biodiversity in grasslands, which threatens the sustainability of pastoral, subsistence and commercial livestock grazing (Rappole et al., 1986; Noble, 1997). Some savanna landscapes are completely encroached by woody species, while in others savanna areas the process is in progress (Archer et al., 2001). Factors that contribute to bush encroachment are poorly understood. However, overgrazing (Moleele and Perkins, 1998), anthropogenic reductions in fire regimes (Heinl et al., 2008; Lehmann et al., 2008), frequent droughts (Cole and Brown, 1976; Scholes and Archer, 1997; Smith and Smith, 2001), and climate change (Fensham et al., 2005) are suspected to facilitate the process.

The increase in bushy vegetation in rangelands threatens livestock production in the savannas because encroaching woody species supress palatable grasses and herbs (Scholes and Archer, 1997) through competition for soil moisture and nutrients. However, encroaching leguminous woody plants such as *Acacia mellifera* may enrich nutrient poor sandy soils in dry savannas through nitrogen fixation (Hagos and Smit, 2005). Research has shown that soils under the canopy of tree species such as *A. mellifera* have higher levels of nitrogen, organic matter and calcium than soils distant to trees (Hagos and Smit, 2005). Some woody plants are also an important fodder resource especially during dry periods (Moleele, 1998). Therefore, a management aim of bush encroached rangelands could be selective thinning of woody vegetation to reduce the grass-tree competition, whilst retaining the beneficial effects of soil enrichment from leguminous tree and shrub species (Hagos and Smit, 2005).

Bush encroachment is an environmental problem in both ranches and communal grazing land despite the difference in grazing intensity between the two grazing management systems (Oba et al., 2000; Kgosikoma et al., 2012). Meanwhile, Wigley et al. (2009) reported that bush encroachment was a slow process in communal grazing lands than ranches due to high utilization rate of woody plants for firewood by the community. High grazing pressure can also reduce fuel loads through consumption and compaction and consequently prevent fire ignition (Douglass et al., 2011). In the absence of fire, bush encroachment may occur (van Langevelde et al., 2003; Sharp and Whittaker, 2003; Savadogo et al., 2007) which decreases rangeland's productivity.

SUSTAINABILITY OF SAVANNA ECOSYSTEMS

The current grazing policies in southern Africa (for example, TGLP of Botswana) were based on the assumptions that ranches would promote sustainable land use and conserve rangeland resources. However, current evidence suggests that rangeland degradation is occurring in both communal and ranching lands (Vanderpost et al., 2011). This had led to others arguing that the current grazing policies had failed to address the land degradation problem and had instead exacerbated it (Rohde et al., 2006). In addition, the communal grazing land which supports the large population of livestock continues to shrink in size as more land is demarcated into ranches (Boone, 2005). Subsequently, grazing pressures have intensified in communal grazing land (Bennett et al., 2010) especially since owners of private ranches continue to use communal grazing land in addition to their ranches (Tsimako, 1991). Considering the limited land currently available for grazing by pastoral communities (Bennett et al., 2010), the existing policies need thorough revision and dual grazing rights to the farmers allocated ranches should be eliminated. This would address the current inequity of land distribution, which could ultimately threaten the sustainability of the entire savanna ecosystem (Eriksen and Watson, 2009a).

Sustainable agro-ecosystem management depend on understanding the effects of different land use and environmental factors on ecosystems dynamics (Wallgren et al., 2009). Savanna ecosystems are complex and therefore management policies should rely on research-based understanding of whole ecosystems. Given that long term ecological data is often missing, the local pastoral community could provide the long term perspective on changes in savanna ecosystem. The social, ecological and economic factors need to be taken into account and participatory management involving pastoral communities should be considered.

CONCLUSION

The review of scientific literature shows that overgrazing and prolonged poorly managed rangelands led to removal of desirable plant species, decrease water infiltration into soil, increase soil erosion, reduce soil nutrients and alter the plant community composition to a less desirable state. These changes compromise both the short and long-term productivity of rangelands in savanna ecosystems. Grazing policies adopted by countries in arid zones, particularly those in Africa have failed to reduce land degradation because they were based on wrong

assumptions or models (Abel, 1997). Most of these policies need to be revised embracing indigenous knowledge systems, if rangeland resources in the savannas are to be used sustainably to benefit the future generation.

REFERENCES

Abel N (1997). Mis-measurement of the productivity and sustainability of African communal rangelands: A case study and some principles from Botswana. Ecol. Econ. 23:113-133.

Abdel-Magid AH, Trlica MJ, Hart RH (1987). Soil and vegetation responses to simulated trampling, J. Range Manage. 40(4):303-306.

Abule E, Smit GN, Snyman HA (2005). The influence of woody plants and livestock grazing on grass species composition, yield and soil nutrients in the Middle Awash Valley of Ethiopia. J. Arid. Environ. 60:343–358.

Asner GP, Elmore AJ, Olander LP, Martin RE, Harris AT (2004). Grazing systems, ecosystem response, and global change. Ann. Rev. Environ. Resour. 29:261-299.

Archer S, Boutton TW, Hibbard KA (2001). Tree in grasslands: biogeochemical consequences of woody plant expansion. In: Schulze E-D, Harrison SP, Heimann M, Holland EA, Lloyd J, Prentice IC, Schimel D (Eds.). Global Biogeochemical Cycles in the Climate Systems. Academic Press, San Diego, pp. 115-137.

Bari F, Wood MK, Murray L (1995). Livestock grazing impacts on interill ersoin in Pakistan. J. Range Manage. 48(3):251-257.

Beeskow AM, Elissalde NO, Rostagno CM (1995). Ecosystem changes associated with grazing intensity on the Punta Ninfas rangelands of Patagonia, Argentina. J. Range Manage. 48(6):517-522.

Belsky AJ, Blumenthal DM (1997). Effects of livestock grazing on stand dynamics and soil in upland forest of the interior west. Conserv. Biol. 11(2):315-327.

Bennett J, Ainslie A, Davis J (2010). Fenced in: Common property struggles in the management of communal rangelands in central Eastern Cape Province, South Africa. Land Use Pol. 27:340-350.

Bilotta GS, Brazier RE, Haygarth PM, Donald LS (2007). The impacts of grazing animals on the quality of soils, vegetation, and surface waters in intensively managed grasslands. Adv. Agron. Academic Press, pp. 237-280.

Bohn CC, Buckhouse JC (1985). Some responses of riparian soils to grazing management in northeastern Oregon. J. Range Manage. 38(4):378–381.

Boone RB (2005). Quantifying changes in vegetation in shrinking grazing areas in Africa. Conse& Soc 3.

Botswana Goverment (1975). National Policy on Tribal Grazing Land. Government printer, Gaborone, Botswana.

Brandy NC, Weil RR (2007). The Nature and Properties of Soils, 14th Edition. Prentice Hall, New York, p. 980.

Cole MM, Brown RC (1976). The Vegetation of the Ghanzi Area of Western Botswana. J. Biogeogr. 3:169-196.

Dahdough-Guebas F, Kairo JG, Jayatissa LP, Cannicci S, Koedam N (2002). An ordination study to view vegetation structure dynamics in disturbed and undisturbed mangrove forests in Kenya and Sri Lanka. Plant Ecol. 161:123-135.

Dahlberg AC (2000) Vegetation diversity and change in relation to land use, soil and rainfall - A case study from North-East District, Botswana. J. Arid Environ. 44:19-40.

Darkoh MBK (2009). An overview of environmental issues in Southern Africa. Afr. J. Ecol. 47:93-98.

Dougill AJ, Thomas DSG, Heathwaite AL (1999). Environmental Change in the Kalahari: Integrated Land Degradation Studies for Non equilibrium Dryland Environments. Ann. Assoc. Am. Geogr. 89:420-442.

Douglass LL, Possingham HP, Carwardine J, Klein CJ, Roxburgh SH, Russell-Smith J, Wilson KA (2011). The effect of carbon credits on savanna land management and priorities for biodiversity conservation. Plos ONE 6(9):e23843.doi:10.131/journal.prone. 0023843.

58 Ecology, Environment and Conservation

du Plessis WP, Bredenkamp GJ, Trollope WSW (1998). Development of a technique for assessing veld condition in Etosha National Park, Namibia, using key herbaceous species. Koedoe 41:19-29.

Du Toit GV, Synaman HA, Malan PJ (2009). Physical impact of grazing by sheep on soil properties in the Nama Karoo subshrub/grassland of South Africa. J. Arid Environ. 73:804–810.

Ellis JE, Swift DM (1988). Stability of African Pastoral Ecosystems: Alternate Paradigms and Implications for Development. J. Range Manage. 41:450-459.

Eriksen SEH, Watson HK (2009a). The dynamic context of southern African savannas: investigating emerging threats and opportunities to sustainability. Environ. Sci. Pol. 12:5-22.

Eriksen SH, Watson HK (2009b). The sustainability of southern African savannas. Environ. Sci, Pol. 12:1-4.

Fatunbi AO, Dube S (2008). Land degradation in a game reserve in Eastern Cape of South Africa: Soil properties and vegetation cover. Sci. Res. Essays 3:111-119.

Fensham RJ, Fairfax RJ, Dwyer JM (2010). Vegetation responses to the first 20 years of cattle grazing in an Australian desert. Ecology. 91:681-692.

Fleischer TL (1994). Ecological costs of livestock grazing in western North America. Cons. Biol. 8(3):629-644.

Geissen V, Sánchez-Hernández R, Kampichler C, Ramos-Reyes R, Sepulveda-Lozada A, Ochoa-Goana S, de Jong BHJ, Huerta-Lwanga E, Hernández-Daumas S (2009). Effects of land-use change on some properties of tropical soils - An example from Southeast Mexico. Geoderma 151:87-97.

Groffman PM, Pouyat RV, Cadenasso ML, Zipperer WC, Szlavecz K, Yesilonis ID, Band LE, Brush GS (2007). Land use context and natural soil controls on plant community composition and soil nitrogen and carbon dynamics in urban and rural forests. For. Ecol. Manag. 246:296-297.

Hagos MG, Smit GN (2005). Soil enrichment by Acacia mellifera subsp detinens on nutrient poor sandy soil in a semi-arid southern African savanna. J. Arid. Environ. 61:47-59.

Heinl M, Sliva J, Tacheba B, Murray-Hudson M (2008). The relevance of fire frequency for the floodplain vegetation of the Okavango Delta, Botswana. Afr. J. Ecol. 46:350-358.

Hendricks HH, Bond WJ, Midgley JJ, Novellie PA (2007). Biodiversity conservation and pastoralism--reducing herd size in a communal livestock production system in Richtersveld National Park. J. Arid Environ. 70:718-727.

Herlocker DJ (1999). Rangeland Ecology and Resource Development in Eastern Africa. Deutsche Gesellschaft für Technische Zusammenarbeit (GTZ), Nairobi, Kenya. p.393.

Hopmans P, Bauhus J, Khanna P, Weston C (2005). Carbon and nitrogen in forest soils: Potential indicators for sustainable management of eucalypt forests in south-eastern Australia. For. Ecol. Manag. 220:75-87.

Ingram LJ, Stahl PD, Schuman GE, Buyer JS, Vance GF, Ganjegunte GK, Welker JM, Derner JD (2008). Grazing impacts on soil carbon and microbial communities in a mixed-grass ecosystem. Soil Sci. Am. J. 72(4):939-948.

Kauffman JB, Krueger WC (1984). Livestock impacts on riparian ecosystems and streamside management implications: A review. J. Range Manage. 37(5):430-438.

Kgosikoma OE (2011). Understanding the savanna dynamics in relation to rangeland management systems and environmental conditions in semi-arid Botswana. PhD thesis. Univesity of Edinburgh.

Kgosikoma OE, Harvie BA, Mojeremane W (2012). Bush encroachment in relation to rangeland management systems and environmental conditions in Kalahari ecosystem of Botswana. Afr. J. Agric. Res. 7:2312-2319.

Kioko J, Kiringe JW, Seno SO (2012). Impacts of livestock grazing on a savanna grassland in Kenya. J. Arid Land 4(1):29-35.

Klumpp K, Fontaine S, Attard E, Le Roux X, Gleixner G, Sousssana JF (2009). Grazing triggers soil carbon loss by altering plant roots and their control on soil microbial community. J. Ecol. 97:876-885.

Lal R (2002). Soil carbon dynamics in cropland and rangeland. Environ. Pollution 116:353-362.

Liebig MA, Gross JR, Kronberg SL, Hanson JD, Frank AB, Phillips RL (2006). Soil response to long-term grazing in the northern Great Plains of North America. Agric. Ecosyst. Environ. 115:270-276.

Lehmann CER, Prior LD, Williams RJ, Bowman DMJS (2008). Spatio-temporal trends in tree cover of a tropical mesic savanna are driven by landscape disturbance. J Appl. Ecol. 45:1304-1311.

Masike S, Urich P (2008). Vulnerability of traditional beef sector to drought and the challenges of climate change: The case of Kgatleng District, Botswana. J. Geogr Reg. Plan. 1:012-018.

Mohammed AS, Bekele T (2010). Forage production and plant diversity in two managed rangelands in the Main Ethiopian Rift. Afr. J. Ecol. 48:13-20.

Moleele NM (1998). Encroacher woody plant browse as feed for cattle. Cattle diet composition for three seasons at Olifants Drift, south-east Botswana. J. Arid Environ. 40:255-268.

Moleele NM, Perkins JS (1998). Encroaching woody plant species and boreholes: is cattle density the main driving factor in the Olifants Drift communal grazing lands, south-eastern Botswana? J. Arid Environ. 40:245-253.

Moleele NM, Ringrose S, Matheson W, Vanderpost C (2002). More woody plants?: The status of bush encroachment in Botswana's grazing areas. J. Environ. Manag. 64:3-11.

Moussa AS, Van Rensburg L, Kellner K, Bationo A (2008). Soil indicators of rangeland degradation in a semi-arid communal district in South Africa. Fut Drylands, pp. 383-393.

Mphinyane WN, Tacheba G, Mangope S, Makore J (2008). Influence of stocking rate on herbage production, steers livemass gain and carcass price on semi-arid sweet bushveld in Southeren Botswana. Afr. J. Agric. Res. 3:084-090.

Muhumuza M, Byarugaba D (2009). Impact of land use on the ecology of uncultivated plant species in the Rwenzori mountain range, mid western Uganda. Afr. J. Ecol. 47:614-621.

Neff JC, Reynolds RL, Belnap J, Lamothe P (2005). Multi-decadal of grazing on soil physical and biochemical properties in southeast Utah. Ecol. Appl. 15(1):87-95.

Nobble JC (1997). The delicate and noxious scrub: Studies on native tree and shrub proliferation in semi-arid woodlands of Australia. CSIRO Division of Wildlife and Ecology, Canberra.

Northup BK, Brown JR, Ash AJ (2005). Grazing impacts on spatial distribution of soil and herbaceous characteristics in an Australian tropical woodland. Agrofor. Syst. 65:137-150.

Oba G, Post E, Syverstssen PO, Stenseth NC (2000). Bush cover and range condition assessments in relation to landscape and grazing in southern Ethiopia. Landsc. Ecol. 15(6):535-546.

O' Connor TG (2005). Influence of land use on plant community composition and diversity in Highland Sourveld grassland in the southern Drakensberg, South Africa. J. Appl. Ecol. 42:975-988.

Owen-Smith N (1999). The annual factor in veld management: Implications of selective patterns of grazing. In:Taintan ND (eds).Veld Management in South Africa. University of Natal Press, Pietrmaritzburg, pp. 129-130.

Parsons DAB, Shackleton CM, Scholes RJ (1997). Changes in herbaceous layer condition under contrasting land use systems in the semi-arid lowveld, South Africa. J. Arid Environ. 37:319-329.

Pluhar JJ, Knight RW, Heitschmidt RK (1987). Infiltration rates and sediments production as influenced by grazing systems in the Texas rolling plains. J. Range Manage. 40(3):240-243.

Popolizio CA, Coetz H, Chapman PL (1994). Short-term response of riparian vegetation to 4 grazing treatments. J. Range Manage. 47(1):48–53.

Rappole JH, Russell CE, Fulbright TE (1986). Anthropogenic pressure and impacts on marginal, neotropical, semiarid ecosystems: The case of south Texas. Sci. Total Environ. 55:91-99.

Reed MS, Dougill AJ, Baker TR (2008). Participatory indicator development: What can ecologists and local communities learn from each other? Ecol. Appl. 18:1253-1269.

Robertson E (1996). Impacts of livestock grazing on soils and recommendations of management. <http://www.cnps.org/cnps/archive/letters/soils.pdf>.

Rohde RF, Moleele NM, Mphale M, Allsopp N, Chanda R, Hoffman MT, Magole L, Young E (2006). Dynamics of grazing policy and practice: Environmental and social impacts in three communal areas of southern Africa. Environ. Sci. Pol. 9:302-316.

Roose E, Barthes B (2001). Organic matter management for soil conservation and productivity restoration in Africa: A contribution from Francophone research. Nutr. Cycl. Agroecosyst. 61:159-170.

Sankaran M (2009). Diversity patterns in savanna grassland communities: Implications for conservation strategies in a biodiversity hotspot. Biod. Conser. 18:1099-1115.

Sankaran M, Hanan NP, Scholes RJ, Ratnam J, Augustine DJ, Cade BS, Gignoux J, Higgins SI, Le Roux X, Ludwig F, Ardo J, Banyikwa F, Bronn A, Bucini G, Caylor KK, Coughenour MB, Diouf A, Ekaya W, Feral CJ, February EC, Frost PGH, Hiernaux P, Hrabar H, Metzger KL, Prins HHT, Ringrose S, Sea W, Tews J, Worden J, Zambatis N (2005). Determinants of woody cover in African savannas. Nature, 438:846-849.

Sasaki T, Okubo S, Okayasu T, Jamsran U, Ohkuro T, Takeuchi K (2010). Indicator species and functional groups as predictors of proximity to ecological thresholds in Mongolian rangelands. Plant Ecol. pp. 1-16.

Savadogo P, Sawadogo L, Tiveau D (2007). Effects of grazing intensity and prescribed fire on soil physical and hydrological properties and pasture yield in the savanna woodlands of Burkina Faso. Agric. Ecosyst. Environ. 118:80-92.

Scholes RJ (1990). The Influence of Soil Fertility on the Ecology of Southern African Dry Savannas. J. Biogeogr. 17:415-419.

Scholes RJ, Archer SR (1997). Tree-grass interactions in savannas. Ann. Rev. Ecol. Syst. 28:517-544.

Sharp BR, Whittaker RJ (2003). The irreversible cattle-driven transformation of a seasonally flooded Australian savanna. J. Biogeogr. 30:783-802.

Skarpe C (1992). Dynamics of savanna ecosystems. J. Veget. Sci. 3(3):293-300.

Smet M, Ward D (2005). A comparison of the effects of different rangeland management systems on plant species composition, diversity and vegetation structure in a semi-arid savanna. Afri. J. Rang. For. Sci. 22:59-71.

Smith RL, Smith TM (2001). Ecology & Field Biology. Benjamin Cummings, New York.

Snyman HA, du Preez CC (2005). Rangeland degradation in a semi-arid South Africa - II: influence on soil quality. J. Arid Environ. 60:483-507.

Sternberg M, Gutman M, Perevolotsky A, Ungar ED, Kigel J (2000). Vegetation response to grazing management in a Mediterranean herbaceous community: a functional group approach. J. Appl. Ecol. 37:224-237.

Steffens M, Kölbl A, Totsche KU, Kögel-Knsbner I (2008). Grazing effects on soil chemical and physical properties in a semi-arid steppe of Inner Mangolia (P. R. China). Geordoma 143:63-72.

Stephenson GR, Veigel A (1987). Recovery of compacted soil on pasture used for winter cattle feeding. J. Range Manage. 40(1):46-48.

Tefera SB (2011). Seed dynamics in relation to land degradation management and soil types in the semi-arid savannas of Swaziland. Afr. J. Agric. Res. 6(11):2494-5505.

Tefera S, Dlamini BJ, Dlamini AM (2010). Changes in soil characteristics and grass layer condition in relation to land management systems in the semi-arid savannas of Swaziland. J. Arid Environ. 74:675-684.

Tefera S, Dlamini BJ, Dlamini AM, Mlambo V (2008b). Current range condition in relation to land management systems in semi-arid savannas of Swaziland. Afr. J. Ecol. 46:158-167.

Tefera S, Snyman HA, Smit GN (2007). Rangeland dynamics in southern Ethiopia: (1) Botanical composition of grasses and soil characteristics in relation to land-use and distance from water in semi-arid Borana rangelands. J. Environ. Manag. 85:429-442.

Thomas DSG, Sporton D (1997). Understanding the dynamics of social and environmental variability: The impacts of structural land use change on the environment and peoples of the Kalahari, Botswana. Appl. Geogr. 17:11-27.

Thurow TL, Blackburn WH, Taylor Jr, CA (1986). Hydrologic characterstsics of vegetation types as affected by livestock grazing systems, Edwards Plateau, Texas. J. Range Manage. 39(6):505-508.

Trollope WSW, Potgieter ALF, Zambatis N (1989). Assessing veld condition in the Kruger National Park South Africa using key grass species. Koedoe 32:67-94.

Tsimako B (1991). The Tribal Grazing Land Policy (TGLP) Ranches Performance to Date. Agricultural Planning and Statistics, Gaborone.

van Auken OW (2009). Causes and consequences of woody plant encroachment into western North American grasslands. J. Environ. Manag. 90:2931-2942.

van Langevelde F, van de Vijver C, Kumar L, van de Koppel J, de Ridder N, van Andel J, Skidmore AK, Hearne JW, Stroosnijder L, Bond WJ, Prins HHT, Rietkerk M (2003). Effects of fire and herbivory on the stability of savanna ecosystems. Ecology 84:337-350.

van der Heijden GMF, Phillips OL (2009). Environmental effects on Neotropical liana species richness. J. Biogeogr. 36:1561-1572.

Vanderpost C, Ringrose S, Matheson W, Arntzen J (2011). Satellite based long-term assessment of rangeland condition in semi-arid areas: An example from Botswana. J. Arid Environ. 75:383-389.

van Vegten JA (1984). Thornbush Invasion in a Savanna Ecosystem in Eastern Botswana. Vegetatio, 56:3-7.

Vavra M, Parks CG, Wisdom MJ (2007). Biodiversity, exotic plant species, and herbivory: The good, the bad, and the ungulate. For. Ecol. Manage. 246:66-72.

Vetter S (2005) Rangelands at equilibrium and non-equilibrium: Recent developments in the debate. J. Arid Environ. 62:321-341.

Walker SM, Desanker PV (2004). The impact of land use on soil carbon in miombo woodland of Malawi. For. Ecol. Manage. 203:345-360.

Wallgren M, Skarpe C, Bergström R, Danell K, Bergström A, Jakobsson T, Karlsson K, Strand T (2009). Influence of land use on the abundance of wildlife and livestock in the Kalahari, Botswana. J. Arid Environ. 73:314-321.

Warren SD, Nevill MB, Blackburn WH, Garza NE (1986). Soil response to trampling under intensive rotation grazing. Soil Sci. Soc. Am. J. 50(5):1336-1341.

Weimer B (1977). The Tribal Grazing Land Policy in the Context of Botswana's Rural Development Policy - A Documentary overview. In: Weimer, B. (Ed.), A policy for rural development: the case of Botswana's national policy on tribal grazing land. National Institute for Research in Development and African studies, Gaborone.

Westoby M, Walker B, Noymeir I (1989). Opportunistic Management for Rangelands Not at Equilibrium. J. Range Manage. 42:266-274.

Wiegand K, David W, David S (2005). Multi-scale patterns and bush encroachment in an arid savanna with a shallow soil layer. J. Veget. Sci. 16:311-320.

Wigley BJ, Bond WJ, Hoffman MT (2009). Bush encroachment under three contrasting land-use practices in a mesic South African savanna. Afr. J. Ecol. 47:62-70.

Wigley BJ, Bond WJ, Hoffman MT (2010). Thicket expansion in a South African savanna under divergent land use: Local vs. global drivers? Glob. Chang. Biol. 16:964-976.

10

Distribution, diversity and abundance of copepod zooplankton of Wular Lake, Kashmir Himalaya

Javaid Ahmad Shah[1]*, Ashok K. Pandit[1] and G. Mustafa Shah[2]

[1]Centre of Research for Development (CORD), University of Kashmir, Srinagar-190006, J & K, India.
[2]Department of Zoology, University of Kashmir, Srinagar-190006, J & K, India.

Wular Lake, the largest freshwater lake of India, plays a significant role in the hydrography of the Kashmir valley by acting as a huge absorption basin for floodwaters. Although rich in biodiversity, no published report is available on Copepoda diversity, distribution and abundance in such an important Ramsar Site of Kashmir Himalaya. Copepod samples, collected over a period of 12 months from September 2010 to August 2011 at five sampling sites revealed the occurrence of about 16 species belonging to 3 families namely cyclopoids with 12 species, calanoids and harpacticoids being represented by two species each. The dominant species seen were *Cyclops bicolor, Eucyclops agilis, Bryocamptus nivalis* and *Diaptomus virginiensis*. Various diversity indices like Shannon-Weaver, Margalef and Fisher_alpha were used to assess the Copepoda diversity in the lake. Further, discernible differences were observed among the studied sites showing 15 species at site III, 13 each at sites IV and II, 9 at site V and 8 at site I.

Key words: Copepods, cyclopoids, composition, diversity, Ramsar Site.

INTRODUCTION

Copepods are claimed to be numerically the most abundant metazoans on earth and conservative estimations revealed that they likely outnumber the abundance of insects (Schminke, 2007; Chang et al., 2010, Hwang et al., 2004, 2010; Kâ and Hwang, 2011), representing one of the biggest sources of animal protein in the world and play a central role in the transfer of carbon from producers to higher trophic levels in most aquatic ecosystems (Jerling and Wooldridge, 1995). Copepoda species which are one of the most important elements of the food chain in the aquatic environments form the inevitable food for fishes and show wide dispersion in all kinds of aquatic ecosystems. Further, copepods are also known to consume large quantities of bacteria (Wroblewski, 1980) and phytoplankton (Calbet et al., 2000) and are the main prey items of larval and juvenile fishes that link pelagic food webs (Tseng et al., 2008; 2009; Vandromme et al., 2010; Wu et al., 2010).

Their abundance and distribution are known to be influenced by hydrographic conditions and they have been suggested as indicator species for waters of different qualities and origins (Bonnet and Frid, 2004; Hwang and Wong, 2005; Thor et al., 2005; Hwang et al., 2006, 2009, 2010).

MATERIALS AND METHODS

Study area

Wular lake, an ox-bow type lake located in the north-west of Kashmir about 55 km from Srinagar city, is situated at an altitude of 1,580 m (a.m.s.l), and lies between 34°16′-34°20′N and 74°33′-74°44′E geographical co-ordinates, being formed by the meandering of River Jhelum which is the main feeding channel besides other tributaries. The lake is drained in the north-east by the only single outlet in the form of River Jhelum. The catchment of the lake is comprised of slopping hills of the Zanaskar ranges of the western Himalaya on the north eastern side and arable land around being used for agricultural purposes (Figure 1). The lake has been declared as the wetland of national importance (1986) under wetland programme of Ministry of Environment and Forests (MoEF), Gov-

---

*Corresponding author. E-mail: javaidshah31@gmail.com.

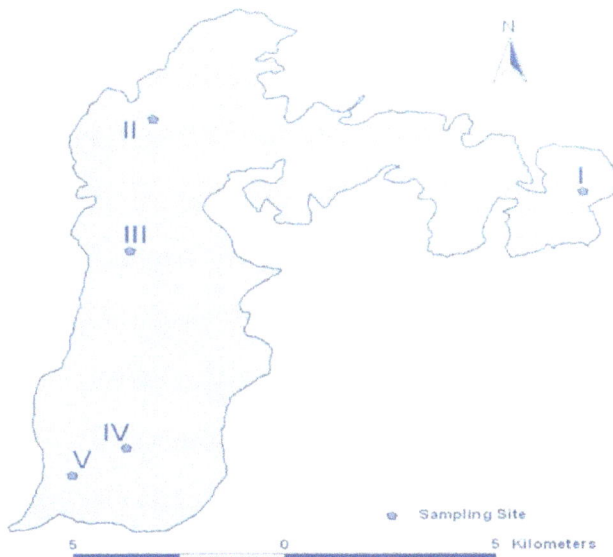

Figure 1. Map of Wular Lake with five sampling sites.

Table 1. Location of five study sites in Wular Lake.

Site	Latitude (N)	Longitude (E)	Distinguishing feature
I	34°21′ 51	74°39′42	Anthropogenic pressures
II	34°24′ 15	74°32 ′35	Good macrophytic growth
III	34°21′ 29	74°31′ 48	Profuse growth of macrophytes
IV	34°17′43	74°31′30	Centre of lake basin
V	34°17′ 16	74°30 ′25	Near outlet of the lake

Government of India and is designated as Ramsar Site (a Wetland of International Importance under Ramsar Convention, 1990).

Study sites

Five sampling sites were chosen to collect the copepods in Wular Lake (Table 1 and Figure 1).

Plankton samples were randomly collected between September 2010 to August 2011 from five sampling sites once in every month. At each site, samples were collected by filtering 100 L surface water through a Nylobolt plankton net (mesh size 52 μm). The plankton samples were fixed in 5% formalin added with 2 to 3 ml of glycerol which acts as a good preservative (Dussart and Defay, 1995). The qualitative and quantitative analyses were carried out in the laboratory with the help of binocular microscope and counting was done by a Sedgewick-Rafter counting cell. Identification of the specimens was carried out with the help of standard woks of Pennak (1978), Edmondson (1992) and Battish (1992). Various diversity indices (Shannon-Weaver, Margalef and Fisher_alpha) were calculated by software PAST.

RESULTS AND DISCUSSION

A total of 16 species of copepods belonging to three families (Cyclopoida with 12 species, Calanoida and Harpac-

ticoida with two species each) were identified (Table 2). Cyclopoids predominated over calanoids in the present study. This is quite expected as the wetland un-der study is weed infested. Rundle and Ormerod (1992) also found the abundance and richness of cyclopoids in wetlands with high weed infestation. Boxshall and Jaume (2000) opined that Cyclopoids are one of the most con-spicuous and diverse group of freshwater copepods and tend to have wide distributional patterns with many spe-cies being cosmopolitan in nature (Reid, 1998). Many studies suggest that species of the family Cyclopoida tend to increase stronger with eutrophication than spe-cies of Calanoida (Gliwicz, 1969; Patalas, 1972; Straile and Geller, 1998; Anneville et al. (2007). However, this pattern might not be consistent among lakes since Jep-pesen et al. (2005) reported reduced numbers of cyclo-poids under eutrophic conditions in eight Danish lakes.

The most abundant taxa in order of dominance were *Cyclops bicolor*, *Bryocamptus nivalis*, *Eucyclops agilis* and *Cyclops latipes* and other species observed were *Acanthocyclops bicuspidatus*, *C. bicolor*, *Cyclops biseto-sus*, *Cyclops bicuspidatus*, *Cyclops panamensis*, *Cyclops scutifer*, *Cyclops vicinus*, *C. latipes*, *E. agilis*, *Bryocamp-tus minutus*, *B. nivalis*, *Diaptomus* sp. and *Diaptomus virginiensis*. Among the recorded copepods some spe-cies dominated in certain seasons while some disap-peared in the other seasons (*Megacyclops viridis* and *Diaptomus* sp. were completely absent in winter and the latter was also absent in spring). Likewise *Cyclops scuti-fer* and *C. vicinus* were absent in summer and the latter was again absent in spring season) indicating the differ-ent growing patterns of the species, as a result, bimodal growth peaks were observed (Table 3). At sites I and II, dominant peaks of copepods were observed in autumn, while as sub-dominant, peaks were evinced in spring. In contrast to the above, sites III, IV and V exhibited the dominant peaks in spring; while the sub-dominant peaks were registered during autumn (for site III) and summer for the remaining sites (Figure 2). This changing beha-vioral patterns of the copepods (e.g. *Cyclops scutifer*- a typical cold water species, *C. vicinus* maintaining higher population in autumn and spring while *Megacyclops viridis* being more abundant in spring) can be attributed to the habitat preferences, environmental conditions or the dormancy period (Alekseev et al., 2007). The same diver-se reasons have also been attributed to varying growth seasons of the copepods during the study period. Many species among the copepods are known to bridge seasonally unfavorable conditions with a dormancy or diapause period (Einsle, 1993). Diapause performance, however, also varies among populations of the same species (Santer, 1998).

This variability in diapause strategies has been obser-ved in a variety of species and is suggested to be indu-ced by environmental conditions, e.g., predation (Hair-ston, 1987) or food availability (Santer and Lampert, 1995). Copepod life cycle strategies differed in the

Table 2. Distribution of Copepoda at five study sites during September 2010 to August 2011.

Cyclopoida	Site I	Site II	Site III	Site IV	Site V
Acanthocyclops bicuspidatus	-	+	+ +	+	+ +
Cyclops bicolor	+ + +	+ +	+ + +	+	+
C. bisetosus	-	+	+	+	-
C. bicuspidatus	-	-	+ + +	-	-
C. scutifer	+	+	+	+	+
C. vicinus	+	+	+	+	+
C. latipes	-	+ + +	+ +	+ + +	-
C. panamensis	-	-	.	+	-
Eucyclops agilis	-	+ + +	+ + +	+	+ ++
Macrocyclops fuscus	+ +	-	+	-	-
Megacyclops viridis	+	+	+	-	-
Paracyclops affinis	-	+	+	+++	-
Calanoida					
Bryocamptus minutus	-	+ +	+ +	+ + +	++
Bryocamptus nivalis	+ +	+++	+	+ + +	++
Harpacticoida					
Diaptomus sp.	+	+	+	+ +	+
Diaptomus virginiensis	+ +	+ +	+	+	+
Grand total = 16	08	13	15	13	09

+++ = Most abundant; ++ = fairly present; + = present; - = absent.

Table 3. Seasonal variations in the population density* of copepods (ind./L.) at different sites from September 2010 to August 2011.

Site		Autumn	Winter	Spring	Summer
I					
01	*Bryocamptus nivalis*	44.0	21.3	45.7	37.0
02	*Cyclops bicolor*	63.0	68.3	23.7	51.7
03	*C. scutifer*	43.7	55.0	17.7	N.D
04	*C. vicinus*	66.3	16.3	N.D	N.D
05	*Diaptomus* sp.	46.7	N.D	N.D	46.7
06	*Diaptomus virginiensis*	34.0	25.3	64.7	41.3
07	*Macrocyclops fuscus*	30.0	20.3	60.7	37.0
08	*Megacyclops viridis*	17.3	N.D	47.7	32.5
II					
01	*Acanthocyclops bicuspidatus*	31.3	32.7	33.3	32.4
02	*Bryocamptus minutus*	55.3	32.7	43.3	43.8
03	*Bryocamptus nivalis*	43.0	34.3	67.3	48.2
04	*Cyclops bicolor*	47.7	62.0	25.3	45.0
05	*C. bisetosus*	50.0	12.3	37.0	33.1
06	*C. scutifer*	40.3	91.7	34.7	N.D
07	*C. vicinus*	70.3	33.3	N.D	N.D
08	*C. latipes*	60.3	33.7	53.3	49.1
09	*Diaptomus* sp.	32.7	N.D	N.D	32.7
10	*Diaptomus virginiensis*	38.3	27.7	68.7	44.9
11	*Eucyclops agilis*	51.3	43.7	70.0	55.0
12	*Megacyclops viridis*	21.7	0.0	60.0	27.2
13	*Paracyclops affinis*	26.0	24.0	46.7	32.2

Table 3. Contd.

III					
01	*Acanthocyclops bicuspidatus*	52.3	33.0	43.7	43.0
02	*Bryocamptus minutus*	49.0	26.0	43.0	39.3
03	*Bryocamptus nivalis*	49.3	40.3	83.7	57.8
04	*Cyclops bicolor*	24.0	68.7	19.7	37.4
05	*C. bisetosus*	37.3	17.0	48.7	34.3
06	*C. bicuspidatus*	56.0	27.7	62.3	48.7
07	*C. scutifer*	35.3	79.7	30.7	N.D
08	*C. vicinus*	59.0	29.0	N.D	N.D
09	*C. latipes*	67.7	21.0	37.3	42.0
10	*Diaptomus* sp.	35.0	N.D	N.D	35.0
11	*Diaptomus virginiensis*	30.3	20.7	57.0	36.0
12	*Eucyclops agilis*	53.3	30.7	55.3	46.4
13	*Macrocyclops fuscus*	20.0	10.3	55.0	28.4
14	*Megacyclops viridis*	14.3	N.D	37.0	17.1
15	*Paracyclops affinis*	22.3	29.3	55.7	35.8
IV					
01	*Acanthocyclops bicuspidatus*	44.0	32.3	44.0	40.1
02	*Bryocamptus minutus*	41.0	36.0	55.3	44.1
03	*Bryocamptus nivalis*	31.0	34.3	68.7	44.7
04	*Cyclops bicolor*	21.7	60.3	14.0	32.0
05	*C. bisetosus*	26.0	13.7	55.7	31.8
06	*C. panamensis*	29.0	46.0	45.0	40.0
07	*C. scutifer*	32.0	64.7	23.0	N.D
08	*C. vicinus*	44.0	31.7	N.D	N.D
09	*C. latipes*	51.0	19.7	53.7	41.4
10	*Diaptomus* sp.	30.7	N.D	N.D	30.7
11	*Diaptomus virginiensis*	34.0	18.7	61.3	38.0
12	*Eucyclops agilis*	18.0	22.3	63.3	34.6
13	*Paracyclops affinis*	31.3	29.3	85.0	48.6
V					
01	*Acanthocyclops bicuspidatus*	39.0	37.3	43.7	40.0
02	*Bryocamptus minutus*	29.0	39.0	55.3	41.1
03	*Bryocamptus nivalis*	25.7	29.3	70.7	41.9
04	*Cyclops bicolor*	23.0	72.7	32.3	42.7
05	*C. scutifer*	30.7	73.3	26.0	N.D
06	*C. vicinus*	30.0	43.3	N.D	N.D
07	*Diaptomus* sp.	45.7	N.D	N.D	45.7
08	*Diaptomus virginiensis*	17.0	24.3	75.0	38.8
09	*Eucyclops agilis*	85.3	13.0	61.0	53.1

N.D = Not detected; *average results based on three analyses.

presence and timing of a seasonal diapause, that is, diapause in winter, autumn, or none at all, resulting in a distinct seasonality of community structure. Further, zooplankton use a variety of environmental parameters such as temperature (Marcus, 1982), photoperiod (Hairston and Kearns, 1995; Alekseev et al., 2007) and food quality to optimize the timing of diapause initiation and/or termination. Furthermore, maternal effects might also be important for the production of diapausing stages (LaMontagne and McCauley, 2001). However, high predation pressure during winter can be ultimate cause of lower population density at all biotopes.

Diversity indices are important tools for ecologists to understand community structure in terms of richness, evenness or total number of existing individuals underlying the basis of diversity indices (Wilhm and Dorris,

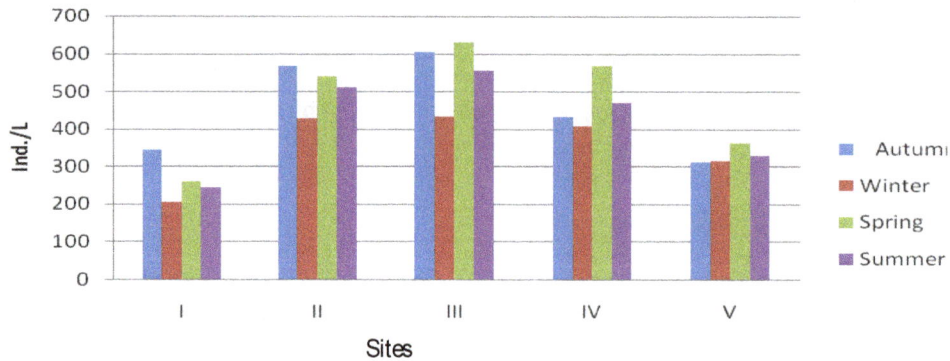

Figure 2. Seasonal variations in the population density of copepods at the different study sites.

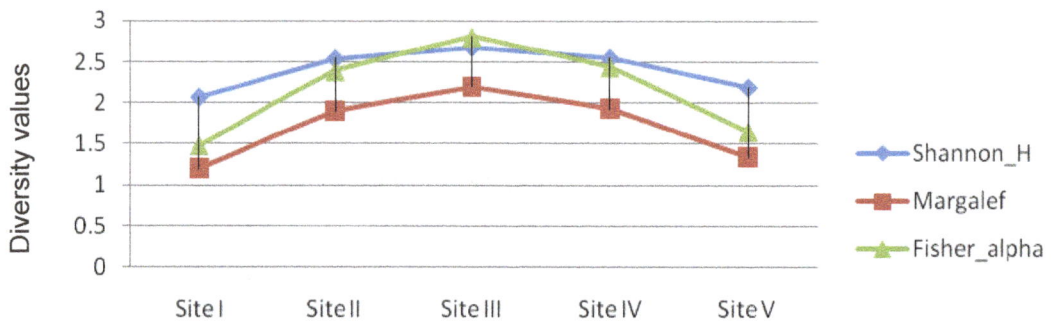

Figure 3. Diversity indices of copepods in Wular Lake during September 2010 to August 2011.

1968; Allan, 1975). Thus, any change in any of these three features will affect the whole population. So, the diversity indices depending upon these features are used effectively to determine the changes in a population (Mandaville, 2002; Dügel, 1995). During the study period, the Shannon index varies from 2.07 at site I to 2.67 at site III, Margalef diversity index ranged from the lowest, 1.20 at site I to the highest 2.20 at site III. Fisher_alpha index varied from 1.48 to 2.81 (Figure 3) reflecting the stress condition of site I which has higher anthropogenic activity. In contrast to the above, site III showed high diversity and number of different species characterized by undisturbed habitats.

During the study period, discernible differences among the species were observed at different studied sites, registering 15 species at site III, 13 species each at sites IV and II, 9 species at site V and 8 species at site I, respectively. Thus, copepods were found to be dominant at sites which were densely infested by macrophytes (site III). Accordingly, sites I and V, being least infested with macrophytes, maintained the lowest population of copepods. Similar conclusion was drawn by many workers in other studies (Blindow et al., 1993; Kuczyńska-Kippen, 2007). Gliwicz and Rybak (1976) further suggested that the more biologically and spatially complicated the habitat is, the more the available niches. It was also seen that most of the copepod species had a more or less positive associations among themselves. A positive association

between two species is indicative of the similar requirement in environmental gradients, while a negative association is indicative of different environmental requirements or active competition between the species involved. Thus, it may be inferred that most of the copepod species in the present study perhaps have overlapping ecological niche to some extent. Moreover, food particle size and food quality appears to play a definite role in their niche separation (Maly and Maly, 1974). Lakes rich in organic matter support higher number of cyclopoids, thus suggesting their preponderance in higher trophic state of waterbody (Subbamma, 1992), a fact also revealed by Pejaver and Somani (2004) in Lake Masunda.

ACKNOWLEDGEMENTS

Thanks are due to the Director, Centre of Research for Development and Head, Environmental Science, University of Kashmir for providing necessary laboratory facilities.

REFERENCES

Alekseev VR, Destasio BT, Gilbert J (2007). Diapause in Aquatic Invertebrates. Springer, New York.
Allan JD (1975). The distributional ecology and diversity of benthic insects in cement creek, Colorado. Ecol. 56: 1040-1053.
Anneville O, Molinero JC, Souissi S, Balvay G, Gerdeaux D (2007).

Long-term changes in the copepod community of Lake Geneva. Journal of Plankton Research, 29:49-59.

Battish SK (1992). Freshwater Zooplankton of India. Oxford and IBH Publishing Co., New Delhi.

Blindow J, Andersson G, Hargeby A, Johansson S (1993). Long-term patterns of alternative stable states in two shallow eutrophic lakes. Freshwat. Biol. 30: 159–167.

Bonnet D, Frid C (2004). Seven copepod species considered as indicators of water mass influence and changes: Results from a Northumberland coastal station. S. Afr. J. Mar. Sci. 61: 485-491.

Boxshall GA, Jaume D (2000). Making waves: The repeated colonization of fresh water by copepod crustaceans. Advances in Ecological Research, 31: 61–79.

Calbet A, Landry MR, Scheinberg RD (2000). Copepod grazing in a subtropical bay: Species-specific responses to a midsummer increase in nanoplankton standing stock. Mar. Ecol. Prog. Ser. 193: 75-84.

Chang WB, Tseng LC, Dahms HU (2010). Abundance, distribution and community structure of planktonic copepods in surface waters of a semi-enclosed embayment of Taiwan during monsoon transition. Zool. Stud. 49: 735-748.

Dügel M (1995). Köyceğiz Gölü'ne dökülen akarsuların su kalitelerinin fiziko-kimyasal ve biyolojik parametrelerle belirlenmesi, Bilim Uzmanlığı Tezi, Hacettepe Üniversitesi, Fenilimleri Enstitüsü,Ankara, 88s.

Dussart BH, Defay D (1995). Copepoda: Introduction to Copepoda. Guides to the Identification of the Microinvertebrates of the Continental Waters of the World (H.J. Dumont, ed.) Vol. 7 SPB Academic Publication.

Edmondson WT (1992). Ward and Whiple Freshwater Biology. 2nd ed. Intern. Books and Periodicals Supply Service, New Delhi.

Einsle U (1993). Crustacea: Calanoida und Cyclopoida - Süsswasserfauna von Mitteleuropa. Gustav Fischer Verlag, Stuttgart.

Gliwicz ZM (1969). Studies on the feeding of pelagic zooplankton in lakes of varying trophy. Ekologia Polska 17: 663-708.

Gliwicz ZM, Rybak JI (1976). Zooplankton. p. 69–96. In: Selected Problems of Lake Littoral Ecology (E. Pieczyńska, ed.). Uniwersytet Warszawski Press, Warszawa.

Hairston NG (1987). Diapause as a predator-avoidance adaptation. p. 281–290. In Predation: Direct and Indirect Impacts on Aquatic Communities: Kerfoot W C, Sih A (eds). University Press of New England,London,

Hairston NG, Kearns CM (1995). The interaction of photoperiod and temperature in Diapause Timing - A Copepod Example. Biological Bulletin, 189:42-48.

Hwang JS, Kumar R, Kuo CS (2009). Impact of predation by the copepod, Mesocyclops pehpeiensis, on life table demography and population dynamics of four cladoceran species: A comparative laboratory study. Zool. Stud. 48: 738-752.

Hwang JS, Kumar R, Dahms HU, Tseng LC, Chen QC (2010). Interannual, seasonal, and diurnal variation in vertical and horizontal distribution patterns of six Oithona sp. (Copepoda: Cyclopoida) in the South China Sea. Zool. Stud. 49: 220-229.

Hwang JS, Kumar R, Hsieh CW, Kuo AY, Souissi S, Hsu, MH, Wu JT, Cheng LW (2010). Patterns of zooplankton distribution along the marine, estuarine and riverine portions of the Danshuei ecosystem in Northern Taiwan. Zool. Stud. 49: 335-352.

Hwang JS, Souissi S, Tseng LC, Seuront L, Schmitt FG, Fang LS, Peng SH, Wu CH, Hsiao SH, Twan WH, Wei T, Kumar R, Fang TS, Chen Q, Wong CK (2006). A 5-year study of the influence of the northeast and southwest monsoons on copepod assemblages in the boundary coastal waters between the East China Sea and the Taiwan Strait. J. Plankt. Res. 28:943-958.

Hwang JS, Tu YY, Tseng LC, Fang LS, Souissi S, Fang TH, Lo WT, Twan WH, Hsaio SH, Wu CH, Peng SH, Wei TP, Chen QC (2004). Taxonomic composition and seasonal distribution of copepod assemblages from waters adjacent to nuclear power plant I and II in Northern Taiwan. J. Mar. Sci. Tech. 12(5): 380-391.

Hwang JS, Wong CK (2005). The China coastal current as a driving force for transporting Calanus sinicus (Copepoda: Calanoida) from its population centers to waters off Taiwan and Hong Kong during the winter NE monsoon period. J. Plankt. Res. 27: 205-210.

Jeppesen E, Jensen JP, Sondergaard M, Lauridsen TL (2005). Response of fish and plankton to nutrient loading reduction in eight shallow Danish lakes with special emphasis on seasonal dynamics. Freshwater Biol.50:1616-1627.

Jerling HL, Wooldridge TH (1995). Plankton distribution and abundance in the Sundays River estuary, South Africa, with comments on potential feeding interactions. S. Afr. J. Mar. Sci. 15: 169-184.

Kâ S, Hwang JS (2011). Mesozooplankton distribution and composition on the northeastern coast of Taiwan during autumn: effects of the Kuroshio Current and hydrothermal vents. Zool. Stud. 50: 155-163.

Kuczyńska-Kippen N (2007). Habitat choice in Rotifera communities of three shallow lakes: Impact of macrophyte substratum and season. Hydrobiol. 593: 190-198.

LaMontagne JM, McCauley E (2001). Maternal effects in Daphnia: What mothers are telling their offspring and do they listen? Ecology Lett. 4:64-71.

Maly EJ, Maly MP (1974). Dietary differences between two co-occurring calanoid copepod species. Oecolog. 17: 325-333.

Mandaville SM (2002) Benthic Macroinvertebrates in Freshwater – Taxa Tolerance Values, Metrics, and Protocols. Project H - 1. Nova Scotia: Soil & Water Conservation Society of Metro Halifax.

Marcus HN (1982). Photoperiodic and temperature regulation of diapause in labidocera aestiva (Copepoda: Calanoida). Biol. Bull. 162(1): 45-52.

Patalas K (1972). Crustacean plankton and the eutrophication of St. Lawrence Great Lakes. J. Fish. Res. Bd. Can. 29: 1451-1462.

Pejaver MK, Somani V (2004). Crustacean zooplankton of Lake Masunda, Thane. Journal of Aquatic Biol. 19 (1): 57-60.

Pennak RW (1978). Freshwater Invertebrates of United States of America. Wiley Interscience Pub., N. Y.

Reid JW (1998). How 'cosmopolitan' are the continental cyclopoid copepods? Comparison of the North American and Eurasian faunas with description of Acanthocyclops parasensitivus sp. (Copepoda: Cyclopoida) from the U.S.A. Zoologischer Anzeiger, 236: 109–118.

Rundle SD, Ormerod SJ (1992).The influence of chemistry and habitat features on some microcrustacea of some upland welsh streams. Freshwater Biol. 26(3): 439 – 452.

Santer B (1998). Life cycle strategies of free-living copepods in fresh waters. J. of Marine Syst.15:327 336.

Santer B, Lampert W (1995). Summer diapause in cyclopoid copepods: Adaptive response to a food bottleneck? Journal of Animal Ecol. 64:600-613.

Schminke HK (2007). Entomology for the copepodologist. J. Plankt. Res. 29: 149-162.

Straile D, Geller W (1998). Crustacean zooplankton in Lake Constance from 1920 to 1995: Response to eutrophication and re-oligotrophication. Archiv für Hydrobiol. (special issue Advances in Limnology) 53:255-274.

Subbamma DV (1992). Plankton of a temple pond near Machili Patnam, Andhra Pradesh. J. Aqua. Biol. 7(1 & 2): 17-21.

Thor P, Nielson TG, Tiselius P, Juul-Pederson T, Michel C, Møller EF, Dahl K, Selander E, Gooding S (2005). Post-spring bloom community structure of pelagic copepods in the Disko Bay, Western Greenland. J. Plankt. Res. 27: 341-356.

Tseng LC, Kumar R, Dahms HU, Chen QC, Hwang JS (2008). Copepod gut contents, ingestion rates and feeding impact in relation to their size structure in the southeastern Taiwan Strait. Zool. Stud.47: 402-416.

Tseng LC, Dahms HU, Chen QC, Hwang JS (2009). Copepod feeding study in the upper layer of the tropical South China Sea. Helg. Mar. Res. 63: 327-337.

Vandromme P, Schmitt FG, Souissi S, Buskey EJ, Strickler JR, Wu CH, lang S, Hwang JS(2010). Symbolic analysis of plankton swimming trajectories: Case study of Strobilidium sp. (Protista) helical walking under various food conditions. Zool. Stud. 49: 289-303.

Wilhm JL, Dorris TC (1968). Biological Parameters for Water Quality Criteria, Bioscience, 18(6): 477-481.

Wroblewski JS (1980). A simulation of distribution of Acartia clause during the Oregon upwelling, August 1973. J. Plank. Res. 2:43-68

Wu CH, Dahms HU, Buskey EJ, Strickler JR, Hwang JS (2010). Behavioral interactions of Temora turbinata with potential ciliate prey. Zool. Stud. 49: 157-168.

11

Evaluating the influence of open cast mining of solid minerals on soil, landuse and livelihood systems in selected areas of Nasarawa State, North-Central Nigeria

Ezeaku, Peter Ikemefuna

Department of Soil Science, Faculty of Agriculture, University of Nigeria, Nsukka, Nigeria. E-mail: ezeakup@yahoo.com or peter.ezeaku@unn.edu.ng.

The study was conducted on soils of selected mining areas in Nasarawa State, North Central Nigeria to assess the environmental impact of open cast mining of coal and Baryte minerals. These entailed a survey of twenty four farm households, while soils, minespoils and water were collected and subjected to standard laboratory analyses. The results showed that soils around the mine sites were coarse textured, acidic with pH of 4.8, have high bulk density as evidenced on the hardness of the soil and load content of basic cations. Minespoils were found to be strongly acidic with a pH of 3.5, coarse textured, high organic matter content and low contents of basic cations. The high organic matter content if properly managed will enhance soil aggregation. However, the high contents of iron, copper, cyanide, and sulfate may have caused adverse effect on the soil ecosystem hence unsustainability of plants life. Water resources are polluted with high contents of acidity (pH 4.20), lead (0.23 mg/L), iron (0.6 mg/L), cyanide (0.7 mg/L), and nitrate (32 mg/L). The effluents from the mines caused water hardness (33.67 mg/L). The cumulative effects of the pollution loads on the soil and water resources affected the landuses in the host communities. Consequently, soil reclamation for farming purposes, establishment of water shed protection programme and provision of portable pipe-borne water, among others, were recommended as intervention measures.

Key words: Livelihood systems, landuse, mining, Nigeria.

INTRODUCTION

Mining may be described as an activity and occupation concerned with the extraction of minerals. These minerals include coal, petroleum oil, Baryte-limestone, quartz, lignite etc. Since pre-historic times, mining has been integral and essential to mans' existence (Madigan, 1981), hence the striking of stone flints by early man to make fire. The ever-increasing demand for minerals and energy as well as advancement in extraction techniques has increased the mining of minerals (Whiteman, 1982; Schobert, 1987). In line with this demand, the past four decades has witnessed an unprecedented exploration of solid mineral resources in Nasarawa State of central Nigeria often tagged the "Home of solid Minerals". The major increases in aggregate mineral resource production in this period have been associated with different kinds of minerals such as coal and Baryte. The soils of the communities of Obi LGA have their parent

materials as Awgu formation, which is generally fossiferous (Obaje et al., 2005). Furthermore, the communities are characterized to have sedimentary rocks of cretaceous Tertiary ages. The sedimentary rocks in these areas are the host to high rank of coal (seam) and Baryte deposits, which are principally mined at Agwantashi, Jangwa, Shankodi and Agaza communities in Obi Local Government Area (LGA) of Nasarawa State of Nigeria. The quantity of coal and Baryte in these communities makes them attractive to individuals and prospecting companies such as Steel Raw Materials Exploration Agency that mine coal and Baryte minerals under the supervision of Nigerian Mining Corporation. Reserve of good quality coal (of high to medium volatile bitumen) is usually extracted by way of striping away the soil to expose the mineral. This is the activity case of the communities under study.

The solid minerals of coal and Baryte are mined for a variety of reasons. For instance, coal has been assessed to be suitable for coke in steel manufacture because it has high calorific values and heat rising (Obaje et al., 2005). Baryte as the chief constituent of lithopone paint is used extensively as an inert volume and weight filler in drilling mud, glass, paper and in the chemical industry (Whiteman, 1982). The general importance of the mining sub-sector has been documented to include foreign exchange, employment and economic development (Obaje, 1996, 2005; Nwajiuba, 2000). Surface mining of solid mineral resources are necessary in regions where the solid minerals do not lie deep beneath the earth. This has been the practice more particularly when the mineral seems to lie fairly close to the surface and the rock above them may not be solidly consolidated (Schobert, 1987). The usual way is by stripping away the soil to expose the mineral extraction. This is the case of the communities under study. Adverse environmental consequences of surface mining is land degradation arising from vegetation destruction (extensive deforestation), exposure of the soil to run-off and even burden spoils as well as dumps that have been confirmed as having harmful minerals and chemicals that pollute the soil environment (Lawal et al., 1981).

Based on these facts, mining activities in the study areas may, consequently, result in the pollution and destruction of the natural environment. The effects may also have serious consequences for livelihood particularly in agrarian communities such as Obi LGA, whose residents are predominantly arable crop farmers with some livestock mostly cattle, sheep, goat, poultry, etc. The hypothesis for this study, therefore, is that negative externalities of open solid mineral mining impacts on the biophysical and agricultural resources of host communities. Information on soil is highly needed to enhance soil management techniques. More so, to restore our devastated lands for optimal productivity, the precise detection and identification of soil conditions and erosion features constitute core elements for monitoring our environmental ecosystems. This is with a view to conserving and managing the soils so as to produce more food and ensure food security for the populace. The main objective of the study is to assess the negative externalities of coal mining on the biophysical environment and agricultural resources of the communities in Obi LGA, Nasarawa State, Nigeria, where open cast mining predominate. Of interest is also to understand the relationship between the host communities and the mineral prospecting companies.

MATERIALS AND METHODS

Study area

The study was conducted in Agwantashi, Jangwa, Azara and Shankodi communities in Obi Local Government Area (LGA) of Nasarawa State, North Central Nigeria. Nasarawa State is in the southern Guinea savanna agro-ecological zone of Nigeria and characterized by abundant solid mineral reserves of coal, baryte, lead-zinc, lime stone, glass sand, tin, Thorium, Niobum, Monazite, precious metals, Titanium, gemstones, Uranium, manganese, Tungsten, etc, hence tagged the "Home of solid minerals" (Obaje et al., 2005). The predominant landuses is agriculture, mining and tourism. The aggregate spatiality of the mining areas is about 12 ha in an undulating landscape with upland and lowland features. The upland has two special features – stoniness and concretions (mine sites), while lowlands are unmined farming areas.

Sources of data

Biophysical surveys were conducted and involved taking soil samples from the earth surface around the sites and minespoil. Samples of water from adjourning streams into which seepage from the mine effluents flow were collected. Farmers were interviewed to ascertain the level of impacts, on the livelihood and ecosystems, of the mining activities as a major land use in the communities. Primary data was collected during the exploratory survey with the objective of making field observations. During the visit, twenty soil auger samples were randomly collected from mined and unmined sites (total of 40 soil samples) at the depth of 0 to 40 cm, a depth considered to be the agricultural plough layer that contains most essential macro-nutrients for plant growth and the maximum rooting depth for most arable crops (Ezeaku et al., 2002). Also, ten undisturbed core samples (15 cm diameter by 15 to 20 cm length) were randomly collected from each site at the depth of 0 to 15 cm for bulk density and total porosity determinations.

Nine water samples each from mine effluents and polluted stream water were collected (total = 18). The water collection was done up- , mid- and down streams (3 samples each) and thereafter bulked, respectively, to get composite samples.

The soil and water samples were collected in a sterilized polyethylene bags and bottles, respectively, and analyzed for physico-chemical properties in a standard laboratory.

Laboratory determinations

The soils were air-dried, crushed to pass 2 mm screen and analysed for physical properties: particle size distribution according to Gee and Bauder (1986) and bulk density was estimated by core method as described by Blake and Hartge (1986). Total porosity was obtained by Bouma's (1991) method.

Chemically, soil pH was distilled in both water and 0.1 NaCl solution using soil/liquid ratio of 1:2.5 (McLean, 1982), total N was obtained by micro Khajeldal method of Bremmer and Mulvaney (1982), organic carbon by Nelson and Sommers (1982) method. Organic carbon (OC) was multiplied by a factor of 1.724 to get organic matter. Available phosphorus (P) was determined by the method of Olsen and Sommers (1982). Exchangeable Ca and Mg by complexometric titration method, while K and Na were obtained by flame photometric method. Water samples were analyzed according to the standard procedure of APHA-AWWA-WPCF (1980).

Farmer sampling and data collection technique

Farmers within the spatial boundary of Obi LGA were considered in the study. Sampling was based on purposive non-probability method and only willing famers were interviewed. Twenty four (24) old and experienced farmers (20 males and 4 female headed households) indicated willingness and therefore formed the sample. The farmers were interviewed on the major land uses especially mining activities and how it impacts on their environment (land and water), agriculture and socio-economic life. These farmers had

Table 1. The mean effect of physical properties of soils analysed (n=40).

Nature of soil	Clay (%)	Silt (%)	Cs (%)	Fs (%)	Bd (g/cm^3)	Tp (%)
Unmined	13.75	10.25	22.5	53.50	1.355	33.98
Minespoil	8.75	7.00	28.2	55.75	1.502	32.63
F-LSD (0.05)	1.229	2.387	NS	NS	NS	3.162

% = Percentage; CS = coarse sand; Fs = fine sand; Bd = bulk density in g/cm^3; Tp (%) = percentage total porosity.

been involved in farming activities for over 40 years. Formal and informal in depth interviews were done using a structured questionnaire, and through informal discussions with the experienced farmers. Structured questionnaire containing objective and subjective questions were given to farmers. Data collected were through the farmers' responses as guided by Agricultural extension agent in the locality.

Statistics

The soil physical and chemical property data were analyzed using Genstat Discovery Edition 3 and SPSS 16.0 Version. Data were also analyzed through descriptive statistics and computation was done for percentages.

RESULTS AND DISCUSSION

Reconnaissance visits to the mining communities showed that exploration of solid minerals was by surface exploitation through excavation of the earth over-burden using heavy equipment. This resulted to farm land and forest/vegetation destruction. About 80% of the people indicated that minimal compensation was made for economic trees lost in the process but not for land. Furthermore, households affected were not relocated as prescribed by the Nigerian land use Law of 1978.

Effects of open cast on land, forest, and water resources

Land resources

The inhabitants of Obi LGA communities, where mining activities were done, already perceive mining as taking substantial area of land as indicated by 85% of the respondents. They were also aware that mining could lead to poor soil quality and infertility, and therefore declining crop yield. Reconnaissance observation of the sites showed that plant growth on excavated land was not vigorous in growth. The soils were characterized by scanty vegetation after some years. The poor performance of crops on mined soils may be related to compactness and low nutrient status of the soils. Crops were grown on unmined soils but to which upturned earth materials were washed, especially through water induced erosion were also affected. Growths of plants were observed to be stunted with yellow leaves, suggesting that the soils are poor soil quality. Eroded materials

(sediment load) cover farm land with fine sand which does not allow water to percolate into the soil sub-surface. This may also have contributed to the observed poor plant development and growth.

Results of physical analyses of soils (Table 1) show that unmined soils have coarse texture with mean sand, clay and silt percentages (%) of 76, 14 and 10, respectively. Minespoils (Coal remains) was texturally coarse with 9% clay, 7% silt and 82% sand. Though sand particles dominate those of silt and clay in both soils, it is not significant. Silt and clay values were significant (P≤ 0.05). Bulk density of mined soil (1.50 Mgm^{-3}) was higher than that of unmined soil (1.36 Mgm^{-3}). The high value observed in mined sites may be due to removal of vegetation arising from mining activities. Similar reports have been made relating high soil bulk density to poor vegetal cover (Ezeaku et al., 2005), soil surface crusting and compaction by raindrop and machine impacts (Neil et al., 1997), and suficial erosion (Lal, 1994). The bulk density of unmined soils may be considered ideal for plant root growth following the threshold classification values of soil bulk density by Arshad et al. (1996) which shows that a value less than 1.40 is ideal for root growth. However, bulk density values of both sites (Table 1) showed non-significance. The values of total porosity in both sites were significant (P≤ 0.05) although mined soil had slightly lower value (32.6 %) than that of unmined soils (33.9 %) as can be seen in Table 1.

In terms of chemical characteristics, the mean soil pH (H$_2$O) values of both soils was generally high; an indication of high acidic load. Even though unmined soil reaction value (4.9) was lower than those of the minespoils (4.1), both values were significant at P≤ 0.05 (Table 2). The pH variations of these two soils could be due to natural variation in soils from one location to the other. Soil with a pH value as that of the minespoil has been rated very acidic (Fokken, 1970) and is not likely to favor the entire soil environment including impairment of the functions of microbial organisms. High soil acidity has been shown to hinder the release of available essential plant nutrients and reduce crop yield (Ezeaku et al., 2003). Opara-Nnadi (1988) related yellowish coloration of plant leaves and suppressed vegetative growth, as observed in the mined sites, to high acidity. The results in Table 2 also show that the mean organic matter content (OMC) in the minespoils was higher (13.82 gkg^{-1}) than those of unmined soil (22.93 gkg^{-1}) even though both

Table 2. The mean effect of chemical properties of soils analysed (n=40).

Nature of soil	Soil pH	OM (g/kg)	TN (g/kg)	Ca[1]	Mg[1]	K[1]	Na[1]	CEC[1]	Base sat (g/kg)	Av. P (Mg/kg)
Unmined	4.9	22.93	0.838	1.08	0.500	0.1550	0.135	12.58	11.65	24.68
Minespoil	4.1	13.82	0.287	0.80	0.325	0.1075	0.145	9.27	8.62	20.25
FLSD (0.05)	0.42	2.728	0.2647	NS	NS	0.035	NS	NS	2.052	1.924

OM = Organic matter; TN = total nitrogen; Ca =calcium; Mg = magnesium; K = potassium; Na = sodium; CEC= cation exchange capacity; Av.p = available phosphorus; NS= non significance.

Table 3. Summary of the mean micronutrient values (range and % CV) of the minespoils.

Micronutrients	Range	Min	Max	Mean	SD	CV (%)
Boron	2.5	11.7	14.2	13.008	1.2915	9.93
Iron	0.97	3.22	4.19	3.873	0.4515	11.66
Copper	0.36	3.72	4.08	3.930	0.1588	4.04
Zinc	1.22	3.07	4.29	3.950	0.5878	14.88
Cyanide	0.51	3.79	4.30	4.120	0.2359	5.73
Sulphate	0.35	3.92	4.27	4.083	0.1457	3.57

Min = Minimum; Max = maximum; SD = standard deviation; %CV = percentage coefficient of variation.

values were significant ($P \leq 0.05$). Reverse trend was the case for total nitrogen, which also showed significance at the same probability level. The high OM content observed in minespoil may be associated to the vegetation origin of coal seams. The values of exchangeable bases such as calcium (1.08 Cmolkg^{-1}) and magnesium (0.500 Cmolkg^{-1}) as well as CEC (12.50 Cmolkg^{-1}) were higher in unmined soils relative to minespoils. Sodium value (0.145 Cmolkg^{-1}) was higher in minespoil (Table 2). All the values were non significant. But the values of potassium (0.1550 Cmolkg^{-1}) and base saturation (11.65 Cmolkg^{-1}) were significantly ($P \leq 0.05$) higher in unmined soils than minespoils.

Mean values of available phosphorus (P) concentration in both soils were significant ($P \leq 0.05$) but higher (24.68 mgkg^{-1}) in unmined soils than minespoils with 20.1 mgkg^{-1} as shown in Table 2. The phosphorus values can allow crop growth within the environment without additional P fertilizer application. Vogel (1975) reported similar P value as been sufficient crop plant in mined soils. Data shown in Table 3 indicate the average value ranges and coefficient of variation of micronutrients as observed in minespoils. The value of Boron ranges from 11.7 to 14.4 mg/L with a mean of 13.0 mg/L. This mean value is higher than the rest of other micronutrients. Mean values of iron, copper and zinc were almost similar (mean of 3.9 mg/L). Sulphate (4.08 mg/L) and cyanide (4.12 mg/L) values were almost same (Table 3). In terms of percentage coefficient of variation, zinc varied most (Cv% = 14.88) and sulphate least (Cv% = 3.57) varied. The implication of these variations is that sulphate, copper and cyanide would have more effect on human and livestock than boron, iron and zinc.

Forest resources

Vegetation destruction is a major consequence of surface mining and this applies to the mine sites. Vegetation destruction is due to deforestation activities which usually results from land use pressure. It leads to soil erosion due to run-off, loss of farm lands and plummeting of crop yields as experienced by farmers in the study areas. The mined soils were observed by farmers to be very hard, an evidence of compactness reflecting high bulk density recorded in Table 1. Reduced growth and yields of planted crops by farmers may be related to increased bulk density that inhibits water conduction and availability as well as oxygen availability to root zone. All of these corroborate with earlier observations that compaction, typical of some agricultural systems, decrease aggregate stability (Caron et al., 1992) leading to collapse of soil pores (decreased macro-porosity)(Swartz et al., 2003). These specifically affect the transmission and drainage pores which impair plant performance (Caron et al., 1992). Thus, the net consequences of vegetation destruction are ecological degradation including loss of aquatic life, biodiversity and forest resources (Isirimah, 2004; Ezeaku and Alaci, 2008).

Water resources

During reconnaissance visits, small streams and rivulets, which pass through the mining communities, including neighboring towns were observed. The streams and rivulets were observed to be linked, and both flow into the River Benue (the second largest river after River Niger in

Table 4. Some characteristics of mine effluents and water samples in the study areas.

Property	Mine effluents seeping into streams (a)	Polluted stream water (b)
Colour	Yellow	Yellowish brown
Total dissolved solid	-	50.7 mg/L
pH	2.02	4.20
Conductivity at 28°C	216	285 uskm
Total hardness	43.67	33.60 mg/L
Organic matter	12.91	8.57 mg/L
Iron (Fe)	0.68	0.60 mg/L
Cadmium (Cd)	0.36	0.60 mg/L
Lead (Pb)	0.27	0.23 mg/L
Copper (Cu)	3.8	2.4 mg/L
Zinc (Zn)	9.3	8.0 mg/L
Manganese (Mn)	2.72	2.05 mg/L
Cyanide	0.93	0.72 mg/L
Nitrate (No_3^-)	36.6	31.8 mg/L
Chloride (cl)	312.9	82.7 mg/L
Sulfate (So_4^-)	213.6	103.2 mg/L

Nigeria). The streams are the main source of water supply for households use in the communities. The study however concentrated on the Jangwa and Dep streams because they are the closest to the mine sites and seepage from the mines flow into them. There are indications of pollution of these streams. The indications, as observed by 95% of the respondents, include change in water taste and sometimes odour, water hardness as evidenced by difficulty of soap/detergent to lather or foam when used for washing. Other effects are fish mortality as sometimes farmers observe floating dead fishes on strean surfaces. Itching and whitening cover on the skin were also sometime observed by the farmers when the stream water is used for bathing. Also when the water is used for cooking the local palm oil rarely gives the desired red color. In order to have a safe drink, the communities now largely use alum (calcium carbonate; $CaCO_3$) for treating water from these streams. Isirimah (2004) noted low species diversity and impairment of water recreational use as other effects of stream pollution.

The results in Table 4(a) show an analysis of pollutants and seepage from the mine sites, which flow into the local Jangwa and Dep streams. The effluent had a pH of 2.02, which indicate extreme acidity. Cyanide content at 0.93 mg/L was at high level relative to established standards (Table 5). Chloride at 312.9 mg/L was high and unsafe level. Nitrate ion at 36.6 mg/L was moderately high. Isirimah (2004) earlier reported that any water that has nitrate content in excess of 35mg/L is dangerously unsafe for human health. The sulfate (SO_4^-) value at 213.6 mg/L was very high. It can therefore be inferred that the seepage of this effluent that contain common pollution loads, into the streams is related to the

observed characteristics of the water from the streams. The inferences compare well with other studies (Hill, 1979; Paker, 2000). Samples of the mixtures of the mine effluent with local Jangwa and Dep streams, and allowing for a flow of 25 m were analyzed and the results shown in Table 4(b). It however should be expected that the concentration of the effluent in the water downstream, which is available for use by the communities, decreases with distance. Analysis of the water sample shows that the water was still acidic with a pH of 4.20, though this might decrease further downstream.

Drinking water standards as established by the United States Public Health Service (USPHS) and the World Health Organization (Renn, 1970; Isirimah, 2004) is presented in Table 5 and are used to compare with the observed values in Table 4(b). The result show that the stream water was acidic (pH = 4.20), while the values of iron (Fe) (0.60 mg/L), lead (Pb) (0.23 mg/L), copper (Cu) (2.4 mg/L), zinc (Zn)(9.0 mg/L), manganese (Mn) (2.1 mg/L), among others, are unwholesome for human consumption following the USPHS and WHO standards (Table 5). Water pollution is a vital problem and the communities expect improvement in present water supply situation, which is the local streams and which are affected by mining. This is considered as one of the most pressing problem of the host communities.

Agricultural landuses

The traditional land tenure system is through communal access to farmland. This is indicative of low population density and subsistence-oriented agriculture with undeveloped input and product market (Doppler, 1991).

Table 5. Characteristics of drinking water standards (in milligrams per litter-mgL-1 except pH values).

Property	Values
PH	7.0, 9.0
Calcium (Ca)	200.00
Magnesium (Mg)	150.3
Manganese (Mn)	0.05 DL
Iron (Fe)	0.3 DL
Lead (Pb)	0.05 ML
Nitrate (NO_3^-)	35 DL; 40 ML
Chloride (Cl^-)	250.0 DL
Cyanide	0.01 DL; 0.2 ML
Sulfate (SO_4^-)	200.0 DL
Copper (Cu)	1.0DL
Zinc (Zn)	3.0 DL
Total Dissolved Solids	400.0 DL

DL = Desirable limit; ML = maximum limit. Source: Renn (1970) and Isirimah (2004).

Table 6. Characteristics of farmer respondents in the communities (n = 24).

Age	Average	35 years
	Average household size of which	8
Labour	Males	42%
	Females	58%
Occupation	Full time farmers	75%
	Part – time farmers	25%
	Average number of plots per household	6
Land	Average area of land (ha)	2.6
	Average area used (ha)	1.4

Table 6 shows that average age of farmers was 35 years and in an average household of 8. Average number of plots owned per household was 6. Average area owned by each household was 2.6 ha, which means an average plot size of 0.41 ha. However, each household used only 1.4 ha during the study period, suggesting that about 40% of the land holdings were under fallow. Fallow is the main means of soil fertility regeneration. The rotation value for farming systems in the communities studied was 25% (Box 1). This is in tandem with the report by Rutheriberg (1980).

The main form of land utilization was arable cropping of yams, cassava, maize, millet, sorghum, Beniseed (Table 7) under rainfed situation. Mixed cropping was dominant in the area. The major combinations of crops were cassava/maize/yam and cassava/maize. The cropping calender commences in March/April with land clearing and tillage when the rains starts. Usually, planting of yam and maize with vegetables starts from April to June. Cassava planting is virtually done throughout the year except from September to December when the moisture stress is so much that plants wither. In July and August, there is planting of Guinea corn and Pigeon pea. The harvesting of early maize and yam is also in July. Late maize planted between August and September is harvested in November/December.

Community livelihood

Negative environmental externalities of mining may constrain livelihood to host communities by hampering agricultural resource uses. Such possible consequences as soil quality have clear economic importance. The

$$R = \frac{V}{A + B} \times \frac{100}{1}$$

where R = Rotation value;

V = Actually weighted vegetation period (cultivation period);

A = Years of cultivation;

B = years of fallow.

Therefore,

$$R= \frac{1}{3 + 1} \times \frac{100}{1}$$

$$= \frac{1}{4} \times \frac{100}{1} = 25\%$$

Box 1. Rotation value for farming systems in the communities of study.

Table 7. Calendar of cropping system in the communities studied.

Period	Operation
March – April	Land clearing – tillage when rain comes
April – May-June	Planting of yams, maize, cassava
July – August	Planting Guinea corn, Pigeon pea, cassava
July	Harvest of early maize and yam
August – September	Planting of late maize
November –December	Harvest of late maize.

possibility of long-term soil productivity degradation has potential significant implications for economic welfare (Kim et al., 2001). Further, contaminated ground water can have negative impact when consumed or it can contribute to surface water degradation by moving laterally into streams (Parker, 2000; Ezeaku and Anikwe, 2005). The specific positive impact of mining in Obi communities as identified by respondents include employment generation (60% of the population), increased income (80% of the population showed to have earn higher income than realized from farm produce sales), increased economic activities (more petty businesses especially for artisan women) and increased infrastructure (graded roads) and transportation facilities. On the negative side, mining is said to exert pressure on the village livelihood systems through deforestation (vegetation destruction leading to fewer pastures for livestock grazing, low wood harvesting for home use and sale) and declining arable land area. Seepage into the village streams also has adverse effects on the quality of

water. Some of the specific negative effects have already been discussed.

Labour and employment

Mining offer direct employment opportunities to some indigenes and the neighbouring inhabitants. Indirectly, some other employment generating activities are stimulated by the mining operations. These include petty trading and transportation, which service the workers at the mines. With respect to labour use for farming, due to the activities of transport an average of 2 km road has been graded by the local government authority. Money realized from taxes/royalties are used for the renovation and rehabilitation exercises.

Income and living standard

Workers were employed at the mines and they reside

around the towns surrounding the mine sites and earn their income. This was an important contribution to rural liquidity and therefore living standards. Community Development Association (CDA) obtained money from mineral transport, taxes/royalties. This was used for development activities and was considered by respondents as a positive effect of mining on the areas. The CDA uses the money realised to repair and build schools, cottage hospitals and maintain roads.

Conclusion

The study revealed that the soils have been degraded, while the water resources were polluted through accelerated anthropogenic (human) activities of mining and others. In this case, liming and fertilization of the soils is important for continuous cropping but should be done under good soil conservation practices. Further revelation of the study shows that there is a high level awareness and interest on mining by the communities. They have used the activities of mining to their advantages in terms of income generation to improve their living standard. However, the communities desire development and establishment of related industries such as State Mineral Resource Development Agency, as well as Medium and Small-Scale coal Briquetting Enterprises, so that coal briquettes can be supplied for domestic cooking and to generate more income and gain employment.

Furthermore, the communities expect aid from the government in the following areas: improvement of health centre, educational opportunities, drinking water through establishment of watershed protection programme and extension of electricity supply. All these will reduce poverty level and bring positive social transformation. The communities also expect the government, at local, state and federal levels, to engage in some form of land reclamation, so that they can use the areas already mined for farming and this is the desire of all the respondents (100%). This is not surprising considering that the communities are agrarian with all the inhabitants dependent on the land. Despite the provision of the Federal Government of Nigeria decree (Landuse decree, 1978) on the landuse and reclamation, the decree is silent on these mined soils. Therefore, to restore the land for optimal productivity, mapping and monitoring of the degradation process should be the basic means for the restoration of the degraded areas, while the use of soil amendment is imperative.

REFERENCES

APHA – AWWA – WPCF (1980). Standard methods for the examination of water and water use. 13th edition. Am. Pub. Health Assoc. Washington.

Arshad MA, Lowery B, Grossman B (1996). Physical tests for monitoring soil quality. In: Doran, J.W., Jones, A.S (eds) Methods of Assessing Soil Quality. SSSA Special Publication No. 49 SSSA, Madison, W. I., pp. 123-141.

Bouma J (1991). Influence of Soil Microporosity on Environmental Quality. Adv. Agron., 46: 1-38

Blake GR, Hartge KH (1986). Bulk density. In: Methods of Soil Analysis, part 1, A. Klute (ed) Agron. No 9. USA Madison W. I., pp. 370-373.

Bremmer JM, Mulvaney CS (1982). Total N, In: Page et al.,(eds) Methods of Soil Analysis. Part 2. 2nd ed. Agron. Monog. 9. ASA and SSSA. Madison, W.I., pp. 626-895.

Caron J, Kay BO, Stone JA (1992). Improvement of structural stability of a clay loam with drying tendency. Soil Sci. Soc. Am. J., 56: 1583-1590.

Doppler W (1991). Land wirttschaftliche betriebssysteme in den Tropen and subtropen. Verlag Eugen ulmer, Stuttgart. Germany.

Ezeaku PI, Akamigbo FOR, Asadu CLA (2002). Maize yield predictions based on soil fertility parameters in southeastern Nigeria. J. Agro-Tech. Ext., 2 (2): 30-38.

Ezeaku PI, Agbede OO, Olimah JA (2003). Nutritional Aspects of Plant Production in an acrisol (acidic) soil. A review. Savannah J Sci. Agric., 3: 23-33.

Ezeaku PI, Anikwe MAN (2005). A model for description of water and solute movement in soil-water restrictive horizons across two landscapes in south east Nigeria. J. Res. Agric., 2(2): 47-53.

Ezeaku PI, Alaci D (2008). Analytical situations of land degradation and sustainable management strategies in Africa. Pakistan J Agric. Soc Sci., 4: 42-52 (http://www.fspublishers.org)

Fokken A (1970). In Agricultural compendium (1981). For rural development in the tropics and subtropics. Elsevier scientific publishing company. Australia, Oxford, New York, p. 157.

Gee GW, Bauder JW (1986). Particle size analysis. In: Klute, (A.ed) Methods of Soil Analysis. Part 2. 2nd ed. Agron. Monog 9. ASA and SSSA. Madison, W.I., pp. 383-411

GENSTAT (2003). GENSTAT 5.0 Release 4.23 DE, Discovery Edition_1, Lawes Agric. Trust, Rothamsted Experimental Station Press, UK.

Hill FR (1979). Barriers to conservation. In: Pelofsky, Alt (ed), Coal Conservation Technology: Problems and solutions. ACS symposium series 10. Am. Chem. Soc. Washington, D.C., pp. 207-217.

Isirimah NO (2004). Soils and Environmental pollution management. Nuton publishers Oweji, p. 76.

Kim KB, Barham B, Coxheed I (2001). Measuring Soil quality Dynamics. A role for Economists, and implications for Economic Analysis. Agric. Econ., 25(1): 13-26.

Land Use Decree (1978). Promulgation of the Federal Government of Nigeria, 1978.

Lal R (1994). Methods and guidelines for assessing sustainability use of soil and water resources in the tropics. Soil Management Support Services technical monograph #21: 1-78.

Lawal FA, Singh K (1981). Analytical studies of Kano and Sokoto States for possible environmental pollution. Bul. Chem. Soc. Nigeria, 6:68-77.

Madigan RT (1981). Of minerals and man. Austalasian lust. Mag and Wetportville, Australia, p. 138.

Mclean EO (1982). Soil pH and Lime Requirement. P. 199-224. In Page et al (eds) Methods of Soil Analysis, part 2. Chemical and Microbial Properties. 2nd ed. Agron. 9. ASA, SSSA, Madison, W. I.

Nwajiuba CU (2000). Socio-Economic Impact of Solid Minerals Prospecting on host communities. A study of Okaba, Kogi State, Nigeria. Technical Report to the Nigerian Coal corporation, Enugu, Nigeria, p. 59.

Neil C, Melillo JM, Steudler PA, Lerri CC, Demoraes JFL, Picolo MC, Brito M (1997). Soil carbon and nitrogen stock following forest clearing for pasture in the southwestern Brazilian Amazon. Ecol. Applic., 7(4): 1216-1225.

Nelson DW, Sommers LE (1982). Total carbon, organic carbon, organic matter. p 539-579. In Page et al. (eds) Methods of Soil Analysis. Part 2 Agron. 9. ASA, SSSA. Madison, W.I.

Obaje NG, Abaa SI (1996). Potential for coal-derived gaseous hydrocarbons in the Benue Trough of Nigeria. Petroleum Geol., 19: 77-94.

Obaje NG, Nzegbuna AI, Moumouni A, Ukaonu CE (2005). Geology and Mineral Resources of Nasarawa State, Nigeria. A preliminary Investigation. Paper presented at the 4th Nasarawa State Agricultural and Solid Minerals Exposition held at Agyaragu, Nasarawa State,

Nigeria, 21-23 March, pp. 1-27.

Olsen SR, Sommers LE (1982). Phosphorus.In: Page *et al* (eds) Methods of Soil Analysis. Part 2. Agron 9. ASA, SSSA. Madison, W.I., pp. 403-434.

Opara-Nadi OA (1998). Liming and organic matter interactions in two Nigerian ultisols on soil pH, OC and early growth of maize (*zea mays. L*). In: Uzoho, B.U, Oti, N.N and A. Ngwuta, 2007. Fertility rates under land use types on soils of similar lithology. J. Am. Sci., 3(4): 20-29.

Renn CE (1970). Investigating Water Problems: A water analysis manual. La Motta, Chestertown, p. 55.

Rutheriberg H (1980). Farming systems in the Tropics. Clarodon ublishers, London, U.K, p. 230.

Schobert HU (1987). Coal: the energy source of the past and Future. Am. Chem.Soc. Washington, D.C.

Swartz RC, Unger PW, Evelt SR (2003). Land use effects on soil hydraulic properties. Cons. and Production Res. Lab. USDA-ARS.

Vogel WG (1975). Requirement and the use of fertilizers, Lime and mulch for vegetating acid mine soils. Austoalasian Inst. Mag and Wetportville, Australia, pp. 12-20.

Whiteman AJ (1982). Nigeria: Its petroleum Geology, Resources and Potentials. Graham Trotman Ltd, Sterling, London, 1: 165.

12

Status of the Sondu-Miriu River fish species diversity and fisheries: Sondu-Miriu Hydro-Power Project (SMHPP) operations

Dickson Otieno Owiti[1,3] , Raphael Achola Kapiyo[2] and Esna Kerubo Bosire[2]

[1]Department of Zoology, Maseno University, P.O Box 333-40105, Maseno, Kenya.
[2]Department of Environmental Science, School of Environment and Earth Science, Maseno University, P.O Box 333-40105, Maseno, Kenya.
[3]Department of Fisheries and Natural Resources, Maseno University, PO Box 333-40105, Maseno, Kenya.

The Sondu-Miriu Hydroelectric Power Project (SMHPP) is a run-of-the-river hydro-power project on the Sondu-Miriu River, Kenya. The part of the river studied: between the Sondu Bridge upstream, and Osodo Bay on L. Victoria, was divided into three zones; 1. Upstream of the dam, artificial lake approximately 5 km long; 15 – 2 km wide, 2. Depleted section and, 3. Lower section. SMHPP caused part of river below the dam (depleted section) to reduce in volume during power generation. Since inception, construction and implementation of the project, concerns were raised that the project would result in loss of *Labeo* and *Synodontis* fishes which were singled out as most vulnerable. This investigation set to determine the effects of SMHPP on river's fishes; and focused on species diversity, abundance and distribution. An electrofisher was used in sampling. The data obtained was compared with those reported earlier, particularly a single previous report based on electrofishing. Fish biomasses were: Upper zone 4,583 g, depleted zone, 10,666 g and lower zone, 22,004 g respectively, with a total biomass of 37,253 g from fishing activity lasting for 429 min. Catch-effort data for each zone; types, numbers and weights are given, compared and discussed. Diversity was highest, Simpson's Diversity Index was 12.945 in the upper zone.

Key words: Sondu-Miriu, hydropower project, electrofishing, fish diversity, river zones, fishing.

INTRODUCTION

From the inception of the ideas and feasibility activities of the Sondu-Miriu Hydroelectric Power Project (SMHPP) in Western part of Kenya, Lake Victoria Region of East Africa, the community of the Sondu-Miriu River area raised a number of concerns with the fear that the project would affect the local environment adversely in certain ways. High among this concern, was a possible negative effect on the fish and fisheries activities associated with the river. Although, the residents were divided in their complaints and the complaints were variously described as petty (Ouma, 2009), such concerns persisted (Ashe et al., 2009). According to Ashe et al. (2009), one of the complaints was that the community members had suffered loss of most sought after local fishes such as *Labeo* spp. and *Synodontis* spp. According to the community of the area, these fish species, (*Labeo* and *Synodontis*), were once abundant in the river unlike presently that they have become scarce. This scarcity has been caused by untreated waters released back to the river after running the turbines at the SMHPP Power

House. Willoughby et al. (2003) and Kapiyo et al. (2003) had investigated and reported on the status of the fisheries and socioeconomic activities of the area to coincide with the initial stages of the SMHPP construction. The SMHPP was commissioned in July 2009 and soon after, it became operational. However, continued social and public protests and upheavals over the socioeconomics and environmental impacts by the project on the river fisheries, called for explanations based on properly carried out research and analyzed data. The present research was undertaken to provide information based on facts to both the locals and the international forums conscious of fisheries and environmental issues. The investigations aimed at determining the changes, if any, that occurred in the various aspects of the fisheries of the Sondu-Miriu River, particularly fish biodiversity, abundance and distribution which could be associated with the establishment and operations of SMHPP. Such information would benefit the SMHPP project contractors, and implementers, who "would be in the know" concerning the past and present status of fish and fishery of the Sondu-Miriu River. The studies were conducted during the March-April 2010 period.

Sondu-Miriu and other Western Kenya Rivers fisheries

Documented information on Sondu-Miriu River fisheries is scanty, and such information is almost non existence for the river upstream of Odino Falls. The scarce literature available addresses mainly the Lower Zone of the river up to Odino Falls, 10 km river course from the river mouth upstream. The available literature is also at variance with each other concerning the dominance and abundance of the fish species involved. However, from this scanty literature, it is noted that the Lower portion of the river and its floodplain has supported fisheries activities for a long period of time.

Amongst the first documented information on the Sondu-Miriu River fishery is that of Whitehead (1959) who reported that the most important migrant fishes of Sondu-Miriu River were as follows: the large barbus- *Barbus altianalis radicliffi* (fuani), *Labeo victorianus* (ningu), *Schilbe mystus* (sire) and *Tilapia vaiabilis* (Mbiru). Other less abundant species according to White head (1959) were; *Alestes– (Brycinus)* (osoga), *Synodontis* spp. (okoko), *Clarias* (mumi) *Bagrus* (sewu) and *Protopterus* (kamongo). Later, Manyala and Ochumba (1992, 2008) found that the five most important species in the lower Sondu-Miriu were *Schilbe mystus, Clarias gariepinus, Synodontis afrofischeri, Lates niloticus* and *Oreochromis variabilis*. Later, Muli and Ojwang (1998) reported that the most abundant species in the lower part of the river were *Schilbe mystus, Synodontis victoriae, Synodontis afrofischeri, L. niloticus* and *Labeo victorianus*. According to Gichuki et al. (2001), *Clarias gariepinus, Protopterus*

aethiopicus, Schilbe intermedicus, L. victorianus, Oreochromis niloticus, Synodontis victoriae and *Xenoclarias eupogon* were the most dominant species in the same lower part of Sondu-Miriu River.

Indeed, many reports on the Sondu-Miriu River fisheries are concerned mainly with the lower part of the river, from the river's mouth up to the Odino Falls. Muli and Ojwang (1996) and Ochumba and Manyala (1992) had noted a significant decline of fish population in Sondu-Miriu River. Some researchers had reported on the general decline in river fisheries within the Lake Victoria basin including the Sondu-Miriu River. These reports include Cadwalladr (1964, 1965, 1969), Whitehead (1958, 1959), Kibaara (1981) and Awange and Ong'ang'a (2006).

Ochumba and Manyala (1992) noted that fish yield in the Sondu-Miriu River had decreased from 668 tons in 1959 to 108 tons in 1992; and that species such as *Clarias gariepinus, Schilbe mystus* and *S. afrofischeri* had replaced the more favorite traditional fisheries that had hitherto been dominated by *Barbus altianalis* and *L. victorianus*. Ochumba and Manyala (1992) and Willoughby (2006) had independently recommended fish species transfers and re-stocking of the river as a management measure in reviving the river fisheries. Fish relocation was later undertaken by Willoughby (2003), when 350 individuals of a mixture of *L. victorianus* and the large barbus (*B. altianalis*) were transferred into the depleted zone between the downstream of the Apoko dam and the upstream of the Odino Falls.

Like the other rivers of Lake Victoria Basin such as Nzoia, Nyando, Yala, Mara and the others. Particularly, the Lower part of the Sondu-Miriu River fisheries must have played important role as source of fish for the riparian community in the first half of the last century. But the significant river fisheries as an important source of fish for the local people gradually waned as a result of reducing fish stocks.

The reasons for the decreased fish populations in the rivers are varied and include increased fishing intensity and pressure and poor fishing methods, as well as the effects of climate change with changes in water quality and erratic river flow regimes. For the case of the Sondu-Miriu River, Mainga (1981) noted that river pollution by the agric-based chemicals washed down into the river by surface runoffs caused by the decline in fish stocks. The decline in the catches of such species such as *Labeo, Alestes (Brycinus)* and the *Barbus* species had been attributed to predation by the Nile perch and intensive fishing that include use of prohibited gears (Manyala and Ochumba, 1990). Mwashote and Shimbira (1994) also noted that agricultural practices that are accompanied by use of agrobased chemicals played a big role in the decline of fish in the Sondu-Miriu River. Witehead (1959), Ochumba and Manyala (1992), Ogutu-Ohwayo (1990) and Muli (1990) stated that the main causes of decline in the catches of riverine species in the region include over-

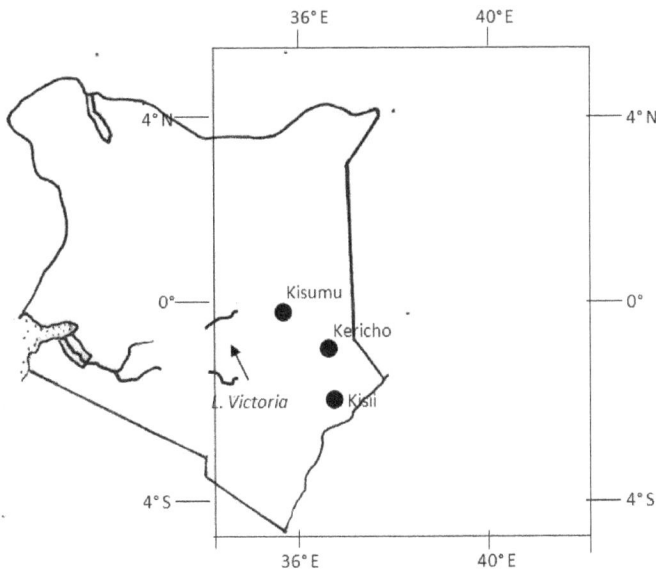

Figure 1. Map of Kenya showing the position of the Sondu-Miriu River and the portion of the river in the present study.

fishing, use of destructive fishing methods, habitat destruction, poor fisheries management and pollution.

Various fishing methods are used to catch fish in the Sondu-Miriu River. These include the use of gill-nets, barrier traps, long lines (hooks and rods), seine nets and mosquito nets. The use of the various traditional fishing baskets known by the locals as sienyo, ounga and traps are becoming rare. Also, traditional spear fishing is fading away. The main fisher folk are males, both the adults and the youth. In the past, women used to fish using baskets, but this has since faded away. Only two previous studies (Willoughby et al., 2003 and Kapiyo et al., 2003), of the Sondu-Miriu River have so far include the upstream of the Odino Falls; and in the course of these studies, the river course about 40 km from the Sondu Bridge to the river mouth had been categorized into the three zones; upper, middle and lower zones.

MATERIALS AND METHODS

Study area and the Sondu-Miriu River

The study area is depicted in Figures 1 and 2. The Sondu-Miriu River is one of the several main rivers that feed Lake Victoria on the Kenyan side of the lake; the Nyanza Gulf. It is one of the five major rivers of the Kenya's Lake Victoria, basin; the others are Nzoia, Yala, Migori-Kuja, and the Nyando Rivers. As shown in Figure 1, Sondu-Miriu has its source on the western slope of the Mau Escarpments of Kenya with approximately 3,470 km² of catchment area of the highlands within and adjacent to western part of Kenya. In its lower course, the river flows through narrow gorges in the southern escapements of the Nyakach plateau; it then meanders into the Odino Falls and to the flood-plains of Nyakwere and Sango as it pours its waters into the Lake Victoria at the Osodo Bay.

The Odino Falls on the river is a major and the most conspicuous and significant physical feature on the Sondu-Miriu River course. The falls are located in the Nyakach plateau and the escapement. The Odino (locally called barrier) falls (Figure 2), situated 10 km upstream of the Sondu-Miriu River mouth, actually forms an effective barrier on the path of upstream migration for any aquatic organisms, particularly fishes that might attempt to take such migrations.

A significant portion of the Sondu-Miriu River waters is, dammed and diverted for use in the generation of hydroelectric power by the Sondu-Miriu Hydroelectric Power Project (Figure 2). The SMHPP power house is located at Kolweny, approximately 400 km from Kenya's capital, Nairobi. The water is dammed and diverted from the river at Apoko, and directed via underground tunnel (Figure 2) to the Power House, approximately 10 km from the river. The dam at Apoko has created an artificial lake in the upstream of the river (Figure 2). The name of the newly created lake varies from Lake Apoko to Pond Area.

The studied portion of the river (Figure 1) was divided into three sections as described below:

1. Upper zone (UZ) which partly contains the pond area (the newly created artificial lake), the impoundment zone and the river course upstream of the lake to Sondu Bridge. The Upper Zone has been identified in 5 sites (Table 1)
2. Depleted zone (DZ) where much water is intermittently withdrawn to run the turbines of the SMHP project at Kolweny, approximately 10 km away from the dam intake, Depleted Zone has been identified in 6 sites (Table 1)
3. Lower zone (LZ) part of the river into which the water from the turbines renters the river and continue to flow down to the lake. Lower Zone has been identified in 9 sites (Table 1)

In the present study, a total of twenty sampling sites, each corresponding to those that had been sampled in the previous (Willoughby et al., 2003) electro-fishing method was used.

Fishing

Eletro-fishing (Figures 3 and 4) was adopted as the preferred fishing method and this technique had been successfully used by the previous (Willoughby et al., 2003) survey. Twenty stations were sampled along the five stretches (zones) of the river.

For each site, a span of 50 m on either side of the generator and giving a 100 meter stretch of the sampling site. The electrofisher used was a Variable Voltage Pulsator Electro-Shocker Model VVP-15 equipment with a generator machine, producing direct current at 400 V and 10 A; an anode of a 30 cm ring attached by 50 m of heavy cable to the control box and a cathode of 2 m of splayed copper cable. The system was operated by trained and experienced technicians. The time taken for electro-fishing activity at each sampling site was from 14 to 43 min. Each fish caught was identified to species using field guides of fishes of East Africa and of the Lake Victoria region (Greenwood, 1966; Whitehead, 1959).

The average fishing activity (approximately 22 min) resulted in a catch rate of 87 g of fish per minute. This compares closely with that of Willoughby et al. (2003), of average fishing activity of approximately 23 min with catch rate of 53 g of fish per minute.

Length-frequency distribution and length-weight data for each sampling site were recorded. The observed data were compared with the information gathered from the local fishermen, administrative personnel and local residents; and with the available documented reports from previous investigations. The catches from the various zones were compared to show distribution, variety and abundance along the Sondu-Miriu riverine fisheries; and this fishery sub-sector's contributes to the general well being of the riparian populations of the study area.

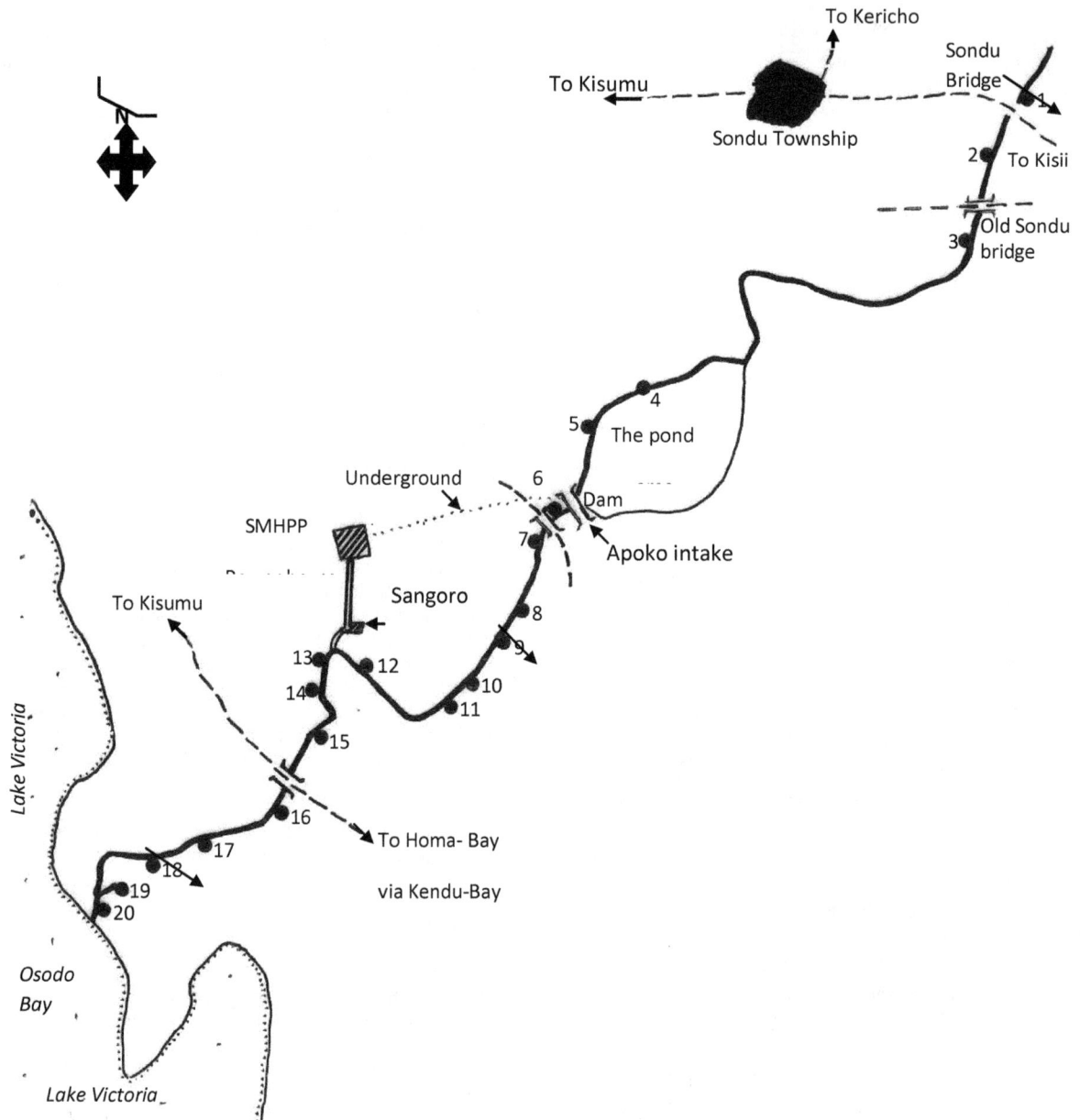

Figure 2. Sampling sites by numbers of the Sondu-Miriu River. This is a longitudinal section of the river from the Sondu-Bridge in the upstream to the river's mouth at Osodo Bay on Lake Victoria.

Table 1. Sampling sites identified in each zone of the studied river and named accordingly.

Upper zone (5 sites)	Depleted zone (6 sites)	Lower zone (9 sites)
1) Sondu Bridge	6) Intake Bridge	12) Wadhlango Rapids
2) Sondu Gauge	7) Intake Downstream	13) Wadhlango Beach
3) Sondu Old Bridge	8) Kut Pool	14) Sang'oro Gauge
4) Mamboleo Gauge	9) Kut Pool Downstream	15) Asera Oyster Bed
5) Mamboleo Downstream	10) Kuoyo Sand Site Up	16) Nyakwere Bridge
	11) Kuoyo Sand Site Down	17) Nyamnacha Banks
		18) Nyandho Ferry
		19) Chuowe Stream
		20) Osodo Beach

Figure 3. Electrofishing in progress in the pond area (the newly created artificial lake).

Figure 4. Bringing the catch ashore.

RESULTS AND DISCUSSION

Representatives of the fish caught are given on Table 2. According to Reynolds (1983) it is appropriate and more meaningful to record catch rates by electrofishing in terms of weight (g) per 100 m or per minute. A number of reporters have, however, reported their catch rates in terms of kg/h or kg/km. It was noted in the present study that the electrofishing in Sondu-Miriu River was more convenient to report the catches in terms of weight (g) per minute or weight (g) per 100 m.

During the present survey, a total of 25 species were caught in the river (Table 2). A number of reports have variously reported the number of fish species in the Sondu-Miriu River at between 14 and 30 species. Some of these reports include that of Whitehead (1959), 14 species; Willoughby (2006), 24 species; Manyala and Gichuki (1992), 30 species; Mugo and Twaddle (1999), 18 species; and Gichuki et al. (2001), 28 species. Many of these reports are for the lower portion of the river (from river-mouth to Odino Falls) only.

Comparison in the present electrofishing survey was made with that of the previous survey (Willoughby et al., 2003) and the following species; which had been reported by Willoughby et al (2003) were: *Protopterus aethiopicus, Bagrus docmack, Brycinus salderi;* from the Lower Zone of the river and *Xystichromis* spp. reported from the Upper Zone. Conversely, the following species which were captured during the present survey were absent from the previous (2003) survey; *Synodontis victoriae* and *Schilbe mystus* (Lower Zone) and Haplochromine species which were caught in the various zones of the river. Other species such as *Mormyrus* (Mormyridae), and black bass *Macropterus salmoides* which have been frequently reported from the Sondu-Miriu River by some other workers were not caught; neither during the present nor in the previous electrofishing surveys.

Species diversity

Catch-effort data collected were used to determine the diversity indices for each of the three zones. Simpson's Diversity Index was highest (12.945) in the Upper Zone.

The main fishes of Sondu-Miriu caught during the present survey ranked in order of biomass (total weights landed): *B. altianalis, L. victorianus, C. gariepinus, S.e mystus, O. leucostictus, O. variabilis, T. rendalii, L. niloticus* and *C. theodorei.*

Each of the three "traditional zones" of the river is well described in the present as well as by the two earlier reports (Willoughby et al., 2003; Kapiyo et al., 2003). However, on the establishment of the SMHPP and it becoming operational, five zones of the area are now evident, and there are clear deviations from the "traditional" three zones. The five zones as per the present study, and which are hereby recommended for any future studies are depicted on Table 1 and Figure 2. These suggested that new zones are represented by the sampling sites as follows:

1. Upstream Zone: sampling sites 1, 2 and 3.
2. Pond Area Zone: sampling sites 4 and 5.
3. Depleted Zone up of Odino Falls: sampling sites 6, 7, 8, 9, 10 and 11
4. Depleted Zone Down of Odino Falls: sampling sites 12 and 13.
5. Lower Zone: sampling sites 14, 15, 16, 17, 18, 19 and 20.

These newly created zones can be designated as Zones I – V, described as follows:

Zone I: Upstream of the newly created lake. This is part of the river between Sondu-Bridge and the upper reaches (or upstream-end) of the lake created by the dam at Apoko (Figure 2).

Table 2. Inventory of fishes of the Sondu-Miriu River, a result from the electrofishing survey (2010).

Scientific names (Species)	Common name (English)	Local name (Luo)
Barbus altianalis	Large barb	Fuani
B. jacksonii	Three spot barb	Adel
B. appleugramma	Small barb	Adel
B. kestenii	Orange cheecked barb	Adel
B. neumayeri	Neumayer's barb	Adel
B. nyanzae	Nyanza barb	Adel
B. paludinosis	Pallid barb	Adel
Labeo victorianus	River carp	Ningu
Rastrineobola agentae	Sardine	Omena
Clarias alluaudi	Catfish	Nyapus
C. gariepinus	Catfish	Mumi
C. theodorei	Catfish	Ndhira
Synodontis afrofischeri	Squeaker	Okoko rateng
S. victoriae	Squeaker	Okoko rachar
Schilbe mystus	-	Sire
Apploccheilichthyes bukobanus	Mosquito fish	-
Mastacembelus franatus	Spiny eel /Freswater eel	Luri/Okunga
Lates niloticus	Nile perch	Mbuta
Astatoreochromis alluaudi	Haplo	Fulu
Oreochromis niloticus	Tilapia	Ngege / Nyamami
O. leucostictus	Tilapia	Opat/Ngege
O. variabilis	Tilapia	Mbiru/Ngege
Pseudocrenilubras multicolor	Haplo	Fulu
Tilapia rendallii	Tilapia	Opat/Ngege
T. zillii		Silisili/Ngege
Haplochromine sp.	Haplo	Fulu

Zone II: The Pond Area (Apoko Lake) that holds the dammed waters accumulating upstream of dam.

Zone III: The Depleted Zone from the Apoko dam down-stream to the Odino Falls.

Zone IV: The continued Depleted Zone from the base of Odino Falls down-stream to the Sang'oro Re-entry of outlet. This is the section of the river between the Odino-Falls and the outlet adjacent to "the presently constructed Sang'oro Hydroelectric Power Project (SHPP).

Zone V: The Lower Zone. This is the section of the river from Sang'oro Re-entry of outlet to the river mouth on the Lake Victoria's shores at Osodo-Bay.

It is noted that in the present study, Zones I and II coincide with the Upper Zone (UZ) of Willoughby et al. (2003) and Kapiyo et al. (2003). Depleted Zone (DZ) of Willoughby et al. (2003) coincides with the present Zone III. Lower Zone (LZ) by the earlier authors is presently represented by Zones IV and V. The number of sites sampled in each zone and their descriptions are given in Table 2. The sampling sites, except sites 4 and 5 in UZ were, however, identical to those of Willoughby et al. (2003). It is particularly noted that in the 2003 studies, the present Zone II sampling sites were riverine, while in the present study, these sites are the near shore waters of a lacustrine ecosystem of the newly created artificial lake. The two sites are thus proximate but not identical to those sampled previously (in 2003).

Conclusion

Although, river fisheries in the Lake Basin of Kenya have reported decline in the recent past, the present results reveal increase in the abundance of fish for the case of the Sondu-Miriu River. The relations between this positive occurrence and the SMHPP is evidently due to new hydrological factors that have resulted in five distinct river zones each with unique characteristics. The five zones emerged out of the traditional three zones due to damming of the river by the SMHPP. A new fishery based on lacustrine ecosystem has been created. Besides this, it should be noted, also, that technology can "emancipate" nature by waking up its forces (Kluxen, 1994). In the present case, there is evidence that the establishment of SMHPP resulted in the upsurge in fish abundance in the Sondu-Miriu River.

Fish has become more abundant. The questions to ask now may be about sustenance, conservation and further improvement on the abundance. These results are contrary to the Sondu-Miriu riparian community's claim that

Figure 5. Different sizes of tilapia fish caught by beach seining at the newly created artificial Apoko Lake of the Sondu-Miriu River.

this part of the river (UZ and DZ) has no fish. The community seems ignorant of the present and abundant fish in this part of the river. It is noted, also, that in spite of the newly created and promising fisheries that has, apparently occur due to the creation of a new lake in the area (Lake Apoko/ the Apoko Pond area in the Upper Zone), there are already uncontrolled and destructive irresponsible methods of fish harvesting going on in the area (Figure 5).

The beach seining activities with the use of wrong and unauthorized gears that include the use of mosquito nets destroy even the young fish (Figure 5). According to one fisherman and a key informant interviewed during the present study, "the beach seines used here destroy even the eggs and the new spawn". This irresponsible over-exploitation of a fishery must be stopped, but this can only be an enormous challenge to authorities given the poor understanding, misconceptions and the miscommunications among the riparian community regarding fishes in their sections of the river and the effects of the establishment and operations of the SMHPP.

Although, river fisheries in the Lake Basin region has been declining rapidly, the outcry by the Sondu-Miriu River Riparian Community concerning the loss of the river fish cannot be substantiated. The claim on loss of fish stocks from the river seem to be based on lack of information on this river fishery. The present results of the scientific survey reveal increase in the abundance of fish in the river, contrary, also to the current general trend of fish decline in many rivers.

The possible reasons for increased fish abundance in the Sondu-Miriu River may be due to the following reasons:

1. Reduced predation from the Nile perch

2. NEMA activities which has created awareness against polluting the river by agrochemicals.
3. Since many important riverine species are actually potamodromous, migrating from the lake to the river, the water hyacinth, *Eichhornia crassipes* which has plagued Lake Victoria, particularly parts of the lake adjacent to this river's mouth, has afforded these prey fish protections from the Nile perch predation. When the fish are still within the lake, prior to taking the migration into the rivers they are under cover of the water hyacinth. Increased fish populations are thus available at areas of the lake adjacent to the river mouth at the onset of the migration process to the river. This translates to the current increase in the abundance of fish in the Sondu-Miriu River;
4. The creation of the artificial Apoko Lake above the SMHPP intake Dam at Apoko has provided fish with a conducive environment (the lacustrine ecosystem) in which to live and multiply;
5. The reduced water level in the depleted area made the catching of fish easier during low water levels.

ACKNOWLEDGEMENTS

The Kenya Marine and Fisheries Research Institute, Kisumu provided the "Set of Elecrofisher Equipment" and the technical staff to operate them. The Kenya Electricity Generating Company (Ken Gen) provided suitable means of transportation by which the sampling sites along the river were accessed.

REFERENCES

Ashe B, Greef L, Haya B, Obbo B (2009). The Information Relevant to the Validation of the Sondu- Miriu Hydroproject, Kenya Memo to DNV on CDM Validation for Sondu _Miriu Large-Hydro Kenya, September 11, 2007: Current Status; Negative as of August 2009 International Rivers, People, Water, Life.

Awange JC, Ong'ang'a O (2006). Lake Victoria, Ecology, Resources, Environment Springer, Berlin. p 347.

Cadwalladr DA (1964). An account of a decline of *Labeo victorianus* Boulenger fishery of Lake Victoria and an associated deterioration of indigenous fishing methods in the Nzoia River, Kenya E Afr. Agr. For. J. 30:245–256.

Cadwalladr DA (1965). Notes on the breeding biology and ecology of *Labeo victorianus* Boulenger (Pisces: Cyprinidae) of Lake Victoria Rev Zoo. Bot. Afr. 72(1-2):109-134.

Cadwalladr DA (1969). A discussion of possible management methods to revive the *Labeo victorianus* fishery of Lake Victoria with special reference to Nzoia River, Kenya Uganda Fisheries Department Occasional Paper 2.

Gichuki J, Dahdohuh GF, Mugo J, Rabuor CO, Triest L, Dehairs F (2001). Species inventory and the local uses of the plants and fishes of the Lower Sondu-Miriu wetland of Lake Victoria Hydrobiologia 458: PDF (7282 KB) Free Preview ISSN 0018-8158 (Print) pp.1573-5117.

Gichuki N (2006). Wetland research in Lake Victoria basin, Kenya part Analysis and synthesis report www.iucea.org

Kapiyo RK, Kibwage JJ, Abila RO (2003). The status of fisheries and fishing activities of the Sondu-Miriu River Technical Report to KenGen / KOEI LTD Japan.

Kibaara D (1981).   Endangered fish species of Kenya's inland waters with emphasis on the *Labeo* spp In *Proceedings of the workshop of the Kenya Marine and Fisheries Research Institute on Aquatic Resources of Kenya* Mombasa, Kenya pp.157–164.

Kluxen W (1994). Philosophical aspects of interplay between nature and artificial environment of man. In Odera Oruka (Ed) *Philosophy, human and ecology. Philosophy of Nature and Environmental Ethics*. ACTS Press, Nairobi African Academy of Sciences pp 98-106.

Mainga OM (1981).   Some hydrobiological conservation at the mouths of two affluents rivers of lake Victoria with special emphasis on Synodontis. MSc Thesis University of Nairobi.

Manyala JO, Ochumba PBO (1990). Small scale fishery of lower Sondu-Miriu River In the proceedings of the symposium on swocioeconomics aspects of Lake Victoria Fisheries IFIP Project Symposium, CIFA sub-committee for Lake Victoria April 25–27, Kisumu, Kenya.

Muli JR (1998). An appraisal of stochking and introduction of fish in Lake Victoria (East Africa). In IG Cowx (Ed), Fishing News Books Blackwell Science Ltd. pp.258–266.

Mwashote BM, Shimbira W (1994).   Some limnological characteristics of the lower Sondu-Miriu River, Kenya. In recent trends of research on Lake Victoria Fisheries. Proceedings of the second EEC Regional Seminar on recent trends of research on Lake Victoria Fisheries 25th – 27th Sept 1991, Kisumu, Kenya Eds E Okemwa, EO Wakwabi and Getabu A ICIPE Science Press.

Ochumba PBO, Manyala JO (2008). A description of the fisheries of the Lower Sondu-Miriu River before the construction of the SMHPP Aquaculture Research 23:701-719 Published on line 21st April 2008 Journal Compilation © 2010 Blackwell Publishing Limited.
http://www3interscienceWilleycom/journal/119337227/Abstract?

Ochumba PBO (1994). Conservation plans for Lake Victoria Magazine of East African Wildlife Society March/April 17:36.

Reynolds JB (1983). Electrofishing In Nielsen, L A & Johnson D L (eds) Fisheries Techniques American Fisheries Society Blacksburg, Virginia ISBN 0-913235-00-8. pp.147–163.

Whitehead PJP (1958). Indigenous river fishing methods in Kenya E Afr. Agr. For. J. 24:111.

Whitehead PJP (1959). The anadromous fishes of Lake Victoria Victoria Rev. Zoo. Bot. Afr. 59:329-263.

Wikipedia Free Encyclopedia (2010). Economy of Kenya CIA World Fact Book, Wikimedia Foundation; last modified, 11th February 2010.

Willoughby NG (2006). Fisheries and ecology of Sondu Miriu river, Western   Kenya.   www.thenrgroup.net/theme/Sondu/sondu-fish.htm

Willoughby NG, Kapiyo R, Mugo J, Owiti D (2003). The implication of the Sondu Miriu Hydropower project (Kenya) for fisheries and fishing communities. The Land 7(1):3-20.

13

Effect of crusher dust on floristic composition and biological spectrum in tropical grassland of Odisha, India

B. K. Tripathy[1], R. B. Mohanty[2], N. Mishra[3] and T. Panda[4]

[1]Department of Botany, Dharmasala Mohavidyalaya, Dharmasala,Jajpur, Odisha, India.
[2]Department of Botany, N. C.Autonomous College, Jajpur, Odisha, India.
[3]Department of Zoology, Chandbali College, Chandbali, Bhadrak-756133, Odisha, India.
[4]Department of Botany, Chandbali College, Chandbali, Bhadrak-756133, Odisha, India.

An investigation was carried out to estimate the effect of crusher dust on floristic composition and biological spectrum in grassland of Odisha, India. The vegetation analysis was performed following standard procedures. The floristic survey of the study area, which includes natural as well as crusher unit affected sites, revealed the presence of 24 plant species distributed in 11 families representing a two storied floristic composition in the area. Out of the 24 species, maximum number of species belonged to the life-form class therophytes followed by hemi cryptophytes and chamaephytes and least number of species belonged to class geophytes. Some of the important families like Poaceae, Fabaceae, Cyperaceae and Convovulaceae are noted to be prevalent in the study area. The seasonal variation of relative frequency, relative density and relative dominance of different species fluctuated. The majority was from the species *Aristida*; the next two in order of dominance were *Heteropogon* and *Setaria*. The grass species have more IVI values as compared to non-grass species in the community. The value of Simpson's dominance index varied from 0.188 to 0.536. It showed increasing trend from rainy to summer having highest value in summer. In contrary, the diversity index showed decreasing trend from rainy to summer.

Key words: Diversity indices, floristic composition, Jajpur district.

INTRODUCTION

The demand for basic necessities for survival and sustenance of life like food, water, air, etc, from natural resources is significantly increasing due to exponential growth of human population. Currently, India has taken major initiatives on developing the infrastructure, to meet the requirement of globalization in the construction of buildings and other structures; crusher unit plays a rightful role and is being utilized. In turn, land degradation due to the alteration and destruction of natural habitat is a major concern throughout the globe. The conversion of natural ecosystems into anthropo-ecosystems as a result of human activities is most apparent in and around urbanizing landscapes (Gupta and Narayan, 2010). The major factors responsible for land degradation is rapid urbanization, industrialization, unsustainable developmental projects and spread of mining activity which often results in the loss of natural ecosystems with associated biodiversity (Ezeaku and Davdson, 2008). The most affected part in this process was plants which is most sensitive and delicate in the environment of forest or grassland. In India, some 2500 species of total vascular flora fall in one or other category of threat (Jain, 1991); of these, the most prone are the angiosperms (Daniels and Jayathi, 1996). Grassland, one of the prominent

Ecology, Environment and Conservation

ecosystems in earth's terrestrial surface area (approximately 1/3rd) is highly sensitive to change in environmental factors (Lalrammawia and Paliwal, 2010). Grassland biodiversity is considered as a natural (biomass) resource base of subsistence for the people of tropical countries. Therefore, loss of grassland very often jeopardizes economic progress in tropical regions. Grassland or forest degradation adversely affects soil physico-chemical and microbiological characteristics and thereby alters the nutrient status of the soil that lead to the genesis of barren degraded wasteland (Mroz et al., 1985; Lal, 1989). The widespread existence of different dust producing units (cement, marble, stone, etc.) in the rural-urban fringes of the 3rd world countries was found to be one of the important anthropogenic activities with a potential to alter not only the ecology but also the metabolic activities of flora, fauna of the surrounding soil and vegetation (Raina et al., 2008; Okita et al., 1996; Chowdhary and Rao, 1996; Pandey and Nand, 1995; Mishra et al., 1993; Somashekar et al., 1999). Particulate matter released during stone crushing is usually of relatively large size. The chemical composition of the dust is a homogenous mixture of oxides of calcium, potassium, silica and sodium, which settles into a head mass when it comes in contact with water (Raina et al., 2008). Stone crusher dust is extremely harmful to human health as well as the ecological health of the region (Pandey et al., 2002; Srivastava et al., 2005). Literature data indicated that when these particulate matters are deposited on vegetation, the plant growth is adversely affected (Chatter, 1991; Mishra et al., 1993; Pandey and Nanda, 1995; Kumar, 2000). There have been several reports on the floristic composition of slag dump, abandoned mines and mine tailings (Prach et al., 2001; Roy et al., 2002; Pyseket al., 2003; Remon et al., 2005; Singh, 2006; Padey and Maiti, 2008) and such studies emphasizes the role of particulate matter on vegetation structure. But reports on the floristic composition effected by crusher dusts from Odisha are still lacking. Dust emissions occur from many operations in the dust units viz. cutting, loading and transportation. Therefore, it was necessary to study the effect of dust on floristic composition. In the present study, we analyzed the diversity, distribution and change in species composition structure with particular reference to crusher dusts in tropical grassland of Orissa, India, so as to develop appropriate strategies for their protection and conservation.

MATERIALS AND METHODS

Study site and its characteristics

The present investigation was carried out in tropical grassland (20° 33'-21° 10"N and between 85° 40'-86° 44"E) of Dharmasala locality and its adjoining areas under Darpani and Sukinda Tahasil, situated north-central region of the Jajpur district, Odisha, India, mostly in both sites of NH-5 and NH-5A. Climate of the area

is tropical monsoonic, experiencing three distinct seasons: rainy (mid June to mid October), winter (mid October to February) and summer (March to mid June). The air temperature ranges from 38°C in summer to 13°C in winter, with an annual average rainfall of approximately 1500 mm. Soil of the area is red and laterite. The pH and average moisture content of the soil arwe recorded to be between 5.6 and 6.4 and 3.5 and 13%, respectively.

Sukinda and Daitari are the two important hill ranges of the district. The hilly tracts with Lalitgiri, Udayagiri, Ratnagiri, Alamgiri, Bathuria-Khola, Phulajhar, Dalimbapani, Gobarghati, Chandikhol-Mahavinayak, Olasuni, Langudi, Deuli, etc. rise up to 900 m above the sea level and contains thick and varied vegetation. In and around Dharmasala area, about 100 crusher units are operating in the vicinity of National highway No. 5 and 5 A (Daitay– Pradeep express highway) as the major small scale industries which fulfill the growing demand of urban expansion. These crusher units are established mainly adjacent to agricultural lands, grasslands, rice fields or village wastelands and using abundantly available granites of this area as raw material.

Vegetation analysis

The vegetation analysis was conducted during June 2008-May 2009, to cover all the spectrum of vegetation. Two sites of about five hectare each were selected for the investigation. First one on the crusher dust effected grassland and the second about 5 km away without any crusher units. The life form classes were determined as per Raunkiaer's (1934). Random sampling method was adopted to sample the vegetation. Five separate plots were laid on each study sites. Ecological enumeration of plant species were done according to quadrate sampling method (Mishra, 1968). Each tiller was counted as an individual plant in case of grasses and each forbs was considered as an individual. However, in the case of runners, each node rooted at the base was considered as an independent individual. The size of the quadrate was 50 x 50 cm. Vegetation data were quantitatively analysed for relative values of frequency, density and abundance (Phillips, 1959).

Relative frequency = (number of plots containing a species / number of occurrences of all species) × 100

Relative density = (number of individuals of a species / total number of individuals of all species) × 100

Relative dominance = (basal area of a species / total basal area of all species) × 100

Importance value index (IVI) for individual plant species was determined as the sum of their relative frequency, relative density and relative dominance.

Statistical analysis

The following indices of diversity were calculated based on species level identification (Ludwig and Reynolds, 1988).

Shannon's species diversity (H = -$\sum$ pi ln pi) where pi is the proportion of the total number of individuals / green biomass belonging to its ith species, ln denotes natural logarithm and and H is the Shannon–Wiener index.

Simpson's dominance indices (D = $\sum$(ni/N)^2, where ni is the density of species, that is, total number of individuals/green biomass of the species, N denotes total number of individuals or plant biomass of the same sample and D is the Simpson's index.

Table 1. Floristic composition and respective life-form classes at experimental sites.

S/N	Name of the species	Grass/Forbs/Sedge/Shrub	Family	Life-form classes
1.	*Aristida setacea* Retz.	Grass	Poaceae	Hemicryptophytes
2.	*Bothriochloa pertusa* (L.) A. Camus	Grass	Poaceae	Hemicryptophytes
3.	*Chrysopogon aciculatus* (Retz.) Trin.	Grass	Poaceae	Hemicryptophytes
4.	*Cynodon dactylon* (L.) Pers.	Grass	Poaceae	Hemicryptophytes
5.	*Cyperus rotundus* L.	Sedge	Cyperaceae	Geophytes
6.	*Cyperus nivenus* Retz.	Sedge	Cyperaceae	Geophytes
7.	*Digitaria ciliaris* (Retz.) Koeler	Grass	Poaceae	Therophytes
8.	*Eragrostis gangetica* (Roxb.) Steud.	Grass	Poaceae	Therophytes
9.	*Heteropogon contortus* (L.) P. Beauv.	Grass	Poaceae	Hemicryptophytes
10.	*Panicum repens* L.	Grass	Poaceae	Hemicryptophytes
11.	*Setaria pumila* (Poir.) Roem. & Schult.	Grass	Poaceae	Therophytes
12.	*Sporobolus indicus* (L.) R. Br.	Grass	Poaceae	Therophytes
13.	*Andrographis echioides* (L.) Nees	Grass	Acanthaceae	Therophytes
14.	*Cassia tora* L.	Shrub	Caesalpiniaceae	Therophytes
15.	*Celocia argentea* L.	Forbs	Amaranthaceae	Therophytes
16.	*Crotolaria prostrate* Rottl. ex Willd	Shrub	Fabaceae	Therophytes
17.	*Desmodium triflorum* (L.) DC	Forbs	Fabaceae	Chamaephytes
18.	*Evolvulus alsenoides* (L.) L.	Forbs	Convolvulaceae	Chamaephytes
19.	*Evolvulus numularius* (L.) L.	Forbs	Convolvulaceae	Chamaephytes
20.	*Phyllanthus virgatus* Forst. f.	Forbs	Euphorbiaceae	Chamaephytes
21.	*Sida acuta* Burm. f.	Forbs	Malvaceae	Chamaephytes
22.	*Spermacoce hipsida* L.	Grass	Rubiaceae	Chamaephytes
23.	*Vernonia cinerea* (L.) Less.	Forbs	Asteraceae	Therophytes
24.	*Zornia gibbose* Sponoghe	Forbs	Fabaceae	Therophytes

RESULTS AND DISCUSSION

The floristic survey of the study area, which includes natural as well as crusher unit affected sites, revealed the presence of 24 plant species distributed in 11 families representing a two storied floristic composition in the area (Table 1). The two storey's observed were consisting of shrubs (height range: 2 to 4 m) and herbs of seasonal and perennial type. Out of the 24 species, maximum number of species belonged to the life-form class therophytes (10 species) followed by hemi cryptophytes and chamephytes (6 species each) and least number of species (02 species) belonged to class geophytes. It was observed that therophytes contributes maximally (41.66%) followed by chamaephytes and hemicryptophytes with 25% each and geophytes with least percentage (8.33%) (Figure 1). Since the percentages of chamephytes/hemicryptophytes is next to therophytes, the vegetation may be called thero-chamephytic.

The higher percentage of therophytes in the grassland was due to periodic climate change along with interference of crusher unit activities (Misra, 1978; Barik and Mishra, 1998). When comparing the present findings with Raunkiaer's (1934) normal spectrum, it was observed that the percentage contribution of therophytes was 28.66 times, chamaephytes- 16 times, geophytes- 4.33 times higher and hemicryptopytes was 1% less. However, the biological spectrum of the present study was found to be close to the findings of Pradhan (1994) and Misra and Mishra (1981). Among the grasses, *Aristida setacea, Bothriochloa pertusa, Chrysopogon aciculatus, Heteropogon controtus* and *Andrographis echioides*, among the sedge, *Cassia tora* and *Demodium triflorum* among forbs in either of the sites showed dominance during the study period (Table 2). All the species appeared with the onset of monsoon and completed their life-cycle towards the end of the winter season. The common occurrence of certain grasses, forbs, sedges and shrubs on both sites indicates their adaptability to thrive in adverse conditions of the waste material. Adaptability of some genera like *Cyperus, Eragrostis* and *Evovulus* to survive in coal mine area has already been reported by Ekka and Behera (2010).

Comparative analysis of the percentage contribution of different groups of plant species indicated the highest contribution by grasses, followed by forbs. Some of the important families like Poaceae, Fabaceae, Cyperaceae and Convovulaceae are noted to be prevalent in the study area. Poaceae is the dominant family and have been reported to play positive role as the initial colonizer of the different derelict mine soils (Helm, 1995; Singh,

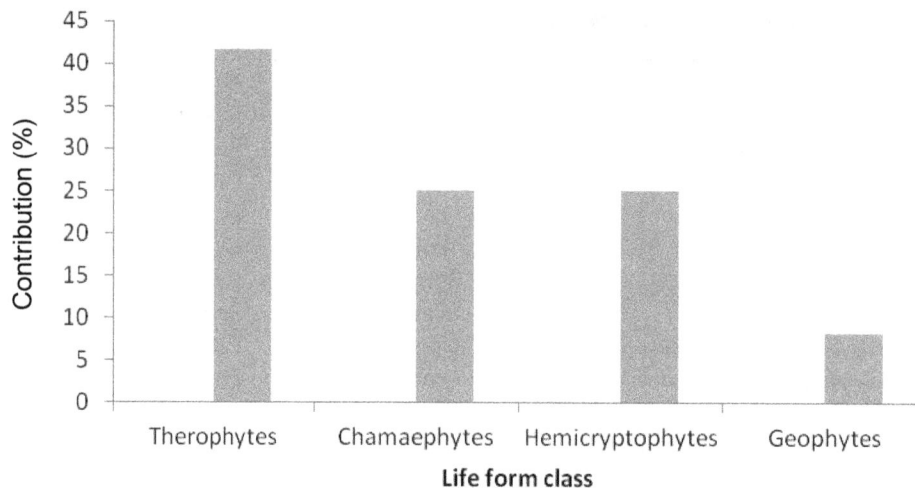

Figure 1. Percentage contribution of different life form class.

Table 2. Seasonal variation of relative density and relative frequency of plant species at the study sites.

Plant species	Relative frequency						Relative density					
	Site1			Site2			Site1			Site2		
	R	W	S	R	W	S	R	W	S	R	W	S
Aristida setacea	5.09	5.55	10.58	5.39	5.93	12.1	16.51	19.97	38.3	18.53	24.22	42.25
Bothriochloa pertusa	4.83	5.28	8.99	5.05	5.64	0.33	4.28	4.7	6.77	4.4	5.57	7.65
Chrysopogon aciculatus	4.33	4.44	7.41	4.26	4.15	7.27	1.79	1.66	1.97	1.75	1.88	2.14
Cynodon dactylon	4.33	4.44	4.76	4.23	4.45	3.64	4.27	4.087	2.07	4.37	4.19	1.22
Cyperus rotundus	4.83	4.722	1.59	5.05	4.75	2.42	3.5	2.76	0.62	2.95	2.41	0.61
Cyperus nivenus	4.58	4.72	2.12	4.25	4.75	2.42	2.84	2.25	1.14	2.75	2.28	0.77
Digitaria ciliaris	3.82	3.61	1.59	3.72	3.56	1.82	2.13	1.35	1.55	2.06	1.61	1.22
Eragrostis gangetica	4.07	3.89	1.06	4.52	5.64	1.21	3.2	3.58	4.72	3.05	3.27	4.16
Heteropogon contortus	4.83	5.00	8.99	5.05	5.34	10.3	18.16	19.71	26.16	20.01	21.49	23.23
Panicum repens	3.56	3.61	-	4.52	4.75	-	4.32	3.45	-	4.41	3.75	-
Setaria pumila	4.58	4.72	6.88	3.72	3.56	4.24	18.38	21.43	11.92	16.68	17.42	9.78
Sporobolus indicus	4.33	4.17	3.17	3.99	3.86	4.85	7.83	5.75	1.3	6.3	4.82	0.61
Andrographis echioides	4.33	4.72	8.47	4.26	5.04	4.85	1.15	1.28	0.73	1.22	1.34	0.48
Cassia tora	4.58	4.72	6.35	4.26	3.56	6.06	0.98	0.49	-	1.07	0.48	-
Celocia argentea	3.56	3.61	5.29	3.99	3.56	4.85	0.78	0.88	0.65	0.74	0.74	0.68
Crotolaria prostrate	4.07	3.61	2.65	4.55	3.56	2.42	0.69	0.77	0.52	0.71	0.8	1.01
Desmodium triflorum	4.83	3.89	5.29	3.99	3.56	6.06	0.77	0.88	0.7	0.61	0.73	0.61
Evolvulus alsenoides	3.82	3.89	3.7	4.26	4.15	3.64	0.25	0.24	0.19	0.26	0.27	0.18
Evolvulus numularis	3.31	3.33	1.59	3.72	3.56	1.82	0.3	0.27	0.05	0.29	0.2	0.21
Phyllanthus virgatus	4.07	3.89	1.59	3.19	2.67	1.21	3.41	1.6	-	3.03	1.88	-
Sida acuta	3.05	3.33	1.59	2.93	3.56	1.21	0.21	0.23	0.39	0.41	0.17	0.09
Spermacoce hipsida	4.07	4.17	5.29	4.26	4.15	6.06	0.36	0.39	0.031	0.43	0.38	0.12
Vernonia cinerea	3.05	2.78	1.06	3.72	2.97	1.21	0.26	0.22	0.13	0.26	0.27	0.21
Zornia gibbose	4.07	3.89	-	3.72	3.26	-	3.62	2.05	-	3.7	2.25	-

R = Rainy, W = winter and S = summer.

2006).The seasonal variation of relative frequency, relative density and relative dominance of different species fluctuated (Tables 2 and 3). The majority was from the species Aristida; the next two in order of dominance were *Heteropogon* and *Setaria*. Considering the individual species, it is clear that species composition

Table 3. Seasonal variation of relative dominance and importance value index of plant species at the study sites.

Plant species	Relative dominance						Important value index					
	Site1			Site2			Site1			Site2		
	R	W	S	R	W	S	R	W	S	R	W	S
Aristida setacea	7.67	8.16	13.70	7.9	8.48	14.6	29.3	33.68	62.63	31.73	38.63	68.98
Bothriochloa pertusa	4.36	4.46	6.69	4.47	4.59	6.96	13.48	14.33	22.46	13.92	15.80	24.91
Chrysopogon aciculatus	4.28	3.89	5.22	3.59	3.79	4.87	10.4	9.99	14.59	9.6	9.82	14.28
Cynodon dactylon	2.05	1.71	2.12	1.93	1.7	1.57	10.65	10.24	8.95	10.55	10.33	6.42
Cyperus rotundus	2.57	2.85	4.73	2.63	2.79	4.35	10.9	10.33	6.94	10.63	9.95	7.38
Cyperus nivenus	2.4	2.47	3.59	2.45	2.5	3.13	9.81	9.44	6.84	9.45	9.92	6.33
Digitaria ciliaris	3.42	3.23	3.0	3.51	2.99	2.2	9.36	8.18	3.14	9.29	7.77	3.04
Eragrostis gangetica	4.02	3.98	5.87	3.94	3.89	5.57	11.29	11.45	11.65	11.51	12.8	10.93
Heteropogon contortus	5.13	5.5	8.48	4.91	5.39	8.17	28.13	24.71	43.64	28.97	32.22	41.71
Panicum repens	6.16	6.45	-	5.61	5.89	-	14.04	13.51	-	14.54	14.39	-
Setaria pumila	4.45	4.55	7.01	4.56	4.69	6.78	27.41	30.71	25.81	25.7	25.67	20.81
Sporobolus indicus	4.88	5.03	4.5	4.91	5.29	4.5	17.03	14.94	4.47	15.2	13.97	5.5
Andrographis echioides	4.28	4.27	6.69	4.38	4.39	6.78	9.75	10.27	15.88	9.86	10.77	12.07
Cassia tora	5.99	5.5	5.4	6.13	5.69	5.35	11.56	10.72	6.35	11.46	9.72	6.06
Celocia argentea	4.28	4.55	6.53	4.38	4.69	6.26	8.62	9.05	12.46	9.11	8.99	11.79
Crotolaria prostrate	5.05	5.12	4.5	5.17	5.19	4.8	9.81	9.51	3.16	10.14	9.55	3.43
Desmodium triflorum	3.68	3.61	3.5	3.77	3.69	3.48	9.29	8.38	5.99	8.37	7.98	6.67
Evolvulus alsenoides	4.36	4.36	6.85	4.38	4.19	6.61	8.43	8.5	10.55	8.9	8.61	10.24
Evolvulus numularis	2.14	1.8	2.45	2.19	1.7	3.83	5.74	5.4	4.08	6.2	5.46	5.64
Phyllanthus virgatus	3.68	3.51	-	3.77	3.49	-	11.16	8.99	1.59	9.99	8.04	1.21
Sida acuta	3.76	3.7	5.71	3.86	3.79	5.22	7.02	7.56	7.68	7.19	7.52	6.51
Spermacoce hipsida	4.28	4.27	6.36	4.47	4.19	5.91	8.71	8.83	11.68	9.13	8.73	11.97
Vernonia cinerea	5.22	5.41	7.99	5.26	5.39	7.65	8.53	8.41	9.18	9.26	8.62	9.08
Zornia gibbose	1.88	1.61	-	1.84	1.6	-	9.58	7.55	-	9.26	7.11	-

R = Rainy, W = winter and S = summer.

Table 4. Dominance and diversity indices of species at the study sites.

Parameter	Site1			Site 2		
	Rainy	Winter	Summer	Rainy	Winter	Summer
Dominance index	0.188	0.212	0.446	0.262	0.28	0.536
Diversity index	1.597	1.578	0.845	2.42	2.397	0.886

in plantation site greatly differed from plantation with crusher units. The grass species have more IVI values as compared to non-grass species in the community. *Aristida setacea* and *Evolvulus numularis* contributed maximum and minimum IVI value for both sites. However, *Phyllanthus virgatus* showed the minimum IVI in summer season. It was also observed that the IVI value in total for all grass species gradually increased from rainy to summer season whereas the non-grasses value showed reverse trend. The value of relative frequency, density, dominance and important value index was found to coincide with the findings of Misra (1978), Pradhan (1994) and Mishra and Mishra (1981). The value

of Simpson's dominance index varied from 0.188 to 0.536. It showed increasing trend from rainy to summer having highest value in summer. In contrast, the diversity index showed decreasing trend from rainy to summer. The highest value of Shannon diversity index (2.42) was in site with crusher units and the lowest value (0.845) was in site without crusher units. The variability of index value in the present study may be due to the action of crusher dust upon the vegetation. The trend of different indices in the present study corroborates with the findings of Pradhan (1994), Misra and Misra (1981). It is concluded that the crusher dusts influence the floristic composition and biological spectrum of the grassland adjacent to crusher units.

REFERENCES

Barik KL, Misra BN (1998). Biological spectrum of a grassland ecosystem of South Orissa. Ecoprint 5(1):73-77.

Chatter, H (1991). Effect of Cement dust on the enzymatic activity in the levels of *Triticum aestivum*. Acta Ecol. 13:113-119.

Chowdhary U, Rao TVR (1996). Effect of cement dust on the enzymatic activity in the levels of *Triticum aestivum*. Acta Ecol. 13:113-119.

Daniels RRJ, Jayanthi M (1996). Biology and conservation of endangered plants: the need to study breeding systems. Trop. Ecol. 37:39-42.

Ekka NJ, Behera N (2011). Under storey plant diversity analysis on different age series coal mine spoil dumps in an open cast coal field in Orissa, India. Trop. Ecol. 52(3):373-343.

Ezeaku PI, Davidson A (2008). Analytical situations of land degradation and sustainable management strategies in Africa. J. Agric. Soc. Sci. 4:42-52.

Gupta S, Narayan R (2010). Brick Kiln industry in Long term impacts biomass and diversity structure of plant communities. Curr. Sci. 99(1):72-77.

Helm, DJ (1995). Native grass cultivars for multiple revegetation goals on a proposed mine site in South Central Alaska. Restor. Ecol. 3(2):111-122.

Jain SK (1991). The problems of endangered species, concepts, problems and solutions. In: Singh KP, Singh JS (eds.).Tropical ecosystems: Ecology and Management. Willey Eastern, New Delhi, India, pp. 69-80.

Kumar RR, Shadaksharasamy N, Srinivas G (2000). Impacts of granite quarrying on environment in Banglore district with reference to socioeconomic status of workers. Pollut. Res. 19:51-54.

Lal R (1989). Soil degradation and conversion of tropical rain forest. In: Botkin, DB, Caswell, MF, Estes, JE, Orio, AA (eds.).Changing the global environment. New York: Academic Press, pp. 135-153.

Lalrammawia C, Paliwal K (2010). Seasonal changes in net ecosystem change of CO_2 and respiration of *Cenchrus ciliaris* L. grassland ecosystem in semiarid tropics: An eddy covariance measurement. Curr. Sci. 98(9):1211-1218.

Ludwig JA, Reynolds JF (1988). Statistical ecology. New York: John Wiley.

Mishra JV, Pandey SN, Singh N, Singh MY, Ahmed KJ (1993). Growth response of *Lycopersicum esculentum* to cement dust treatment. J. Environ. Sci. Health 28:1774-1780.

Mishra MK, Mishra BN (1981). Species diversity and dominance in a tropical grassland community. Folia Geobotanica et Phytotaxonomica 16(3):309-316.

Mishra R (1968): Ecological Work Book. Oxford and I.B.H. Publishing Co., New Delhi.

Misra MK (1978). Phytosociology and primary production of a grassland community of Berhampur. Ph.D. Thesis, Berhampur University, Berhampur, Orissa.

Mroz GD, Jurgensen MF, Federick DJ (1985). Soil nutrient changes following whole tree harvesting on tree -northern hard wood sites. Soil Soc Sci. Am. 49:1552-1557.

Okita T, Hara H, Fukuzaki N (1996). Measurements of atmospheric SO_2 and SO_4^{2-} and determination of the wet scavenging coefficient of Sulfate aerosols for the winter monsoon season over the Sea of Japan. Atmos. Environ. 30:3733-3739.

Pandey DD, Nand S (1995). Effect of stone crushers dust pollution on grain characteristic of Maize. Environ. Ecol. 13:901-903.

Pandey JS, Khan S, Joseph V, Kumar R (2002). Aerosol scavenging: Model application and sensitivity analysis in the Indian context. Environ. Monit. Assess. 74:105-116.

Pandey S, Maiti T (2008). Physicochemical and biological characterization of slag disposal site at Burnpur, West Bengal. Pollut. Res. 27(2):345-348.

Phillips EA (1959): Methods of Vegetation Study. A Hoff Dryden Book, Henry Holt Co. Inc., New York.

Prach K, Bartha S, Joyee CB, Pyšek P, Van Diggelen R, Wiegleb G (2001). Role of spontaneous succession in ecosystem restoration: A perspective. Appl. Veget. Sci. 4:111-114.

Pradhan D (1994). Primary production and phytosociology of a grassland community of Bhubaneswar. Ph.D. Thesis, Berhampur University, Berhampur, Orissa.

Pyšek A, Pyšek P, Jarošík V, Hájek M, Wild J (2003). Diversity of native and alien plant species on rubbish dumps: Effects of dump age, environmental factors and toxicity. Divers .Distrib. 9:177-189.

Raina AK, Rathore V, Sharma A (2008). Effect of stone crusher dust on leaves *Melia azadarach* L. and *Dalbergia sissoo* Roxb. in Jammu (J&K). Nat. Environ. Pollut. Technol. 7:279-282.

Raunkiaer C (1937). The life forms of plants and statistical plant geography.Clarendon Press, Oxford. pp. 639.

Remon E, Bouchardon JL, Cornier B, Guy B, Leclerc JC, Faure O (2005). Soil characteristics, heavy metal availability and vegetation recovery at a former metallurgical landfill: Implications in risk assessment and site restoration. Environ. Pollut. 137:316-323.

Roy A, Basu SK, Singh KP (2002). Modeling ecosystem development on blast furnace slag dumps in a tropical region. Simulation 78(9):531-542.

Singh A (2006). Herbaceous species com position of an age series of naturally revegetated coal mine spoils on Singrauli Coalfields. India. J. Indian Inst. Sci. 86:76-80.

Somashekar RK, Ravi R, Ramesh AM (1999). Impact of granite mining on some plant species around quarries and crusher of Bangalore district. Pollut. Res. 18:445-451.

Srivastava A, Joseph AE, Patil S, More A, Dixit RC, Prakash M (2005). Air toxics in ambient air of Delhi. Atmos. Environ. 39:59-71.

14

Species composition and abundance of small mammals in Chebera-Churchura National Park, Ethiopia

Demeke Datiko and Afework Bekele

Department of Biology, Addis Ababa University, P. O. Box 1176, Addis Ababa, Ethiopia.

In this study, we investigated species composition and abundance of small mammals (rodents and insectivores) in the recently established Chebera Churchura National Park, in Ethiopia between 2010 and 2011. Two study grids were established in seven selected habitats. In each grid, forty-nine Sherman live traps were used to capture the small mammals. A total of 704 individuals were captured in 5488 trap-nights. Among them, 682 (98.3%) were rodents and 12 (1.7%) were insectivores. These comprised 16 species of rodents and 2 species of insectivores. The captured species were: *Mastomys natalensis, Mastomys erythroleucus, Lemniscomys striatus, Arvicanthis nilotiucs, Arvicanthis dembeensis, Acomys cahirinus, Rattus rattus, Mus musculus, Stenocephalemys albipes, Gerbilliscus robusta, Lophuromys flavopunctatus, Pelomys harringtoni, Mus tentellus, Crocidura fumosa* and *Crocidura flavescens. Tachyoryctes splendens, Xerus rutilus* and *Hystrix cristata* were recorded as observed species. The abundance of species varied among habitats and between seasons. *M. natalensis* and *L. striatus* were the most abundant species, whereas *C. flavescens* was the least abundant. The age distribution and trap success of small mammals varied between seasons and among habitats. This study clearly shows that the park has a diverse fauna.

Key words: Abundance, Chebera Churchura, Ethiopia, small mammals.

INTRODUCTION

Small mammals comprise the highest proportion among the mammal species all over the world (Vaughan et al., 2000). A total of 5416 species of mammals are recorded globally, of which more than 2,277 species are rodents and insectivores (Wilson and Reeder, 2005). They account for about 42% of mammalian species (Wilson and Reeder, 2005). In Africa, small mammals are probably the most ubiquitous and numerous (Skinner and Chimimba, 2005). Over 1150 species of mammals are currently listed for Africa, but still more mammalian species, especially rodents, insectivores and bats, still await discovery (Kingdon, 1997). Many studies were carried out in the continent on small mammals (rodents and insectivores). These include, Skinner and Chimimba

(2005) in Southern African Subregion, Avenant and Cavallini (2008) and Avenant (2011) in South Africa, Linzey and Kesner (1997) in Zimbabwe, Oguge (1995) in Kenya, and Leirs et al. (1994) in Tanzania. For instance, one study on the community structure of small mammals (Rodentia and Soricomorpha) from the Gulf of Guinea region of West Africa found 45 species of soricomorphs and 101 of rodents (Amori and Luiselli, 2011). However, ecological studies for small mammals in Africa focused mostly on the western region, with minimal attention on the eastern part of the continent (Habtamu and Bekele, 2008). Faunal exploration is an important component of the study in a given protected area. It helps in understanding the potential of the area in composition and diversity of animals as well as to carryout conservation action for the future. Information on the diversity of small mammals besides reinforcing scientific knowledge will boost the importance of the region to establish protected areas and refugia (Habtamu and Bekele, 2008). Moreover, small mammals are known to have ecological, economical, social and cultural values (Avenant, 2011). They play an important role in natural communities and provide the main supply of fresh food for many predators (Davies, 2002).

Ethiopia's past geological history, unique topography and great variation on climate is home for diverse biological resources (Yalden and Largen, 1992; Bekele, 1996b; Lavrenchenko et al., 1998; Takele et al., 2010). The country has 284 species of mammals of which 39.4% are small mammals (Lavrenchenko et al., 1997; Datiko et al., 2007; Habtamu and Bekele, 2008). The country possesses 84 rodent species of which 15 are endemic (Bekele, 1996a; Lavrenchenko et al., 1998), comprising 25% of the Ethiopian mammal fauna and about 50% of the total endemic species (Bekele, 1996b; Yalden et al., 1996; Datiko et al., 2007). Despite this diversity, only few studies on the taxonomy and population ecology of rodents have been conducted in Ethiopia (Bekele, 1996a).

A number of investigations was carried out to study several aspects of small mammals communities in some of the National Parks of the country including Alatish National Park (Habtamu and Bekele, 2008), Nechisar National Park (Datiko et al., 2007), Bale Mountains National Park (Yalden, 1988; Lavrenchenko et al., 1997) and Simen Mountains National Park (Yalden et al., 1996) and in central Ethiopia by Bekele and Leirs (1997). Many of the endemic mammals are associated with high altitude moorland and grassland habitats (Yalden and Largen, 1992; Bekele, 1996a), although the species diversity is less than many lowlands (Yalden and Largen, 1992). Few areas of the country are extensively surveyed for small mammals (Yalden and Largen, 1992; Bekele, 1996a, b), while composition, habitat use and population dynamics of the small mammal community are poorly known for many regions of Ethiopia. This fact is probably a result of inaccessibility, remoteness and inhospitable conditions of such areas (Bekele, 1996a, b; Habtamu and Bekele, 2008). The same is true for the small mammal of the Chebera-Churchura National Park (CCNP). The National Park is among the recently established ones in the country, and is located in the southwestern extensive lowland area of Ethiopia. This National Park is a key biological resource conservation area based on the diversity of large mammal fauna, birds and the unique ecosystem (Timer, 2005; Weldeyohanes, 2006). The park was established in 2005 (Timer, 2005). So far, no studies have been carried out on small mammal in the area. The current study aimed at investigating the species composition, abundance and seasonal variation of small mammals in the CCNP.

MATERIALS AND METHODS

Study area

CCNP is located along the southwestern part of Ethiopia. It is partly located within Dawro zone and in Konta special district, about 300 and 580 km southwest of Awassa and Addis Ababa, respectively. It covers an area of 1,250 km2 and lies between the coordinates 36°27'00"- 36°57'14"E longitude and 6°56'05"-7°08'02"N latitude (Timer, 2005). The Park is bordered by Konta special district to the north, Omo River to the south, Dawro zone to the east and southeast and Agare high mountains and Omo River to the west (Weldeyohanes, 2006). It lies at the centre of Omo-Gibe River Basin. There are four small crater lakes that are distributed in different parts of the Park area (Figure 1). The altitude of the area ranges from 550 to 1,700 m asl at the volcanic peaks of the western boundary (Timer, 2005; Weldeyohanes, 2006). It is characterized by few flat lands and highly undulating to rolling plains with incised river and perennial streams, valleys and gorges. The mean annual temperature of the area is 17°C. The monthly mean minimum temperature ranges between 10 and 11.4°C (June and August), and the maximum range of between 27 and 29°C (January and February) (Weldeyohanes, 2006). The rainfall distribution is unimodal, with one long rainy season between April and August. The average amount of annual rainfall in the area varies from 1,000 to 3,500 mm.

The natural vegetation of CCNP is highly diverse (Megaze, 2006; Weldeyohanes, 2006). For instance, the ground water forest type of vegetation is dominated by Podocarpus juniperus and broad-leaved tree species. The riverine forests is characterized by mixed vegetation dominated by plant species such as *Albizia grandibracteata, Aspilia mosambicensis, Arundo donax, Chionantus mildobradii, Ehretia cymosa* and Grewia ferrugunea. The grassland has scattered trees and covers the largest part of the Park. It is dominated by elephant grass (Pennisetum purpureum) and few scattered trees. Notable species in woodland are *Acacia brevispica, Combretum colinum, Combretum mole, Maytenus arbutifolia, Terminalia brownii* and *Vitex doniana* (Timer, 2005; Megaze, 2006). In the surroundings of CCNP, farmers commonly cultivate cereals, coffee and root crops for their food source. Land-use patterns include grazing, grass cutting, fire wood collection and timber for construction (Weldeyohanes, 2006). Wild honey harvesting and collection of spices and wild coffee are also carried in the Park (Timer, 2005).

Trapping

In a study of Capture-Mark-Recapture (CMR), non-volant small mammals were trapped using Sherman live traps (16 x 6.5 x 5.5 cm) during the dry and wet seasons between 2010 and 2011. Each grid was sampled once per a season from the same site. Before the start of current data collection, a preliminary survey was conducted on September 2010. This helped us to identify the boundaries of different habitat types and to decide the number of grids/sites. The vegetation of the study area is not markedly differentiated into habitat types. However, based on the classification of vegetation regions of Ethiopia (Yalden and Largen, 1992) and the habitat classification scheme of White and Edwards (2000), seven habitat types: ground water forest (GWF), riverine forest (RF), grassland (GL), wooded grassland (WGL), bushland (BL), lake shore (LS) and agricultural field (AF) were identified. Seven randomly selected sites were identified to represent each habitat type. At each habitat type, two permanent live trapping grids 4900 m2 were established. In each trapping site, a standard square (seven rows by seven columns) trapping grid was established during both seasons. A total of 49 Sherman live traps were set per grid at 10 m intervals between points. The traps were baited with peanut butter mixed

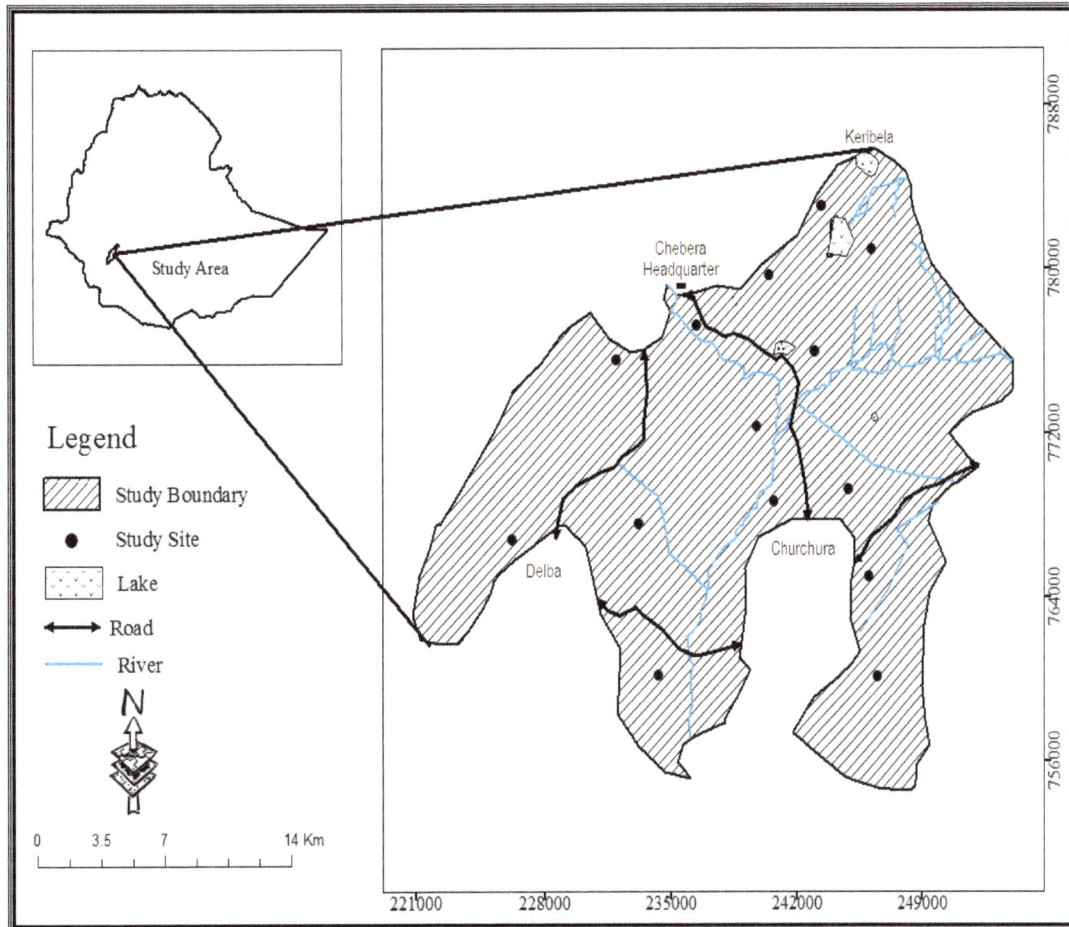

Figure 1. Location of the study sites/grids in Chebera-Churchura National Park, Ethiopia.

with crushed maize. To provide protection against the strong heat, traps were covered with hay and plant leaves during the dry season. The traps were checked twice a day (early morning between 07:00 and 09:00 h a.m., and late afternoon between 17:00 and 19:00 h p.m.) for three or five consecutive days depending on accessibility of the site. Each captured animal was identified, marked by toe clipping, and released at the site where it was trapped. In addition, body mass, sex and approximate age (juvenile, sub-adult and adult) was recorded. Reproductive condition (females: closed or perforated vagina; males: scrotal or abdominal testes) was also recorded (Bekele, 1996a). Age was determined by a compromise between body mass and reproductive condition (Bekele, 1996a, b; Habtamu and Bekele, 2008). Moreover, two or three representative individuals of each species were collected as voucher specimens from snap trapped animals. Thirty-six snap-traps were used along Sherman live-traps at the same time. The aim was to collect voucher specimen for identification. The traps were apart from the live trapping grids by about 200 m at each site to minimize the effect of home range. Skin and skull of the specimens were mounted and deposited in the Zoological Natural History Museum, Addis Ababa University. The specimens were compared and identified at species level by referring to additional reference materials deposited in the Museum. Data collected were analysed using SPSS version 16 computer software programme (SPSS Inc., Chicago, IL, USA) and descriptive statistics to compute the variation in relative abundance, total number of captures between seasons and habitats as well as between age groups.

RESULTS

A total of 18 (16 rodents and two insectivores) species of small mammals were recorded. Of these, 15 species of small mammals were captured during both dry and wet seasons of 5488 trap-nights and the remaining three species were recorded as observed species (Table 1). Out of the 704 small mammals trapped, 682 (98.3%) were rodents and 12 (1.7%) were insectivores. These small mammals were: *Mastomys natalensis*, *Lemniscomys striatus*, *Arvicanthis niloticus*, *Arvicanthis dembeensis*, *Mastomys erythroleucus*, *Acomys cahirinus*, *Rattus rattus*, *Mus musculus*, *Stenocephalemys albipes*, *Gerbilliscus robusta*, *Lophuromys flavopunctatus*, *Pelomys harringtoni*, *Mus tentellus*, *Crocidura fumosa*, *Crocidura flavescens*, *Tachyoryctes splendens*, *Xerus rutilus* and *Hystrix cristata*. Among these, *S. albipes* is a rodent endemic to Ethiopia (Bekele, 1996a, b). The total number of captures differed among species.

M. natalensis was the most abundant species constituting 29.0% of the total number of captures, followed by *L. striatus* (20.2%). *A. niloticus*, *A. dembeensis*, *M. erythroleucus* were recorded as 8.5, 6.7

Table 1. Species composition and relative abundance of live-trapped small mammals in the Chebera-Churchura National Park, Ethiopia (*= non-trapped/observed species).

Species	Total number of captures	Relative abundance (%)
Mastomys natalensis	204	29.0
Lemniscomys striatus	142	20.2
Arvicanthis niloticus	60	8.5
Arvicanthis dembeensis	52	7.4
Mastomys erythroleucus	44	6.2
Acomys cahirinus	40	5.7
Rattus rattus	37	5.3
Mus musculus	35	5.0
Stenocephalemys albipes	20	2.8
Gerbilliscus robusta	19	2.7
Lophuromys flavopunctatus	18	2.6
Pelomys harringtoni	13	1.8
Mus tentellus	8	1.1
Crocidura fumosa	8	1.1
Crocidura flavescens	4	0.6
*Tachyorctes splendens**	*	*
*Xerus rutilus**	*	*
*Hystrix cristata**	*	*
Total	704	100

Table 2. A abundance of trapped small mammal species in each habitat types (- shows absence of trapped individuals; GWF: ground water forest, RF: riverine forest, GL: grassland, WGL: wooded grassland, BL: bushland; LS: LS: lake shore, AF: agricultural field).

Species	Abundance of species in each habitat type						
	GWF	RF	GL	WGL	BL	LS	AF
M. natalensis	3.9	2.9	14.2	12.3	11.3	12.7	42.7
L. striatus	3.5	2.8	17.6	21.1	26.1	17.6	11.3
A. niloticus	-	-	18.3	21.7	16.6	6.7	36.7
A. dembeensis	-	-	7.7	11.5	15.4	5.8	59.6
M. erythroleucus	-	-	13.6	15.9	20.5	11.4	38.6
A. cahirinus	10.0	7.5	15.0	7.5	15.0	10.0	35.0
R. rattus	-	-	18.9	18.9	13.5	-	48.7
M. musculus	5.7	-	11.4	14.3	22.9	5.7	40.0
S. albipes	5.0	-	20.0	30.0	15.0	20.0	10.0
G. robusta	10.5	5.3	10.5	26.3	31.6	-	15.8
L. flavopunctatus	-	-	38.9	44.4	16.7	-	-
P. harringtoni	7.7	-	23.1	23.1	30.8	15.3	-
M. tentellus	-	-	25.0	25.0	37.5	-	12.5
C. fumosa	-	-	12.5	37.5	25.0	12.5	12.5
C. flavescens	-	-	25.0	50.0	-	25.0	-

and 6.2%, respectively. The other species were below 5% of the total number of captures (Table 1). The least abundant species were *M. tentellus, C. fumosa* and *C. flavescens*, with 1.1, 1.1 and 0.6%, respectively. Larger rodents such as porcupine (*H. cristata*), unstriped ground squirrel (*X. rutilus*) and the mole rat (*T. splendens*) were observed throughout the study area.

M. natalensis, L. striatus and *A. cahirinus* were recorded in all habitat types (Table 2), the highest abundance was 42.7% (agricultural field), 26.1% (bushland) and 35.0% (agricultural field) for *M. natalensis, L. striatus* and *A. cahirinus*, respectively. For these species, the lowest abundance was 3.1% (*M. natalensis*), 2.9% (*L. striatus*) and 7.5 (*A. cahirinus*) in

Table 3. Seasonal variation and sex distribution of small mammals trapped by live-trapping (-: absence).

Species	2010 (Dry season)			2011 (Wet season)		
	Sex		Total number of captures	Sex		Total number of captures
	M	F		M	F	
M. natalensis	54	70	124	35	45	80
L. striatus	31	38	69	34	39	73
A. niloticus	20	23	43	8	9	17
A. dembeensis	17	15	32	10	10	20
M. erythroleucus	16	14	30	8	6	14
A. cahirinus	12	16	28	5	7	12
R. rattus	11	11	22	9	6	15
M. musculus	11	13	24	6	5	11
S. albipes	4	4	8	7	5	12
G. robusta	6	3	9	4	6	10
L. flavopunctatus	6	4	10	4	4	8
P. harringtoni	4	3	7	5	1	6
M. tentellus	-	-	-	3	5	8
C. fumosa	2	2	4	3	1	4
C. flavescens	3	1	4	-	-	-
Total	197	217	414	141	149	290

Table 4. Composition of different age groups of live-trapped small mammals.

Seasons	Age groups			Total catch	Relative abundance
Month/year	Adult	Sub-adult	Young (juveniles)		
Oct-Dec 2010 (Dry)	326	69	19	414	58.8%
Jun-Aug 2011 (Wet)	206	30	54	290	41.2%
Total	532	99	73	704	$x^2 = 21.8$, P < 0.05

riverine forest. *M. musculus* and *S. albipes* were captured in all habitats except the riverine forest, with variable abundance. *G. robusta* was not captured in agricultural field. The other species, *A. niloticus*, *A. dembeensis*, *M. erythroleucus*, *M. tentellus* and *C. fumosa* were captured in the same habitat types with variable abundance. Except for *M. tentellus* and *C. flavescens*, all species were captured both during wet and dry seasons (Table 3). The abundance of small mammals in the dry and wet seasons was 58.8 and 41.2%, respectively. The total number of captures was higher during the dry season than the wet season (58.8%; x^2 = 21.8, df = 1, P < 0.05). *Crocidura flavescens* was only captured during the wet season and *M. tentellus* during the dry season. Out of the 704 captured individuals in all trapping occasions, females comprised 366 (52.0%) and males 338 individuals (48.0%; Table 3), which was statistically non-significant (x^2=1.12, df = 1, P>0.05). However, as the number of individuals (abundance) increased the number of females also increased.

Adults comprised 532 (75.5%), sub-adults 99 (14.1%) and juveniles 73 individuals (10.4%; Table 4), each class

statistically different from the others (x^2 = 566.75, df = 2, P<0.05). During the wet season, juveniles comprised 54 (18.6%), sub-adults 30 (10.4%) and adults 206 (71.0%) individuals, whereas during the dry season, juveniles comprised 19 (4.6%), sub-adults 69 (16.7%) and adults 326 (78.7%) individuals. The variations in the number of sub-adults (x^2 =15.36, df = 1, P < 0.05) and adults (x^2 = 27.06, df = 1, P < 0.05) between seasons were significant. Out of the 73 juvenile rodents captured from different habitats during both seasons, 54 (74.0%) were captured during the wet and 19 (26.0%) during the dry season (x^2 = 16.78, d.f. = 1, P <0.05).

Trap success did not differ among habitats (x^2 = 3.72, df = 6, P>0.05; Table 5). The total number of captures did not vary between different habitats (x^2 = 3.72, df = 6, P>0.05). However, the highest abundance was in agricultural field with 152 (21.6%) individuals and the lowest in riverine forest with 6 (0.9%) individuals during the dry and wet seasons, respectively. Maximum trap success from agricultural field (38.8%) and minimum from riverine forest (1.5%) were recorded during the dry and wet seasons, respectively. There were also variations in

Table 5. Trap success and relative abundance of small mammals in different habitats.

Habitat types	No. of grids	Month/year	Season	Total number of captures	Relative abundance	Trap night	Trap success
GWF	2	Oct-Dec/2010	Dry	14	2.0	392	3.6
	2	Jun-Aug/2011	Wet	9	1.3	392	2.3
RF	2	Oct-Dec/2010	Dry	8	1.1	392	2.0
	2	Jun-Aug/2011	Wet	6	0.9	392	1.5
GL	2	Oct-Dec/2010	Dry	64	9.1	392	16.3
	2	Jun-Aug/2011	Wet	48	6.8	392	12.2
WGL	2	Oct-Dec/2010	Dry	66	9.3	392	16.8
	2	Jun-Aug/2011	Wet	59	8.4	392	15.1
BL	2	Oct-Dec/2010	Dry	73	10.4	392	18.6
	2	Jun-Aug/2011	Wet	54	7.7	392	13.8
LS	2	Oct-Dec/2010	Dry	37	5.2	392	9.4
	2	Jun-Aug/2011	Wet	40	5.7	392	10.2
AF	2	Oct-Dec/2010	Dry	152	21.6	392	38.8
	2	Jun-Aug/2011	Wet	74	10.5	392	18.9

GWF: ground water forest, RF: riverine forest, GL: grassland, WGL: wooded grassland, BL: bushland; LS: lake shore, AF: agricultural field.

trap success among habitats and seasons (x^2 = 96.8, df = 13, P<0.05). The overall trap success of the study was 12.8%.

DISCUSSION

The present study revealed the presence of 18 species of small mammals in CCNP, indicating that the park is an ideal habitat for harboring small mammal species (rodents and insectivores). However, the study of Habtamu and Bekele (2008) revealed more species from Alatish National Park, northwestern Ethiopia. Also, great variations in the relative abundance of small mammals species were observed among habitats and between seasons. *M. natalensis* and *L. striatus* were the most abundant and widely distributed species in all habitat types, whereas *L. flavopunctatus* and *C. flavescens* were the least distributed and abundant species. Most small mammals are captured from grassland, wooded grassland and bushland, whereas ground water forest and riverine forest were very poor in species richness and abundance. Habtamu and Bekele (2008) and Marcello et al. (2008) also revealed similar findings. They noted the effect of vegetation cover and availability of resources on the abundance of animal species. This might be due to the homogeneous vegetation that is dominated by few species of tall trees in both forest

types. In addition, the underground/open area under the forest habitat is open or less covered resulting in shortage of cover, food and diversity of microhabitats. Densely covered habitats with high diversity of plant species were preferred by most small mammal species in the study area. Similar results were found in Arbaminch forest (Datiko et al., 2007), and Alatish National Park (Habtamu and Bekele, 2008), both in Ethiopia.

Also, the total number of captures varied between seasons. Seasonality might cause the dynamic changes which occur in the habitats such as cover and food availability as noted by Oguge (1995) and Martin (1998). More individuals were captured from the agricultural field before harvest during the dry season. The trap success declined during the post-harvest period. The cropping system of the area possibly might have contributed to continuous supply and availability of alternative food and shelter for the species before harvest. Habitat complexity, food and cover availability are key factors influencing the overall distribution of rodents (Gebresilassie et al., 2004; Avenant and Cavallini, 2008). Moreover, lack of cover after harvest might have exposed the animals to predators which could force them to migrate to more suitable habitats (Hansson, 1999). Makundi et al. (2009) also noted that population size of small mammal species fluctuates greatly as a result of change in quality and quantity of resources in the environment. Studies also revealed that the availability (quality and quantity) of resources also determine the movement pattern of small mammals (Kasso et al., 2010; Kilgore et al., 2010). The study of Gebresilassie et al. (2004) also revealed that farmlands provide essential resources better than grasslands before harvest.

During the present study, *M. tentellus* was not trapped during the dry season and *C. flavescens* during the wet season. This shows that extended study including both seasons will provide more information on the diversity and distribution of the species. The sex ratio of most species did not vary. However, the present study revealed that as the number of females increases in a population the abundance also increases. The study by Bekele (1996a), Datiko et al. (2007) and Habtamu and Bekele (2008) also found the same trend in central, south and northwestern part of the country, respectively. For some species, variation in age distribution was observed between seasons. Out of the total number of captured individuals, adults comprised the largest number of individuals (75.5%), followed by sub-adults (14.1%) and juveniles (10.4%). This might be related to the relative large home ranges for adults and sub-adults and small home ranges for young individuals of the same species (Shanker, 2001). In the present study, the number of pregnant females and juveniles was high during the wet season and low during the dry season, which was also observed in the study of Datiko et al. (2007). Even for the most abundant species (*M. natalensis* and *L. striatus*), pregnant and young individuals were rarely trapped during the dry season. This shows that breeding in most small mammal species was during the wet season. The wet season is full of more nutritious food which could promote breeding of animals (Jackson et al., 2004; Marcello et al., 2008). Our results are consistent with other studies that showed breeding patterns for many African rodents to be related to rainfall (Habtamu and Bekele, 2008; Takele et al., 2010; Girma et al., 2012).

Trap success differed between habitat types and seasons. In almost all habitat types, lower trap success was observed during the wet season and the abundance of most species increased during the dry season. The wet season was associated with reproduction for most rodent species (Bekele, 1996a; Marcello et al., 2008). The overall trap success varied among the seven habitats. The average total trap success was 12.8%. Similar studies in different parts of Ethiopia obtained higher trap successes: 19% for the high altitude locality of south Goba (Yalden et al., 1996), 18.7% for Harenna Forest (Yalden, 1988) and 17.6% for Arbaminch (Datiko et al., 2007). As compared to previous studies, a lower trap success was obtained in the present study. It might be related to habitat unsuitability and topographic variation of each of the area.

According to Avenant and Cavallini (2008) fire could be beneficial in that it allows new growth of more nutritious vegetation enabling quick recovery of the population of small mammals. However, in this study area the incidence of fire was very frequent which could have severe adverse effects on small mammal populations. In previous studies, fire was shown to lower the species diversity due to destruction of vast areas of their habitat and food resources, which could also lead to changes in the behavior (Haim and Izhaki, 1994; Clausnitzer, 2003). Similar effects might have contributed to the small mammal fauna of CCNP. Further studies focusing on individual species and their ecology are important. The CCNP ecosystem revealed that the inaccessible, remote areas of the country harbored unique and endemic species. Therefore, in order to have a comprehensive understanding of the area, assessing the ecology of each species within the geographic boundary of this newly established park should be a priority.

ACKNOWLEDGEMENTS

We thank Addis Ababa University for providing financial assistance. The help provided by all staff members of Chebera-Churchura National Park is greatly appreciated.

REFERENCES

Amori G, Luiselli L (2011). Small mammal community structure in West Africa: A meta-analysis using null models. Afr. J. Ecol. 49: 418-430.

Avenant NL, Cavallini P (2008). Correlating rodent community structure with ecological integrity, Tussen-die-Riviere Nature Reserve, Free State Province, South Africa. Integr. Zool. 2: 212-219.

Avenant N (2011). The potential utility of rodents and other small mammals as indicators of ecosystem 'integrity' of South African grasslands. Wildl. Res. 38: 626-639.

Bekele A (1996a). Population dynamics of the Ethiopian endemic rodent, *Praomys albipes* in the Menagesha State Forest. J. Zool. Lond. 238: 1-12.

Bekele A (1996b). Rodents of the Mengasha State Forest, Ethiopia, with an emphasis on the endemic *Praomys albipes* Ruppell 1842. Trop. Zool. 9: 201-212.

Bekele A, Leirs H (1997). Population ecology of rodents of maize fields and grasslands in central Ethiopia. Belg. J. Zool. 127:39-48.

Clausnitzer V (2003). Rodents of Mt. Elegon, Uganda: ecology, biogeography, and the significance of fire. Ecotrop. Monogr. 3: 3-184.

Davies G (2002). African Forest Biodiversity: A field Survey Manual for Vertebrates. Cambridge: Earthwatch, pp. 120-126.

Datiko D, Bekele A, Belay G (2007). Species composition, distribution and habitat association of rodents from Arbaminch forest and farmlands, Ethiopia. Afr. J. Ecol. 45: 651-657.

Gebresilassie W, Bekele A, Belay G, Balakrishnan M (2004). Micro-habitat choice and diet of rodents in Maynugus irrigation field, northern Ethiopia. Afr. J. Ecol. 42: 315-321.

Girma Z, Bekele A, Hemson G (2012). Small mammals of Kaka and Hunkolo, southeast Ethiopia. Trop. Ecol. 53: 33-41.

Habtamu T, Bekele A (2008). Habitat association of insectivores and rodents of Alatish National Park, northwestern Ethiopia. Trop. Ecol. 49: 1-11.

Haim A, Izhaki I (1994). Changes in rodent community during recovery from fire: Relevance to conservation. Biodiv. Conserv. 3: 573-685.

Hansson L (1999). Intraspecific variation in dynamics: Small rodents between food and predation in changing landscapes. Oikos 85: 159-169.

Jackson TP, Aarde RJV (2004). Diet quality differentially affects breeding efforts of *Mastomys coucha* and *M. natalensis*: Implications for rodent pests. J. Exp. Zool. 30: 97-108.

Kasso M, Bekele A, Hemson G (2010). Species composition, abundance and habitat association of rodents and insectivores from Chilalo-Galama Mountain range, Arsi, Ethiopia. Afr. J. Ecol. 48: 1105-1114.

Kilgore A, Lambert TD, Adler GH (2010). Lianas influence fruit and seed use by rodents in a tropical forest. Trop. Ecol. 51: 265-271.

Kingdon J (1997). The Kingdon Field Guide to African Mammals. London: Academic Press, p. 476.

Lavrenchenko LA, Milisnikov AN, Aniskin VM, Warhavsky AA, Gebrekidan S (1997). The genetic diversity of small mammals of the Bale Mountains. SINET: Ethiop. J. Sci. 20: 213-233.

Leirs H, Verhagen R, Verheyen W (1994). The basis of reproductive seasonality in *Mastomys* rats (Rodentia: Muridae) in Tanzania. Trop. Ecol. 10: 55-66.

Linzey AV, Kesner MH (1997). Small mammals of a woodland savannah ecosystem in Zimbabwe. I. Density and habitat occupancy patterns. J. Zool. Lond. 243: 137-152.

Makundi RH, Apia W, Massawe W, Mulungu LS, Katakweba A (2009). Diversity and population dynamics of rodents in farm-fallow mosaic fields in Central Tanzania. Afr. J. Ecol. 48: 313-320.

Marcello GJ, Wilder SM, Meikle DB (2008). Population dynamics of a generalist rodent in relation to variability in pulsed food resources in a fragmented landscape. J. Anim. Ecol. 77: 41-46.Martin TE (1998) Are microhabitat preference of coexisting species under selection and adaptive? Ecol. 79: 656-670.

Megaze A (2006). Population status and distribution of African Buffalo (Synceruscaffer sparrman 1779) in Chebera Churcura National Park, Ethiopia. M. Sc. Thesis. Addis Ababa University (Unpublished), p. 117.

Oguge ON (1995). Diet, seasonal abundance and microhabitats of *Praomys* (*Mastomys*) *natalensis* (Rodentia; Muridae) and other small rodents in a Kenyan sub-humid grass land community. Afr. J. Ecol. 33: 211-223.

Shanker K (2001). The role of competition and habitat in structuring small mammals communities in a tropical montane ecosystem in southern India. J. Zool. Lond. 253: 15-24.

Skinner JD, Chimimba CT (2005). The Mammals of the Southern African Subregion (3rd Ed). Cambridge: Cambridge University Press, pp 874.

Takele S, Bekele A Belay G, Balakrishinan M (2010). A comparison of rodent and insectivore communities between sugarcane plantation and natural habitat in Ethiopia. Trop. Ecol. 52: 61-68.

Timer G (2005). Diversity, Abundance, Distribution and Habitat Association of large mammals in the Chebera Churchura National Park, Ethiopia. M.Sc. thesis, Addis Ababa University, Addis Ababa (Unpublished), p. 127.

Vaughan JA, Ryan JM, Czaplewsiki NJ (2000). Mammalogy, 4th eds. Toronto: Saunders College Publishing, p. 565.

White L, Edwards A (2000). Vegetation inventory and description. In: White L, Edwards A (eds) Conservation of Researches in the African Rain Forest: A Technical Hand Book. New York: Wildlife Conservation Society, pp. 118-119.

Wilson DE, Reender DM (2005). Mammal species of the world: A Taxonomic and Geographic Reference. 3rd edn Vol. I. Maryland: Johns Hopkins University Press, pp. 835-847.

Weldeyohanes D (2006). Diversity, distribution and relative abundance of Avian species of Chebera Churhcura National Park, Ethiopia. M. Sc. Thesis. Addis Ababa University (Unpublished), p. 103.

Yalden DW (1988). Small mammals of the Bale Mountains, Ethiopia. Afr. J. Ecol. 26: 282-294.

Yalden DW, Largen MJ (1992). The endemic mammals of Ethiopia. Mammal. Rev. 22: 115-150.

Yalden DW, Largen MJ, Hillman JC (1996). Catalogue of the mammals of Ethiopia and Eritrea. 7. Revised Checklist, Zoogeography and Conservation. Trop. Zool. 9: 73-164.

15

Impacts of enclosures in rehabilitation of degraded rangelands of Turkana County, Kenya

John N. Kigomo and Gabriel M. Muturi

Kenya Forestry Research Institute, P.O. Box 20412-00200, Nairobi, Kenya.

The role of enclosures in range rehabilitation was investigated through a case study at Kalatum in Turkana County where the use of enclosures and establishment of fodder species were demonstrated to the local communities in late 1980's. However, documented information on the impact of enclosures is scanty. Rehabilitation impact was evaluated through ecological sampling in fenced and unfenced areas and comparison of vegetation variables. The technology adoption was evaluated by Geographical Information System (GIS) techniques to determine the extent of enclosure expansion using social fence by local community. Results showed a higher density of fodder species in the fenced than in the unfenced areas ($p = 0.019$). For example, *Acacia tortilis* attained a density of 204 trees ha^{-1} in the fenced areas as compared to 74 trees ha^{-1} in the unfenced area and the average ground cover of *Crysopogon plumulosus* was 36% in the fenced as compared to 4% in the unfenced area. Technology adoption was good, as evidenced from fence expansion from the initial 5-ha established under research trials for 23-ha community enclosures. The findings show the importance of involving local communities during the project cycle and recommends up scaling of this technology among pastoral communities in Kenya.

Key words: Enclosure, rehabilitation, technology adoption, Geographical Information System (GIS) and Turkana.

INTRODUCTION

Range degradation has become a common phenomenon in Kenya's drylands, thereby threatening the survival of indigenous communities who have long depended on range land resources. According to the Government of Kenya (2007), drylands support 30% of human population and 70% of livestock and the bulk of wildlife that support the tourism sector. In the past, pastoralism thrived under traditional management practices that were characterized by delineation of seasonal grazing areas to support nomadism (Sitters et al., 2009). The practices facilitated range resource resilience through periodic relief of grazing pressure between the foraging intervals. Development interventions have discouraged nomadism in favour of sedentarization for effective provision of social amenities without regard to ecological implications. Consequently, these settlements have become degradation nuclei; with resource degradation intensity increasing towards the settlement centres (Okoti et al., 2004; Kariuki et al., 2007). The main factors associated with this trend include over-exploitation of resources due to localized increase in human and livestock populations, climatic changes due to insufficient and unreliable rainfall and poverty. To arrest this alarming degradation, it is important to design economically feasible, socially acceptable and ecologically viable management and conservation strategies for vegetation resources.

Numerous drought incidences have been reported in Turkana County whose impacts have been devastating

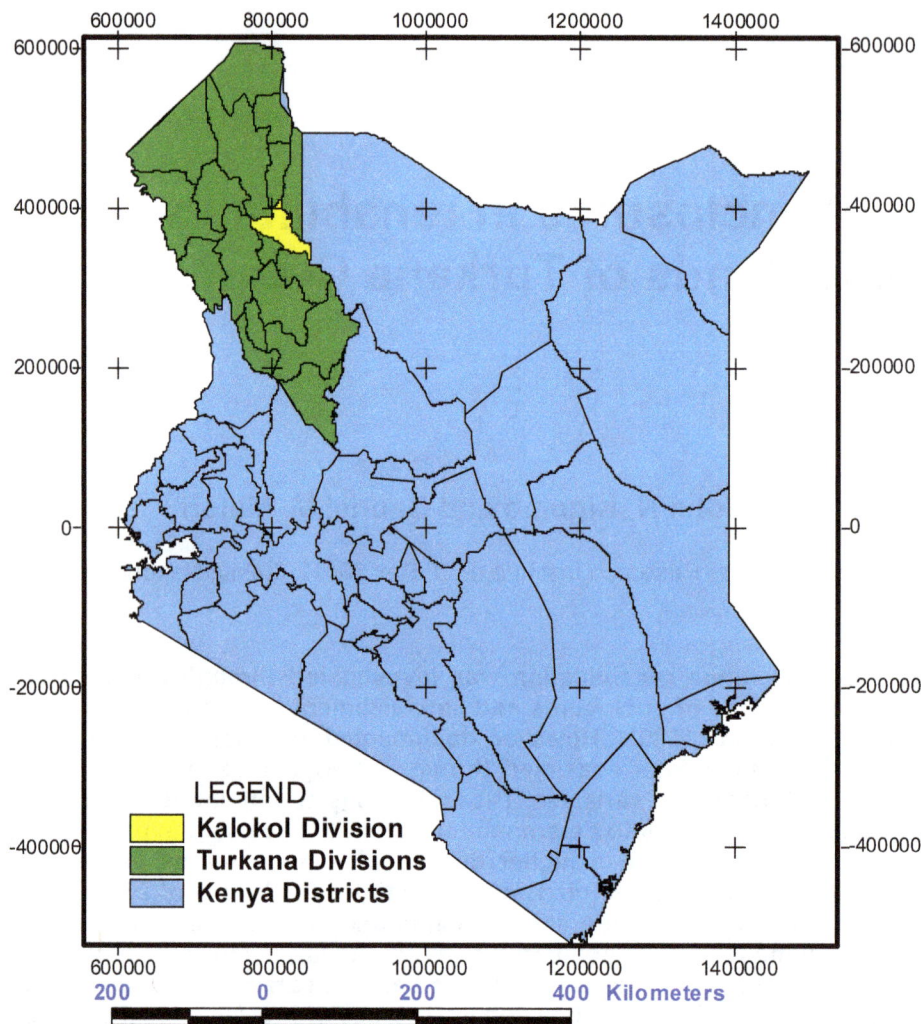

Figure 1. Location of Turkana County in northern Kenya.

(UNEP, 2000). Past reactionary measures included famine relief, sporadic food for work activities and establishment of irrigation schemes. These led to the establishment of settlements resulting into prevalent resource degradation. In the early 1980s, several programmes were initiated to address the emerging degradation problems. Turkana Rural Development Program (TRDP) was one such program that was funded by the Norwegian Agency for Development Co-operation (NORAD). Kenya Forestry Research Institute (KEFRI), the leading implementing agent for this program and other stakeholders were involved in range rehabilitation by use of enclosures, range re-seeding and planting of fodder trees through water harvesting. This involved demonstration of rehabilitation technologies and sensitising the local communities on the potentials of introduced technologies. Although the enclosures have been one of the range rehabilitation technologies in northern Kenya, scientific studies on their socio-economic and ecological importance are scanty. This paper presents an evaluation of about 25 year old

Kalatum rehabilitation site in Kalokol division where a research and development plot was initially established in 1986. The overall objective was to investigate contribution of enclosures as a tool for dryland rehabilitation in northern Kenya. Specifically, the study aimed to assess technology adoption by the local community, investigate and compare quantitative information on vegetation structure in terms of density, spatial distribution and regeneration in enclosures and adjacent open areas and finally, to formulate a baseline for sustainable utilization of enclosures in Turkana County.

MATERIALS AND METHODS

Study site

The study site is located within Kalatum ranges of Kalokol division in Turkana County, North-West Kenya, about 30 km north of Lodwar town (Figure 1). Kalatum rangeland occupies an area of about 112 km^2, which traditionally reserved as wet season grazing (Herlocker et al., 1994). However, this is no longer adhered to due to change of lifestyle.

Soils are predominantly sandy but occasionally intercepted by patches of black cotton soils with high concentrations of carbonates and chlorides of sodium and carbonates (Van Bremen and Kinyanjui, 1992). Rainfall ranges between 150 and 200 mm per annum with peaks in April and October, while vegetation is classified as wooded annual grassland with a general vegetation cover of 41% (Herlocker et al., 1994).

Establishment of research and community plots

A 5-ha browse trial plot was initiated in 1986 at Kalatum ranges to investigate suitability and adaptability of different Australian and indigenous fodder tree species under different water harvesting techniques. Fencing was done using cedar posts and barbed wire. Micro-catchments of 5 x 5 m and 10 x 10 m were constructed and seedlings were planted. The indigenous species were *Acacia mellifera, Acacia tortilis, Balanites aegyptiaca, Zizyphus mauritiana, Cordia sinensis* and *Dobera glabra*. The exotic species were *Acacia aneura, Atriplex aurionformis, Azandrachta indica, Parkinsonia aculeate* and *Acacia horosericea*. Prior to plot establishment, a series of consultative meetings were held to discuss the project objectives with community members and a local resident engaged for surveillance to ensure non-intrusion of livestock into the trial plot.

Contact between the scientists and community members were maintained through collaborative periodic plots assessments and feedback meetings where ideas were freely discussed. During consultations, community members resolved to establish their own plots using available less palatable plant materials and using existing community norms.

Data collection

A line transect was established in 2011 across the fenced and unfenced areas. Plots of 20 x 20 m were systematically laid along the transect which were estimated using a handheld Geographic Positioning System (GPS) according to Mengistu et al. (2005). A total of eighteen plots were established within the fenced areas and seven plots were established in the unfenced areas, as there was minimal variability in the unfenced areas. Smaller sub-plots of 5 x 5 m and 2 x 2 m were nested at the centre of the main plots for saplings and herbaceous layer assessment respectively.

Trees data was collected in 20 x 20 m plots. Trees were classified as the woody plants whose respective height and diameter at breast height (dbh) were above 3 m and 2.5 cm. (dbh is at 1.30 cm over the soil) Saplings data was collected through counting and measuring height in 5 x 5 m, plots, saplings were classified as the woody plants with height between 0.5to 3 m and a dbh of <2.5 cm. Height and dbh were measured with a graduated measuring rod and diameter tape, respectively.

Herbaceous (including grass) and seedlings (woody species of height below 0.5m) were assessed in 2 x 2 m sub-plots through, identification, counting and recording their numbers. Canopy and ground cover was also estimated in main and sub-plots, respecttively. Plant species were identified according to a checklist of plants for Turkana district (Morgan, 1981; Herlocker et al., 1994; Beentje, 2004). For species which could not be identified in the field, herbarium specimens were collected and identified through comparison with specimen in the regional herbarium at KEFRI Lodwar. The nomenclature of the identified species was cross-checked through international plant names index website (www.ipni.org).

Assessment of plots expansion

GPS points were taken along the perimeter of initial research plot and expanded community plots. The data points were entered to Geographic Information Systems (GIS) software (Arch GIS); polygons were drawn by joining the points and polygon areas generated by the software. Focussed group discussions were held with the communities in surrounding villages to document rules and regulations that guide the management and maintenance of community enclosures.

Data analysis

Descriptive synthesis and analysis of data was carried out using Microsoft Excel and SPSS version 19. Average percentage cover, relative abundance and frequency of each herbaceous species was obtained using the following equations (Herlocker, 2005):

Average percentage cover = Total percentage cover of species in the plots/number of plots having species x 100

Relative abundance = Number of individuals of the species recorded/total number of plants in all the plots x 100

Relative frequency = Number of sample plots in which the species occurred/total number of plots x 100

The density of woody plant species was obtained by calculating the number of individuals per hectare and paired sample t-test used to compare between enclosure and open areas. The diversity was obtained using Shannon diversity index (H') as follows (MacDonald, 2003):

$$(H') = -\sum pi \ln pi$$

Where $pi = n_i/N$, the proportion of the *ith* species and ln is the natural logarithm. The index assumes that each representative sample species has an equal chance of being included in each sampling point.

RESULTS

Species composition

A total of 31 plant species within 20 families were identified both in the enclosures and in the open grazing areas. The plant composition consisted of 20% grasses, 30% trees/shrubs and 50% herbs. Eighty percent of identified plant species were palatable to livestock. Fourteen species were recorded both in the enclosures and in the open rangelands while twelve species were found in the enclosures only, in addition, five species occurred only in the open area (Tables 1 and 2).

Woody species density, diversity and population structure

The dominant woody species recorded was *Acacia tortilis*, with a density of 204 trees and 74 trees ha^{-1} inside the enclosure and in open areas-respectively (Table 2). Density was significantly different between enclosure and open areas (t = 3.05 df = 7 p = 0.019). Other species recorded include *Acacia reficiens, Salvadora persica and Zizyphus mauritiana*. Among the indigenous and exotic

Table 1. Herbs and seedlings found in enclosure and open area.

Specie	Enclosure			Open area		
	Mean cover (%)	Relative abundance (%)	Relative Frequency (%)	Mean cover (%)	Relative abundance (%)	Relative Frequency (%)
Indigofera cliffordiana Gillett.	-	-	-	0.14	0.17	14.29
Seddera hirsuta Hall.f.	0.40	0.92	11.11	0.71	0.17	14.29
Barleria eranthamoides C.B.C.L (Cl.)	0.06	0.1	5.56	0.14	0.17	14.29
Evolvulus alsinoides (L) L.plate	1.70	5.75	11.11	0.86	1.89	28.57
Ruella patula Jacq. (Ruellia)	-	-	-	5.71	6.56	28.57
Aristida mutabilis Trin & Rupr.	-	-	-	1.40	7.77	14.29
Heliotropium longiflorum Jaub and Spach.	1.00	6.47	16.67	0.14	0.86	14.29
Acacia tortilis (Forssk) Hayne	0.06	0.10	5.56	-	-	-
Ornithogulum Tennifolium (tennifolium)	0.12	1.13	11.11	-	-	-
Aristida adscensionis Walter	1.20	5.85	16.67	-	-	-
Tribulus cistoides L.	0.06	0.21	11.11	-	-	-
Indigofera swaziensis Bolus	0.20	0.92	22.22	-	-	-
Dactyloctenium aegyptium (L.) K. Ritch. (Willd.)	0.06	5.34	5.57	-	-	-
Ocimum suave Willd	2.50	26.39	11.11	-	-	-
Pupalia lappacea L.	0.40	1.33	5.56	2.14	22.62	28.57
Solanum coagulans Forssk	0.06	0.10	5.56	-	-	-
Acacia reficiens Wawra& Peyr	0.06	0.10	5.56	-	-	-
Acacia nubica Benth	1.20	0.41	5.56	-	-	-
Fersetia stenoptera Hochst	1.70	7.75	11.11			
Eragrostis porosa Nees.	0.17	12.32	5.56	-	-	-
Duosperma eremophilum L. (Willd)	-	-	-	5.71	2.42	14.29
Chrysopogon plumulosus Hochst.	36.30	26.59	50.00	4.14	15.89	28.57
Hyphaene compressa H. Wendl.	0.06	0.1	5.56	-	-	-
Cyperus rotundus L.	-	-	-	5.00	4.45	14.29

Table 2. Density (trees/ha) of dominant tree species.

Specie	Enclosure	Open area
Acacia tortilis (Forssk) Hayne	204	74
Acacia reficiens Wawra& peyr	89	7
Acacia nubica Benth.	76	4
Balanites aegyptica (L.) Del.	60	4
Salvadora persica L.	45	2
Zizyphus mauritiana Lam.	13	0
Acacia mellifera (Vahl) Benth.	4	1
Cadaba rotundifolia Forssk.	0	7

species planted; *Z. mauritiana, Acacia mellifera* and *A. tortilis* has been recruited successfully through micro-catchments. Mean Shannon diversity index (*H*) was 1.6 and 1.0 for enclosure and open areas, respectively. Saplings were only recorded in the enclosures where *A. reficiens, Acacia nubica* and *A. tortilis* were the dominant species. Distribution of dbh classes indicates a higher proportion of lower diameters with 75% of woody species having diameters less than 5 cm (Figure 2). This scenario is absent outside the enclosure since diameter distribution could not be formulated due to low number and diversity of woody species recorded.

Herbaceous and seedlings cover, abundance and frequency

The ground cover was dominated by *Chrysopogon plumulosus* with 36 and 4% within and outside the enclosures, respectively. *Duosperma eremophilum*, a perennial herb was mostly found in open rangelands. The

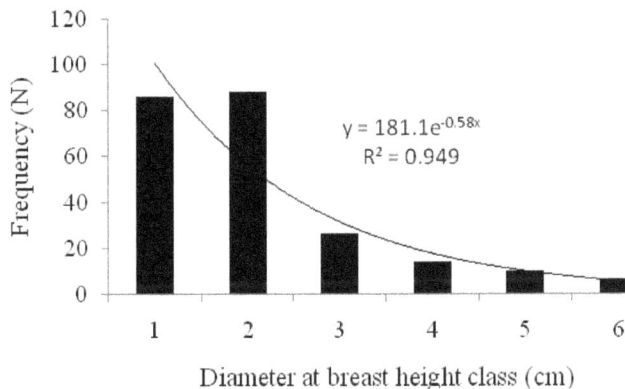

Figure 2. Diameter class distribution of trees in the enclosure (1 = 2.5-4.9, 2 = 5.0-7.4, 3 = 7.5-9.9, 4 = 10-12.4, 5 = 12.5-14.9, 6 = ≥15).

most abundant species in the herbal layer was *C. plumulosus*, *Ocimum suave* and *Eragrostis porosa* with a relative abundance of 27, 26 and 12%, respectively in the enclosure. In the open rangelands, *Pupalea lappaceae* was the most abundant species (22%). The seedlings of *A. tortilis*, *A. nubica*, *A. reficiens* and *Hyphaene compressa* had a relative frequency of 6% within the enclosure whereas *C. plumulosus*, *P. lappaceae*, *Evolvulus alsinoides* and *Ruela patula* dominated in the open.

Technology adoption

The study indicates an increase of the initial fenced area of 5-ha to an extensive area of 23-ha as outlined in Figure 3. Local community adopted the use of locally available materials and traditional rules to keep off the livestock in areas agreed through participatory consultations. Community plot 1 which is an extension of the original plot was established from 1989 to 1993 whereas community plot 2 and 3 were established from 1995 to 2002.

During focussed group discussions, traditional rules and regulations guiding the management and maintenance of enclosures were summarized as follows:

- Decisions are carried out by council of elders selected from surrounding villages;
- Young seedlings were protected from damage during dry season grazing by sick and aged small livestock such as goats and sheep;

- Surveillance of the enclosure was done by sending young men and sometimes the whole community;
- Harvesting of trees is prohibited, but pruning of trees can be done during construction of temporary livestock sheds;
- Harvesting of highly valued *Acacia* pods (*ngitit*) was only allowed under supervision by the council of elders;
- Severe penalties like revoking grazing rights for the enclosure are imposed on community members who violate the established grazing rules.

DISCUSSIONS

The contrasting vegetation patterns between the fenced and unfenced plots referred to in this study (Table 2 and Figure 3) attest for the effectiveness of enclosures in range rehabilitation. The enclosures have marginally higher diversity index than open areas. Although the species diversity is low in enclosures, it is still above minimum threshold of 1.5 in sustainable natural ecosystems (Macdonald, 2003). Although periodical changes of vegetation trends was not factored in this study, similar studies in Northern Ethiopia shown a contrast of area enclosures aged about 10 years and open grazing areas (Mengistu et al., 2005). The high diversity values of enclosures indicate the technologies' significance in conservation of genetic resources of woody species which are under heavy threat of local extermination. Furthermore, they play crucial role in provision of browse during dry spells in this locality.

The higher density of *A. tortilis* in the enclosed areas could be due to accumulation of seed from goat droppings, while foraging along the perimeter of social fence during 'dry season grazing'. This concurs with a previous study that revealed a higher recruitment of the species in abandoned livestock "kraals" (Reid and Ellis, 1995). Goats which form majority of livestock population in Turkana County normally feeds on the *A. tortlis* pods but the seeds are not digested, but are excreted in the droppings and germinate later, although if the young seedlings are not enclosed they are browsed. The contrasting status of *A. tortilis* seedlings and saplings can be associated with seedling protection in the enclosure as opposed to seedling damage through browsing in the unfenced areas.

The presence of *Aristida mutabilis*, *Aristida adscensions* and *Eragrostis porosa* grass in the enclosures only, which are highly valued among the pastoralists, demonstrates the significance of enclosing degraded areas. The low percentage of seedlings of woody species which accounted for 10% could be due to competition from other plants and moisture stress. However, in an area receiving an average rainfall of 150 mm per annum, the seedling proportion is adequate if it can reach maturity. It is worthy to note that no seedlings were found in the open rangelands during the period of study hence enclosing portions of degraded rangeland could greatly

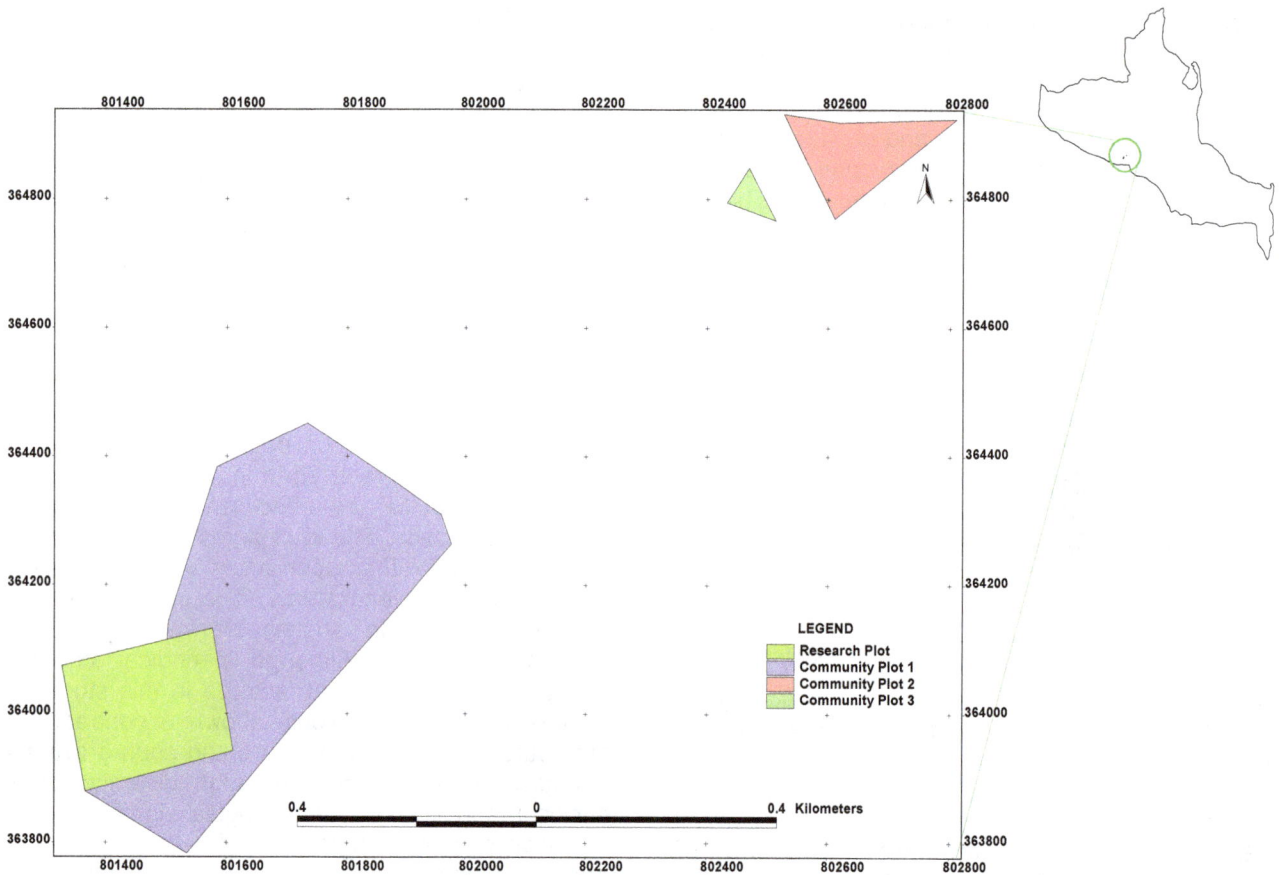

Figure 3. Location and extent of research and community plots in the project area.

enhance regeneration of woody species and consequently hasten the rehabilitation of these sites. High presence of *Pupalea lappaceae* which is an annual herb indicates low preference by the livestock. However *Duosperma eremophilum* is highly preferred by livestock but is mostly found in range lands. It is highly drought tolerant and sprouts quickly after the grazing period. It is conceivable that the species is a poor competitor and this is the likely reason for its low abundance in the enclosed areas.

Diameter distribution revealed a relatively large number of individuals in lower diameter class followed by gradual decline in higher diameter classes (Figure 2). This indicates a sustainable regeneration status within the enclosures. The absence of saplings in the open rangelands can be attributed to the browsing by livestock and human disturbance.

Planting of trees in micro-catchments resulted in successful establishment of indigenous tree species. This shows the importance of water harvesting structures due to limited and unreliable amount of rainfall in this locality. The study also revealed poor long term survival of exotic species which concurs with studies of Olukoye et al. (2003)

Strong traditional systems in management of natural

resources have been demonstrated by the community through expansion of original research plot to an extensive enclosure, which is the only green spot within this locality. The study concurs with studies of Barrow (1996) that documented the evolution of traditional range management practices among the Turkana community and "Sukuma" of Tanzania who practice customary tree resource management system known as "*ngitili*". Collection of woody material was strictly supervised by elders to ensure materials were from dead branches and less palatable species such as *Cadaba rotundifolia*. Jama and Zeila (2005) emphasises the importance of involving local community since they understand the interconnectedness, peculiarities and complexities of dryland ecosystems especially their rangelands and woodlands.

Conclusions and recommendations

The results of this study show that, with proper management and involvement of all stakeholders including local community during project designing, planning and implementation, enclosures can contribute to rehabilitation of degraded areas. Use of traditional rules can be adopted easily as compared to physical fences during establishment of community enclosures. However, there is need to carry out cost benefit analysis to steer up scaling of this technology among different pastoral communities in the dry lands of Kenya.

ACKNOWLEDGEMENTS

The authors thank Government of Kenya through Kenya Forestry Research Institute for financial and logistic support. The contributions of Julius Kaseberi, Jacob Kimani, Giathi Gitehi, Simon Wairungu and Dr. James Kimondo during data collection and preparation of manuscripts are highly appreciated. Special thanks also go to the pastoralists for their hospitality and participation during field data collection.

REFERENCES

Barrow EGC (1996). The drylands of Africa: Local participation in tree management, Initiative Publishers, Nairobi. pp. 268.

Beentje H (1994) Kenya trees, shrubs and lianas, National Museums of Kenya, Nairobi. pp.722.

Government of Kenya (2007). Kenya Vision 2030: A globally competitive and prosperous Kenya, Ministry of Planning and Vision 2030, Government Printers. pp. 136

Herlocker D (1995). Range resource monitoring: field and office guidelines, Range Management Handbook of Kenya vol.III, Signal Press Ltd., Nairobi, p. 60

Herlocker D, Shaaban SB, Wilkes S (1994). Range management handbook of Kenya 2(9) Turkana District, Republic of Kenya, Ministry of Agriculture, Livestock Development and Marketing, Nairobi. pp.189.

Jama B, Zeila A (2005). Agroforestry in the drylands of eastern Africa: a call to action. ICRAF Working Paper no. 1, World Agroforestry Centre, pp.29

Kariuki JG, Machua J, Luvada AM, Kigomo JN, Muindi FK, Macharia EW (2007). Baseline survey of woodland utilization and degradation around Kakuma refugee camp, KEFRI/JOFCA Report. pp. 60.

Macdonald G (2003). Biogeography: Introduction to space, time and life. John Wiley and Sons, Los Angeles. pp. 518.

Mengistu T, Teketay D, Hulten H, Yemshaw Y (2005). The role of enclosures in the recovery of woody vegetation in the degraded dryland hillsides of central and northern Ethiopia. J. Arid Environ. 60: 259-281.

Morgan WTE (1981) Ethno-botany of Turkana: Use of plants by a pastoral people and their livestock in Kenya. Econ. Bot. 35: 96-130

Okoti M, Ngethe JC, Ekaya WN, Mbuvi DM (2004) Land use ecology and socio-economic changes in a pastoral system. J. Hum. Ecol, 16(2): 83-89

Olukoye GA, Wamicha WN, Kinyamario JI (2003). Assessment of the performance of exotic and indigenous tree and shrubs species for rehabilitating saline soils of Northern Kenya. Afr. J. Ecol. 41: 164-170.

Reid RE, Ellis JE (1995). Impacts of pastoralist on woodlands in South Turkana, Kenya Livestock –Mediated Tree Recruitment. Ecol. Appl. 5: 978-992

Sitters J, Heitkoing IMA, Holmgren M, Ojwan GSO (2009). Herded cattle and wild grazers partition water but share forage resources during dry season in East Africa savannas. Biol. Conserv. 142: 738-750.

UNEP (2000). Devastating drought in Kenya: Environmental impacts and Responses, Nairobi, Kenya. pp. 159

Van Bremen H, Kinyanjui HCK (1992). Soils of Lodwar area: An inventory, an evaluation of present land use and recommendations for future land use. Reconnaissance Soils Survey Report No. R17 Soil Survey Nairobi Kenya

16

Dry season herbivore utilization of open grasslands in Lower Zambezi National Park, Zambia

Chansa Chomba[1]*, Ramadhani Senzota[2], Harry Chabwela[3], Jacob Mwitwa[4] and Vincent Nyirenda[5]

[1]Zambia Wildlife Authority, P/B 1 Chilanga, Zambia.
[2]Department of Zoology and Wildlife Conservation, University of Dar es Salaam, P.O. Box 35065 Dar es Salaam Tanzania.
[3]Department of Biological Sciences, University of Zambia, P. O. Box 32379 Lusaka, Zambia.
[4]School of Natural Resources, Copperbelt University, P. O. Box 21692 Kitwe, Zambia.
[5]Office of the Director General, Zambia Wildlife Authority, P/B 1 Chilanga, Zambia.

Utilisation of open grassland plains by large herbivores (≥100 kg) and harvester ants (*Messor capensis*) in Lower Zambezi National Park, Zambia was assessed every September from 1997 to 2007. A point intercept method was used to estimate percent cover for grass, shrub, bare, litter, herbivore droppings and presence of harvester ants in the Jeki open grassland vegetation community as indicators of range condition. Results showed a reduction in grass cover from 35% in 1997 to 10% in 2007 while litter remained stable. Incidence of herbivore droppings and harvester ants showed a decline while shrub cover and species composition of shrubs increased significantly. As grass cover and grass species composition declined, herbivore droppings and incidence of harvester ants also decreased. Unpalatable species such as *Vernonia* spp. were assumed to be signs of poor quality range. Increase in bush encroachment indicated heavy utilization by herbivores in the dry season when quality of range deteriorated as a consequence of over utilization of grass species. Further research is required to determine population estimates and grazing capacity of key herbivores such as buffalo (*Syncerus caffer*).

Key words: Range utilization, overgrazing, flood plains, valley floor, shrub encroachment.

INTRODUCTION

Lower Zambezi National Park and surrounding Game Management Areas (GMAs) are situated in Agro ecological zone I which is the driest ecological zone in Zambia receiving ≤ 400 mm annual rainfall in the valley floor and 800 mm on the plateau. In this ecological zone, droughts occur 61% of the time (Sichingabula, 1998). Drought occurrences particularly in the valley floor reduce availability of open water and forage for herbivores in the dry season (April/May to October/November). At this time of the year, most tributaries of the Zambezi River such as Chongwe, Musigiswa, and others flow intermittently such that from about September every year, they become

sand rivers and any water in them is below surface and not directly available to most herbivores. Elephants *(Loxodonta africana)* are the only species that have the ability to dig down the sand river bed to the water level and in the process making water available to other species. Such water holes made by elephants are limited in number such that by September to October every year, the Zambezi River remains the only main source of water for herbivores (Kajuni, Chansa and Chivumba, 1998). The study assessed dry season grass cover, grass and shrub species composition, incidence of herbivore droppings and harvester ants (*Messor capensis*), (Picker, et al., 2004), in the grassland plains of the Zambezi Valley floor every September during 1997 to 2007. It was reported that most of the valley floor was over grazed (Jarman, 1972) in the dry season, which led to decline of

*Corresponding author. E-mail: chansa.chomba@zawa.org.zm or ritachansa@yahoo.com.

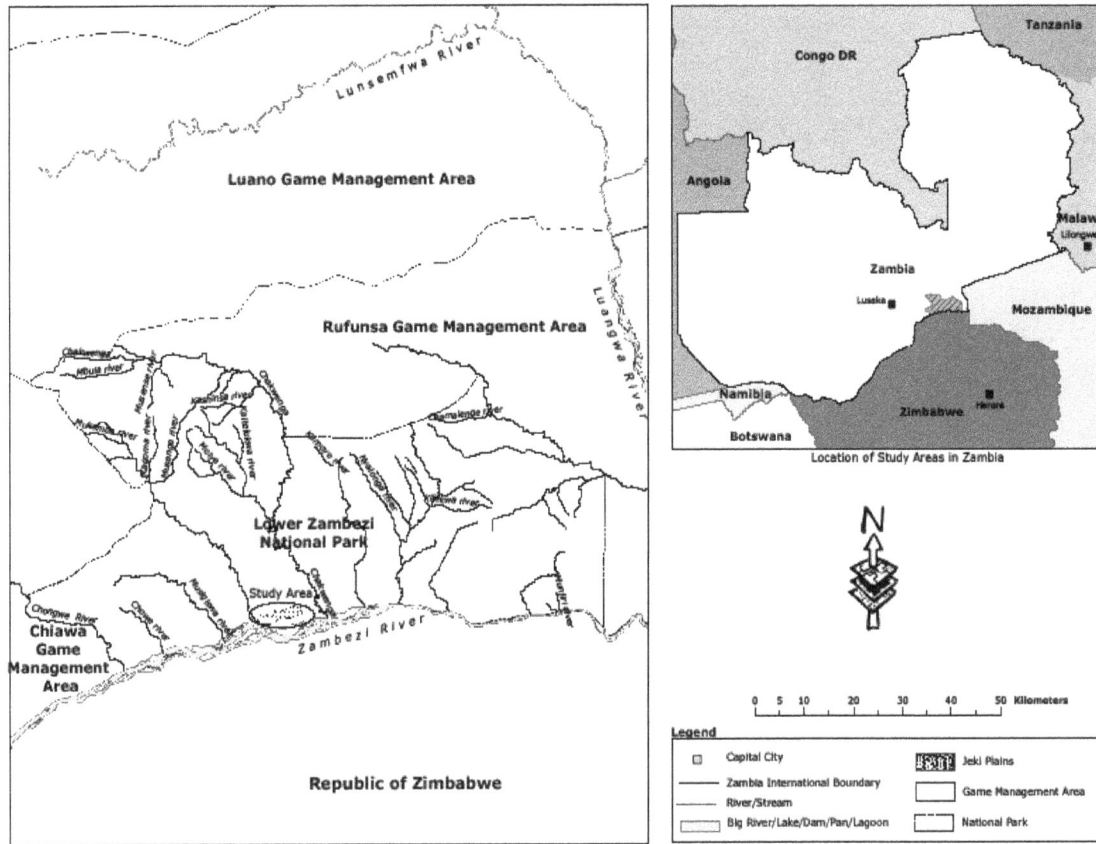

Figure 1. Location of Lower Zambezi National Park, Zambia.

most large herbivores. In light of the potential threats that may arise from global climate change, the suggested research activities were required to monitor range condition and trend for improved management of the Lower Zambezi ecosystem. This study was commissioned to evaluate indicators of range utilization by large herbivores in the dry season when water resources are very limited and suggested ways of ameliorating range deterioration.

MATERIALS AND METHODS

Study area

The Lower Zambezi National Park is approximately 4,092 km^2 in extent located at 15° 7' to 15° 44' South and 29° 10' to 30° 10' East (Figure 1). It is situated in Agro ecological zone I which is a low rainfall region of Zambia receiving about ≤ 800 mm on the plateau and ≤ 400 mm in the valley floor (Sichingabula, 1998). The vegetation communities in the Lower Zambezi National Park comprise riparian forest along the Zambezi River and major tributaries, woodlands which are subdivided into Miombo covering the plateau and escarpment making up over 70% of the area and Mopane on older calcareous alluvial soils along the valley floor and *Acacia* woodlands between Miombo and Mopane woodlands (Kajuni et al., 1998; Chanda , 1991). Grasslands are found mainly in dambos, marshes and on the Zambezi River flood plains which drain the area seasonally. The Jeki plains where the study was

carried out is the largest at about 30 km^2 ha in extent (Kajuni et al., 1998).

Field methods

To determine cover of grass, litter, shrub, percent occurrence of herbivore droppings and harvester ants, a point intercept method was used (Brower and Zar, 1977; Walker, 1974). Shrubs considered in this study were those below 1 m high. A 1-m long and 1 m high wooden frame with 10 slanting wore pins placed 10 cm apart was used. The frame was placed over herbaceous plants at 1,000 randomly selected sites in the five hectare area of the Jeki plains and the pins were lowered vertically one after another, one at a time. Phenology and species of plant litter or bare ground touched by each pin was recorded. The frame was placed 1,000 times representing 10,000 sample points during the period 1997 to 2007.

All the hits were added together to provide the total number of hits for each individual species category. The final number of hits was expressed as percent of total number of pins. Incidences of harvester ants and herbivore droppings were also recorded for each sample point. The total number of harvester ant burrows per hectare was recorded in the five hectares sampled area. The amount of litter accumulated by harvester ants at each burrow encountered was collected and weighed using a digital solar scale with readings calibrated to the nearest 0.5 g. An excavation was made in the ground to collect the litter accumulated in 4, 395 burrows examined. Above ground and below ground litter collected was weighed together and classified as husks, seeds and soldier ant heads.

Table 1. Outcome of Chi-square test for percent cover grass, shrub, litter, bare and percent occurrence of herbivore droppings and harvester ants (figures show probability at which differences are significant: = 95% **, NS = Not significant, n = 8).

Parameter	Year								Percent cover and occurrence		
	1997	1998	2000	2002	2003	2005	2006	2007	χ^2	P-value	Significance
Bare	30	30	15	20	20	18	20	10	36.24	0.001	**
Grass	35	35	30	25	18	15	10	10	34.27	0.001	**
Shrub	10	15	30	32	37	43	45	60	54.20	0.001	**
Litter	25	20	25	23	25	24	25	20	1.39	0.250	NS
Herbivore droppings	65	61	63	45	30	15	13	11	102.36	0.001	**
Harvester ants	95	93	96	70	55	30	22	15	136.15	0.001	**

RESULTS

Grass cover and species composition

There was a reduction in grass cover and species composition (Table 1). In 1997 grass cover was 35% comprising 14 grass species, *Brachiaria* spp., *Cenchurus ciliaris, Chloris guyana, Cynodon dactylon, Dactyloctenium aegyptium, Dactyloctenium* spp., *Digitaria* spp., *Echinochloa* spp., *Eragrostis* spp., *Panicum maximum, Pennisetum* spp., *Phragmites* ssp., *Setaria* spp., and *Sporobolus pyramidalis*. Between 1997 and 2007 percent grass cover declined significantly from 35 to 10% in 2007 (P < 0.001). Number of grass species per site also declined from 14 to 11. Three grass species, *C. ciliaris, P. maximum* and *D. aegyptium* were totally eliminated from all study sites by 2007 (Table 2). Seven grass species, *Brachiaria* spp., *C. guyana, Dactyloctenium* spp., *Echinochloa* spp., *Eragrostis* spp., *Pennisetum* spp. and *Setaria* spp. were reduced to less than 10% incidence. Only four species *Digitaria* spp., *Phragmites* spp., *C. dactylon* and *S. pyramidalis* maintained percent incidence of between 15 and 65% (Table 2). The results also showed that *Chloris, Cenchurus, Dactyloctenium, Eragrostis* and *Panicum* species had decreased and might be eliminated by heavy grazing.

Shrub cover

In 1997, shrub cover was 10% with only eight species recorded, *Faidherbia albida, Acacia tortilis, Balanites aegyptiaca, Cassia obtusifolia, Trichodesma zeylanicum, Acanthospermun hispidum, Vernonia* spp. and *Sonchus* spp. In 2007, the shrub cover increased to 60%, which was higher than any cover category (Figure 2) (Y= 6.4762x + 4.8572; R^2 = 0.9553), with the addition of five species namely; *Adansonia digitata, Borassus aethiopium, Diospyros* spp., *F. albida* and *Trichilia emetica*.

Litter and herbivore droppings

Litter from dry grass and tree leaves remained relatively stable from 25% in 1997 declining slightly to only 20% in 2007 (Figure 2). Herbivore droppings declined during the period 1997 to 2007, from 65% in 1997 to 11% in 2007 (Figure 2).

Harvester ants

The incidence of harvester ants declined from 95% in 1997 to15% in 2007 (Figure 2). The number of burrows per hectare also declined significantly (P< 0.005)from 605 in 1997 to 255 in 2007. The mean per hectare during the period 1997 to 2007 was 400. Each burrow had a mean weight of 300 g of litter giving a mean of 124 kg litter per hectare (Table 3). The mean above ground grass husks during 1997 to 2007 comprised 97% with only 3% seeds. On the other hand below ground, 75% of the litter was grass seeds. The remaining 25% comprised of heads of dead ant soldiers piled up in a separate chamber. The below ground proportion of grass seeds to dead soldier heads remained relatively constant during the period 1997 to 2007 (P>0.005).

DISCUSSION

Shrub encroachment

The results obtained in this study during the period 1997 to 2007 showed that shrubs such as *Vernonia, Sonchus, Ocimum canum* and *Acanthospermum* spp. which are unpalatable to the grazers displaced grasses in many sample points of the study area. This was caused by aggregation of herbivores in the dry season in the open grasslands. Increased frequency of grazing and intensity on palatable grass species reduced plant vigour which became easily out competed by less palatable shrubs as also reported by Western (1975) and Jarman (1972). Unpalatable species of shrubs (Vernon 1983) such as *Acanthospermum* spp, *Sonchus* spp., *Vernonia* spp., *Sida alba* and *Euphorbia* spp., replaced palatable grass species namely *Chloris* spp., *Cynodon* spp., *Dactyloctenium* spp., *Echinocloa colona, Eragrostis* spp., and *Panicum* spp., *Phragmites* spp., *Echinocloa* spp.,

Table 2. Percent of species occurrence of grass and shrub in the study area and the years in which some species were eliminated from the sites and emergence of others.

Species	Years and % of occurrence, disappearance and emergence of species								Rating	Preferred site
	1997	1998	2000	2002	2003	2005	2006	2007		
Grass										
Brachiaria spp	5	5	5	3	3	3	3	3	Heavily grazed	On wet soil under shade
Cenchurus ciliaris	30	28	30	30	15	10	0	0	Heavily grazed,	Near termite mounds, over grazed areas
Chloris guyana	3	3	3	3	3	3	3	3	Heavily grazed	On edges of dry river channels
Cynodon dactylon	47	50	50	50	48	48	48	48	Heavily grazed,	Was present mainly on all soils except on sandy soils. Resisted grazing pressure
Dactyloctenium aegyptium	30	25	15	5	2	1	0	0	Heavily grazed palatable	Common on sandy soils
Dactyloctenium spp	15	15	5	8	7	5	5	3	Heavily grazed	On sandy soils, on over grazed areas
Digitaria spp	15	15	15	15	15	15	15	15	Grazed	On island edges and near water
Echinochloa spp	10	10	10	10	10	10	10	10	Grazed	Common on wet areas only
Eragrostis spp	15	13	10	11	10	8	7	5	Grazed	On shallow soils. Was not usually grazed
Panicum maximum	35	35	3	3	1	0	0	0	Heavily grazed	On old river channels and banks. Heavily grazed
Pennisetum spp	10	10	9	7	10	9	7	7	Heavily Grazed	On edges of permanent water
Phragmites ssp	30	30	30	30	30	30	30	30	Grazed by elephant	On islands.
Setaria spp	15	15	13	10	10	10	7	7	grazed	Near termite mounds. Flower heads catching on clothes
Sporobolus pyramidalis	65	65	65	65	65	65	65	65	Very limited grazing	In many soil types. Generally avoided by grazers
Shrub										
Trichilia emetica	0	0	1	2	5	13	17	25	No record	Particularly on old river channels
Faidhebia albida	10	17	30	35	62	64	66	71	Heavily Browsed	Particularly near termite mounds and ant hills
Diospyros spp	0	0	1	3	3	5	5	5	Browsed	Particularly near termite mounds and ant hills
Borassus aethiopium	0	5	5	10	15	15	15	16	Browsed by elephant	Was non selective but more common near stream banks
Acacia tortilis	3	3	4	4	5	11	15	19	Heavily Browsed	Was non selective, but was more common on dry river channels
Adansonia digitata	0	0	0	2	2	2	2	2	Browsed Fruits eaten	Showed no particular preference
Balanites aegyptiaca	1	1	1	1	1	3	3	3	Browsed	Found mainly on termite mounds and ant hills
Cassia obtusifolia	3	3	0	0	1	1	0	0	No record	Found in wetter areas
Trichodesma zeylanicum	7	7	3	0	0	0	0	0	Not browsed	Dry heavily grazed areas
Acanthospermun hispidum	21	2	18	23	25	13	11	9	Not browsed	Found on heavily grazed areas
Vernonia spp.	55	55	38	45	43	45	45	43	Not browsed	Found on heavily grazed areas
Sonchus spp.	11	8	11	5	5	0	0	3	Not browsed	Was common on dry river channels

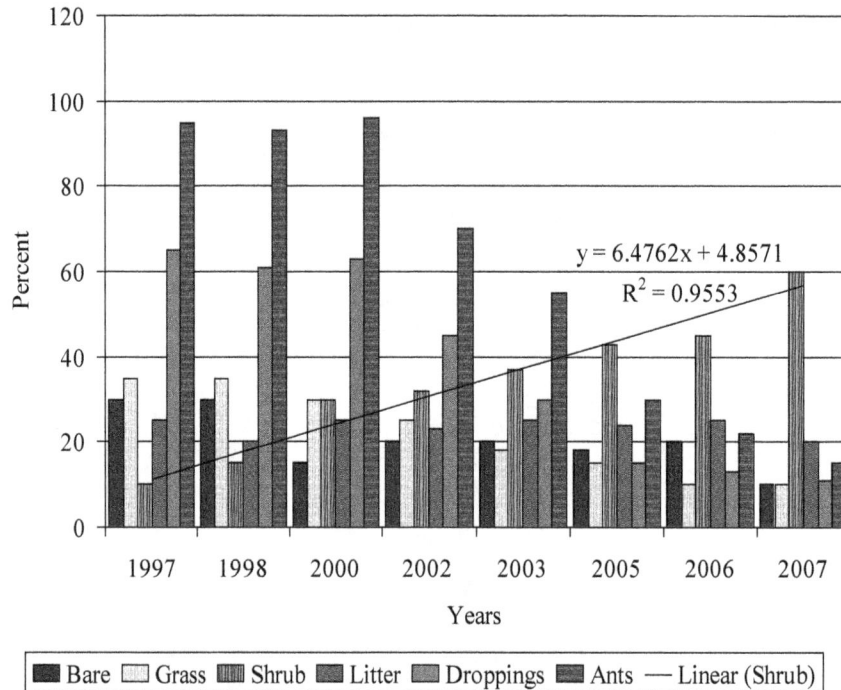

Figure 2. Change in percent cover of grass, bare, litter, shrub and incidence of herbivore droppings and harvester ants, during 1997 to 2007.

Table 3. Number of harvester ant burrows per ha, weight of litter in grammes and percent.

Year	Number of burrows	Mean weight in grammes per burrow	Total weight in grammes per ha.	Percent proportion of grass seeds below ground	Percent proportion of dead soldier heads below ground
1997	605	550	332,750	87	13
1998	650	600	390,000	90	10
1999	500	430	215,000	80	20
2000	355	250	88,750	83	27
2001	400	300	120,000	74	26
2002	550	430	16,500	70	30
2003	320	225	72,000	62	38
2004	255	140	35,700	65	35
2005	250	125	31,250	71	29
2006	255	135	34,425	73	27
2007	255	120	30,600	75	25
Mean	400	300	124,270	75	25

Below ground proportion of grass seeds to soldier heads.

Sporobolus spp., and *Chloris* spp.; however, persisted. Palatable species of grass could not withstand grazing pressure and were replaced by less palatable (increasers) and shrubs. The loss of grass species which were replaced by shrubs as range deteriorated increased species composition and cover of shrubs shown in Figure 2.

The results of this study indicated that open grassland plains were being encroached by shrubs and woody vegetation. We also suggested other factors exacerbating woody encroachment as being; lower numbers of elephants, which previously opened more of the woodlands through their feeding habits of knocking down trees and breaking branches. The numbers of elephants present in the National Park of 1,000 to 3,000 were far much lower than the > 5,000 individuals recorded in the 1970s (Mwima and Yoneda, 1995). The large herds of elephant kept woody vegetation under control. On the other hand, the implications of the removal of fire used by the Nsenga people when the reserve was established in the 1950s eliminated the slash and burn subsistence farming which played a significant role as a factor in the woodland dynamics of the area. This latter factor has not been well researched and understood. It is assumed that the combination of the reduction in elephant numbers, removal of the Nsenga people and the disruption of the natural flooding regime may have collectively accentuated encroachment of woody vegetation on former flood plains.

Incidence of litter and herbivore droppings

The amount of litter remained relatively stable becauseas litter from grass declined it was replaced by litter from shrubs (see Figure 2). Herbivore droppings also declined, which may have affected soil fertility in grassland plains. McNaughton (1985) and McNaughton et al. (1988) explained how herbivores affect primary production and regulate recycling balance through their dung output. Yet little attention is paid to their spatial and temporal variation in the incorporation of dung into the soi (Grimsdell, 1978). As palatable grass species are eliminated from an area, herbivore utilization of the area also declines which also reduces herbivore droppings.

Soil trampling by ungulates

In areas defaced of vegetation cover, trampling of hooves further hardened the soil, reducing water percolation and retention rates and increased water evaporation (Dunnet, 1997). This resulted in open plains being colonized by shallow rooted therophytes, which could not hold the soil together to prevent soil erosion which also contributed to range deterioration. The beneficial effects of large numbers of herbivores and their hooves observed in other natural areas with large numbers of large mammals such as the Serengeti National Park in Tanzania and Masaai Mara National Reserve in Kenya (Bell, 1971) is not felt in the Lower Zambezi area. This is because the herbivores in the Lower Zambezi National Park area, are largely resident except elephants which migrate seasonally otherwise the rest of the herbivores do not leave room for recovery of the range, while the ones in Serengeti migrate giving chance to the range to recover from grazing pressure.

Incidence of harvester ants

Harvester ants mainly feed on grass seeds, which they

store in their nests in the ground. Towards the end of the season, soldiers of the harvester ants are killed by the minor workers and their bodies together with grass seeds are eaten up, but the huge heads are piled up in the chamber of the nest (Bertin, 1967). These heads together with the left over grass materials subsequently decompose releasing nutrients to the soil. Abandoned grass seeds germinate which assist in seed dispersal and maintenance of grass species composition, which later provide food for the ants and large herbivores. Therefore, as grass species and cover declined during the period 1997 to 2007, the incidence of harvester ant burrows and weight of litter buried in them also declined and may have subsequently disturbed the balance as harvester ants depend on grass for their sustenance while the grasses benefit from harvester ants that spread their seeds. In this study it was assumed that the mean number of 400 burrows per hectare was a minimum value for a healthy density of harvester ants in the Lower Zambezi National Park valley floor.

The activities of harvester ants which involves gathering of seeds and leaves for food is important in maintaining the ecological balance of grasslands. *Messor* spp. in particular, collect husk and seeds which they store in their nests and unlike other genera do not grow fungal gardens. Overgrazing by large herbivores on the Jeki plains of the Lower Zambezi National Park therefore, seem to have reduced food available to harvester ants and may have caused the decline in the number of harvester ant burrows and weight of litter per hectare. Fire, particularly late season (September – November) may also be a contributing factor to the overall reduction in grass litter which negatively affect harvester ants. The reduction in the below ground proportion of grass seeds to dead soldiers heads during the period 1997 to 2007 could be attributed to the reduced amount of grass litter available per burrow. It would appear that when grass litter is abundant the proportion of grass seeds to dead soldiers' heads is always skewed in favour of grass seeds and vice versa. Harvester ants are known to ferment grass seeds and eat them completely when the quantity is low. In seasons of food abundance they may eat the fleshy, edible appendages (the fat body or elaiosome) of certain specialized seeds which they also disperse in the process. During the seasons of food shortage, the dispersal of grass seeds may be reduced. In such years, all grass seeds are eaten and all soldiers are killed off by minor workers and their bodies eaten leaving their large heads in a separate chamber. This could be the reason why areas previously occupied by grasses were taken over by shrubs as no grass seeds were left to germinate during the subsequent rainy season.

Further research will be required to determine whether harvester ants select grass seed size, morphology and species in which case, it could be advantageous for a plant species to invest the total productive effort in a large number of very small seeds rather than in a few big ones. Only those species favoured by the harvester ants would succeed by virtue of being spread by harvester ants. Other species may be avoided by virtue of toxic seed constituents and may not benefit from the dispersal by ants. Seed size in its own right may also be a factor to consider. Ecologically, seed size is important in breaking dormancy. Therefore, being a small seed can only sample that part of the environment immediately adjacent to it, which is not necessarily representative of the generally prevailing conditions. In this study, we could not conclude with certainty the major causes of decline in the number of harvester ant burrows and weight of litter. It is therefore, important to investigate further the size, shape, internal structure, life span and number of seeds produced by each decreaser or increaser to determine their vulnerability to harvester ants and vice versa.

Influence of harvester ants on soil fertility

Grazers produce large amounts of dung and insects such as harvester ants also play an important role in nutrient recycling and seed dispersal by burying pieces of grass and seeds in the soil. Such litter later breakdown adding to the soil nutrient status. Such symbiotic relationship between harvester ants and grass has a beneficial effect on soil structure, soil nutrients, soil aeration, water percolation, and seed dispersal. This symbiotic relationship however, was being disrupted in the grassland plains of the Lower Zambezi National Park as shrubs took over most of the sites and the number of harvester ant burrows declined (Table 3). As the number of burrows declined the amount of grass seeds buried under ground also declined which reduced the regeneration potential of many grass species, some of which are grazed by herbivores.

The future of the lower Zambezi grassland plains

Grimsdell (1978) observed that open grasslands and woodland communities were in a state of flux whose major influencing factors were; climate, man, fire and elephants. In the Lower Zambezi area, it will be important to monitor and collect ecological data on a number of ecological factors such as; climate, man, fire, harvester ants, elephants and other large herbivores and many others. Subsequently, a summary of some of the key ecological factors to be considered in long-term ecological monitoring of the Zambezi Valley floor is provided below under recommendations.

RECOMMENDATIONS

Shrub encroachment

The study of encroachment of grassland plains by woody

vegetation is critical in maintaining pasture and sustaining herbivores depending on them. In order to maintain quality pasture for herbivores, Zambia Wildlife Authority should provide guidance on the control of shrubs on open grassland plains.

Monitoring changes in the flood regime

The upstream impoundments of both Zambezi and Kafue Rivers for hydroelectric power generation are of great consequence to the ecology and plant community distribution on the flood plains and islands in the Lower Zambezi National Park (Sichingabula, 1998). Zambia Wildlife Authority should commission a long-term study to assess any adverse impacts not only on plant communities but the gradual loss of river channels which are being colonized by wood plant species.

Disrupted flush floods and possible loss of soil fertility

Current records on rainfall indicate that Lower Zambezi area experiences more drought occurrences than any other area in Zambia (Sichingabula, 1998). The only area of the Zambezi Valley which experiences fewer droughts is around Lake Kariba possibly caused by the formation of a lake breeze system (Hutchinson, 1973) which seems to have minimized the drought vulnerability of this drought prone area. The remaining part of the Zambezi Valley is drought prone. The effects of drought and low rainfall in the Lower Zambezi National Park and the effects of the pre and post impoundments of the Zambezi River at Kariba and Kafue River at Itezhi Tezhi have not been assessed.

A long-term study will be required to determine the impact of these impoundments on the disruption of flush floods and sediment load deposition on the grassland plains as this is likely to reduce soil fertility. It is possible that due to reduced flooding frequency and intensity, a number of areas previously inundated with water are no longer affected by floods and have lost the sediment deposition previously brought about by the regular flooding regimes.

Removal of the Nsenga people and reduction of elephant population

The impacts of the removal of fire originally used by the Nsenga people in their slush and burn agriculture before the area was declared a reserve; the reduction of elephant numbers through poaching evidenced through out the African continent in the late 1970s up to the 1980s have not been critically assessed in order to understand their long term influence on vegetation dynamics in the Lower Zambezi National Park. Understanding the role of elephants in vegetation dyanamics will enable ZAWA to manage the vegetation communities in Lower Zambezi National Park in a manner that would maintain open grasslands for grazers.

Localised seasonal grazing by large herbivores

We noted in this study that shrubs and woody plants showed a significant increase (Figure 2). As a consequence of this change, most herbivores concentrated their grazing on few and smaller plains causing stress on palatable species of grass. The impacts associated with these changes must be studied in more details if management is to understand and manage the National Park based on empirical evidence and sound conservation goals. An effective way of managing the National Park would be to establish a long-term ecological monitoring program. Such a programme would check and document not only changes in vegetation as a result of those major influences introduced on ecosystems but also their effects to animal population dynamics.

Development of tourist accommodation facilities

By 2007 there were nineteen (19) lodges along the riverbank of up to 16-bed capacity. Some of these tourist accommodation facilities were close to each other, in some instances less than 5 km between them. The location of these facilities on grassland plains may have reduced area available to grazers. For instance, during construction phases of the tourist lodges, soil excavation is carried out (Sichingabula, 1998) which because of the fragility of the soils in the flood plain, may have accentuated soil erosion in those sites further reducing the extent of the open grassland plains available to herbivores. The policy of locating tourist facilities or any other infrastructure developments along the riverbank, open plains, islands and near multiple channels should be reviewed within the framework of mitigating their impacts in order to save the open grassland plain habitats and the herbivores which depend on them for their sustenance.

Range condition and trend

Oudtshoorn (1992) and Dunne (1977) outlined major criterion for characterizing range that is in poor condition with signs of erosion. To that effect, poor vegetation cover inevitably leads to accelerated soil erosion because it reduces the capacity of the range to absorb and retain water which results in exhaustion of the soil and hence reduced range ability to recover from herbivore use. In

order to understand the relationship between height and weight of the different grass species in the mid Zambezi area with a view to estimating optimum range utilization, botanical inventories of plant species in the open plains and islands are required.

Additional studies to determine the correlation between height and weight for different grass species must also be undertaken. Such correlation would provide a means to estimate a minimum amount of vegetation cover that could be allowed in the open plains before irreversible range deterioration occurs. Such studies would provide standard forage utilization estimates in respect of height/weight for the different important grass species in the area. Particular attention should be paid to grazing lawns along the river bank and islands, as these are a result of deposited alluvial and as the rate and locality of this deposition is a function of river flow dynamics, monitoring should take place to assess the extent and changes in this key resource. Another area of future study should be the quantification of the abundance and distribution of vegetation communities and cover assessments over time.

ACKNOWLEDGEMENTS

We wish to thank the Warden in Charge Ms. Susan Chimuka, the Senior Warden the late Maxwell Malama, the Chief Warden Dr. Henry Mwima and the entire EDF Team for reviewing the initial draft and for making available transport and other field logistics. Mr. Chaka Kaumba was very helpful in preparing the map.

REFERENCES

Bell RHV (1971). A grazing ecosystem in the Serengeti. Sci. Am., 224: 86-93.
Bertin L (1967). Larousse Encyclopedia of Animal life. Galahad books. New York.
Brower JE, Zar JH (1977). Field and laboratory methods for general ecology. Brown Publishers, Dubuque, Iowa.
Chabwela HN (1985). The Lower Zambezi: the future many times ignored. Black lechwe., 37: 21-24.
Chanda G (1991). Natural Resources Management in Chiawa Middle Zambezi Valley, Zambia. MSc thesis, University of Zimbabwe, Harare.
Dunnet T (1997). Evaluation of erosion conditions and trends. In: Kunkle (Ed) Hydrological techniques for upstream conservation.FAO Conservation guide 1, Rome.
Grimsdell JJR (1978). Ecological monitoring. African Wildlife Foundation, Nairobi.
Hutchinson P (1973). Increase in rainfall due to Lake Kariba. Weather, 28: 499-504.
Jarman PJ (1972). Seasonal distribution of large mammal populations in the unflooded middle Zambezi Valley. J. Appl. Ecol., 9: 325 346.
Kajuni AR, Chansa W, Chivumba R (1998). General Management Plan for Lower Zambezi National Park and Surrounding Game Management Areas, National Parks and Wildlife Service, Chilanga.
McNaughton SJ, Ruess RW, Seagle SW (1988). Large mammals and process dynamics in African ecosystems. Bioscience, 38: 794-800.
McNaughton SJ (1985). Ecology of a grazing ecosystem. The Serengeti.Ecologicol. Monogr., 55: 259-294.

Mwima HK, Yoneda K (1995). Report on the aerial census of large mammals in Lower Zambezi National Park in November 1995, National Parks and Wildlife Service, Chilanga.

Picker M, Griffiths C, Weaving A (2004). A field guide to insects of South Africa, Struik Publishers. Cape Town.

Sichingabula H (1998). Physical characteristics of the Lower Zambezi National Park and adjacent Game Management Areas. Special report for EDF/NPWS Project, Chilanga.

Vernon R (1983). Field guide to important arable weeds of Zambia. M't Makulu Research Station, Chilanga.

Walker B H (1974). Evaluation of eight methods of botanical analysis on grassland in Rhodesia. J. Appl. Ecol., 7: 403-416.

Western D (1975). The distribution of animals in relation to resources. Handbook No. 2 African Wildlife Leadership Foundation, Nairobi.

17

Invasion of the Mozambique tilapia, *Oreochromis mossambicus* (Pisces: Cichlidae; Peters, 1852) in the Yamuna river, Uttar Pradesh, India

Mushtaq Ahmad Ganie[1] , Mehraj Din Bhat [1], Mohd Iqbal Khan[1], Muni Parveen[2], M. H Balkhi [3] and Muneer Ahmad Malla[1]

[1]Department of Zoology, Faculty of Basic Sciences, Bundelkhand University Jhansi–284 128, U. P., India.
[2]Department of Zoology, Faculty of Biological Sciences, University of Kashmir, Srinagar–190 006, J & K, India.
[3]Faculty of Fisheries, Sher-e-Kashmir University of Agricultural Sciences and Technology of Kashmir, Srinagar, India.

Oreochromis mossambicus (Peters, 1852) is a highly successful invader of aquatic ecosystems due to its adaptable life history, tropic flexibility, ability to tolerate extreme and often unfavourable environmental conditions, rapid reproduction and maternal care of offsprings. Upon introduction to areas outside its natural range, these characteristics often give *O. mossambicus* a competitive advantage over indigenous fishes. The present study investigated the population characteristics of non-indigenous Mossambique Tilapia, *O. mossambicus*, for a period of 12-months from August 2009 to July 2010 in the lower stretch of Yamuna River in India. The Mossambique Tilapia, *O. mossambicus,* formed the most abundant fish species in all the catches from the Yamuna River at all the sampling stations. The gonado-somatic index (GSI) and the presence of all six gonadal stages confirmed that *O. mossambicus* has established a breeding population. The GSI for females indicated year-round reproduction with increased spawning intensity in spring (March to April) and monsoon (July to August). Males ranged from 142-280.0 mm total length (TL) and females from 130-265.0 mm TL. Small juvenile fish were collected every month of the year and multiple size classes present in sampling catches suggest successful recruitment of young. Adult *O. mossambicus* consumed primarily detritus and vegetal matter, though the diet of juveniles, collected from the Yamuna River, was found to be carnivorous. We expected Mozambique tilapia to further invade the Yamuna River due to natural dispersal. There is a need for more detailed studies of tilapia abundance, recruitment and local environmental conditions across the country to fully understand the invasion potential and consequences for the endemic aquatic biodiversity.

Key words: Exotic fish, *Oreochromis mossambicus*, invasion, colonization, Yamuna River, U.P.

INTRODUCTION

India is a vast country in terms of natural resources and considered one of the mega-biodiversity countries in the world (Lakra et al., 2011). The indian mainland is drained by 15 major, 45 medium and over 120 minor rivers, besides numerous ephemeral streams (Rao, 1975). The diverse river system in India harbour one of the richest fish germplasm resources in the world (Vass et al., 2009), characterized by many rare and endemic fish species and as much as 166 indigenous fish species have so far been recorded from the rivers of Central India (Sarkar and Lakra, 2007).

Unfortunately, over the last few decades riverine ecosystems of India has suffered from intense human intervention resulting in habitat loss and degradation and as a consequence, many fresh water fish species have become highly endangered, particular in Yamuna basin where heavy demand is placed on fresh waters. A new and potentially serious threat to the indigenous fish faunais the invasion of alien fishes (Singh and Lakra, 2011). Although the negative effects of introduced species are widely recognized (Canonico et al., 2005; Lakra et al., 2008; Singh and Lakra 2006, 2011), many of them are still being released into the aquatic ecosystems of India for production enhancement, without consideration of their potential impact on native fish and fisheries.

The African mouth-brooder cichlid, *Oreochromis mossambicus* (Peters 1852), or the Mozambique tilapia, is native to the eastward flowing rivers of central and southern Africa (Philippart and Ruwet, 1982; Trewavas, 1982). Due to their perceived utility as an aquaculture species, *O. mossambicus* are now widely distributed around the world (Arthington et al., 1984; Philippart and Ruwet, 1982). However, *O. mossambicus* have now fallen out of favour as a preferred aquaculture species because of their propensity to 'stunt' and their general poor quality due to the small size of founder stocks (Pullin, 1988). Invasive populations are now causing environmental and ecological problems in many countries (Canonico et al., 2005) and as such, *O. mossambicus* is listed in the Global Invasive Species Database (2006) as being in the top 100 invasive alien species on the planet.

The species has been described as a 'model invader' due to a number of key biological characteristics including tolerance to wide ranging ecological conditions, generalist dietary requirements, rapid reproduction with maternal care, and the ability to successfully compete with native fish through aggressive behaviour (Pe´rez et al., 2006b). Therefore, given suitable environmental conditions, *O. mossambicus* have become successfully established in almost every region in which they have been cultured or imported (Costa-Pierce, 2003; Cucherousset and Olden, 2011; Diana, 2009; Strecker et al., 2011). Official records show that *O. mossambicus* was first introduced to India from Srilanka in 1952 and thereafter stocked in several reservoirs of southern India for production enhancement (Sugunan, 1995). Tilapia now forms a part of fish fauna in the Godavari, Krishna, Cauvery, Yamuna and Ganga Rivers (Lakra et al., 2008).

In earlier studies, tilapia attracted the attention of scientific communities due to its mouth brooding behaviour (Perez et al., 2006; Russell et al., 2012). Tilapia has remained an objective of astonishment to ethnologists for years but its present behaviour, that is, prolific feeder and prolific breeder changed the scenario.

Tilapia is now known for its invasion to the non-native

water bodies and destruction of their flora and fauna. The reported high incidence of *O. mossambicus* in the catches of artisanal fisheries prompted us to study the population characteristics, that is, the abundance, size range, food and feeding, gonado-somatic index (GSI), maturity and breeding so as to ascertain the abundance and establishment of *O. mossambicus* in the Yamuna River flowing along Etawa to Hamirpur in the state of Uttar Pradesh.

MATERIALS AND METHODS

Study area

Uttar Pradesh (U.P.) is one of the largest states in India, located between 23°52'-31°28'N latitude and 77°04'-84°38'E longitude. Being land-locked, it is endowed with an abundant supply of inland water resources (1165 million ha) that are ideal for fisheries and aquaculture. The availability of 0.72 million ha of running water in the form of rivers and canals enriches the state with plenty of ichthyofaunal diversity (Bilgrami, 1991; Kapoor et al., 2002). Yamuna River, one of the most important Rivers of Indogangetic plains, originates from Yamnotri glacier near Banderpuch peaks of lower Himalayas (38° 59' 78° 27') in the Mussorie range at an elevation of about 6320 m above mean sea level in the Uttarkashi district of Uttrakhand. It is the sub-basin of the Ganga River system. It is 1376 sq. km long basin, covering an area of 320 lakh sq. km of which 61750 sq. Km. lies in U.P.

The study area covered approximately 250 km of the river stretch of the lower Yamuna flowing along the districts of Etawa, Jaluan and Hamirpur in the state of Uttar Pradesh. Etawa (S1), Kalpi (S2) in Jaluan and Hamirpur town (S3) in Hamirpur district were the study sites as demarcated in Figure 1. These landing sites were chosen because they are some of the most active, with high artisanal fisheries landings for the Yamuna stretch in the three districts. Landings from the study sites were therefore considered more appropriate and more representative.

Collection and identification of fish

The data for this study were collected from the commercial catches at the fisheries landing sites of three districts viz, Etawa, Jalaun (Kalpi) and Hamirpur of Uttar-Pradesh state during the period August 2009 to July 2010 on fortnightly basis. The sampling from the selected landing sites was conducted for two consecutive days twice a month in every fifteen days interval in a month from each landing site. Therefore, the monthly sampling frequency represents four days at each landing site. Sampling was conducted in the early mornings or evenings because in these hours all the fresh fish were brought to the landing sites for marketing. Fishermen generally used multi-meshed gill nets of mesh size 8.5 to 50 mm as well as dragnets for fishing. From commercial catches, fishes collected at the landing centres were measured (total length, TL nearest mm), and body mass determined (weighed to the nearest gram) using portable digital balance. Fish identification was confirmed using reference literature (Jayaram, 1981, 1999; Talwar and Jhingran, 1991). In addition data were also collected from fisheries market Jhansi and Lucknow because harvests from all these landing sites (Etawa, Kalpi and Hamirpur) are sold in fish market Jhansi and Lucknow, which maintain landing records (numbers and body mass).

Relative abundance

From the catch, sorting of fish species was done by fishermen for

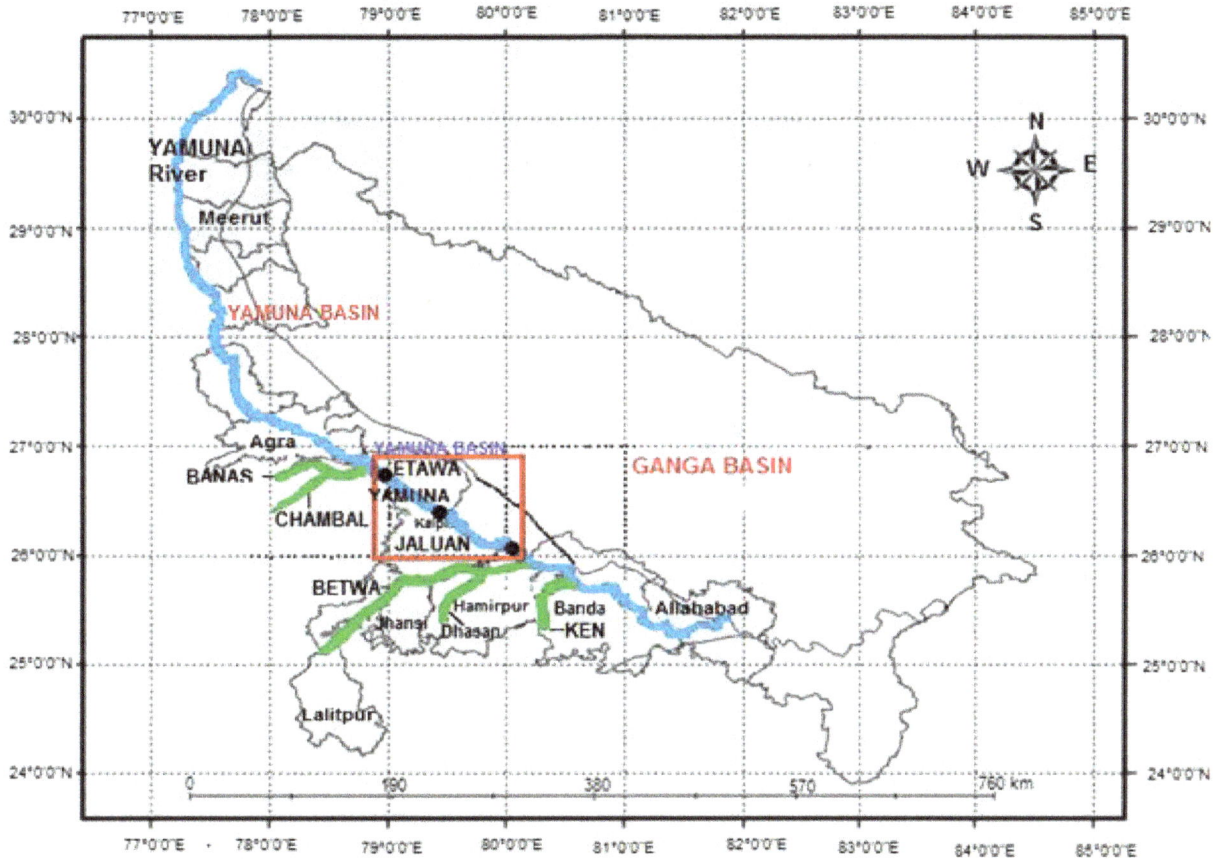

Figure 1. Map of the Yamuna River and the portion of the river in the present study.

marketing and sale. The data from such segregated fish groups were then collected to work out the species contribution. From the total catch, relative abundance (RA) of a individual fish species at each study site was estimated following the formula adopted by Lakra et al. 2010:

$$RA\ (\%) = \frac{Number\ of\ samples\ of\ particular\ fish\ species}{Total\ number\ of\ samples} \times 100$$

Gonado-somatic index (GSI)

From the catch, O. mossambicus was separately counted, sexed as male and female and the gonads of immature, maturing, mature and spent fish were dissected out, weighed and fixed in 10% formalin for microscopic examinations. The GSI was calculated using the formula GSI = GW/EBW x100, With GW = gonad mass/weight (g) and EBW = eviscerated body mass/weight (g). The fecundity of individual females was determined gravimetrically (to the nearest gram), and the gonad maturity stages were determined visually according to reference literature (Nagelkerke and Sibbing, 1996).

Food and feeding habits

The intestines of 150 collected specimens from different sampling sites were cut and fixed in 4% formalin for gut content analysis. The

diet and feeding habits of O. mossambicus were determined based on the contents of the digestive tract and was examined using Binocular Magnus MXL-Bi stereomicroscope. Different taxa of the food items were identified, and counted by numerical methods adopted by Hyslop (1980) and Costal et al. (1992). In the numerical method, the number of each food item was expressed as the percentage of the total number of food items found in the stomach.

RESULTS

Biodiversity of fish species and catch composition

The results of present study showed the occurrence of 21 freshwater fish species belonging to 9 Families (Table 1). The Indian major carps comprising of Catla catla, Cirrhinus mrigala and Labeo rohita constituted 1.10 to 2.15% of total catch and their size ranged from 80 to 500 mm in length and 500 to 7500 g in body mass (Table 2). The minor carps in the total catch were mainly represented by Labeo calbasu, Cirrhinus reba, Puntius sophore, Puntius ticto and Puntius ranga. They constituted 15.23–18.65% with size range of 50 to 350 mm in length and 200–1500 g in body mass (Table 2). Catfishes in general were represented by Channa striatus, Mystus tengra, Rita rita, Notopterus notopterus and constituted 07–10%

Figure 2. Specimen of *O. mossambicus* collected from Yamuna River.

Table 1. Diversity of fish species*, total mean length (TL) and relative abundance (RA) of fish collected from the Yamuna River at three study sites (S1, S2, S3).

Specie	Family	Total length (mm)		RA (%)		
		Max	Min	S1	S2	S3
Oreochromis mossambicus (Peters)	Cichlidae	220	130	24.51	26.24	24.50
Gadusa chapra (Hamilton–Buchanan)	Clupeidae	140	60	8.09	8.87	11.52
Cirrhinus mrigala (Hamilton–Buchanan)	Cyprinidae	450	160	0.15	0.39	0.18
Cirrhinus reba (Hamilton–Buchanan)	Cyprinidae	220	150	4.01	3.61	4.98
Cyprinus carpio	Cyprinidae	400	230	8.32	7.95	7.86
Catla catla (Hamilton)	Cyprinidae	460	280	0.52	0.66	0.35
Labeo rohita (Hamilton–Buchanan)	Cyprinidae	500	80	1.12	0.62	0.61
Labeo calbasu (Hamilton–Buchanan)	Cyprinidae	350	140	0.81	0.93	0.26
Puntius ticto (Hamilton–Buchanan)	Cyprinidae	90	45	6.65	6.34	7.84
Puntius sophore (Hamilton–Buchanan)	Cyprinidae	90	60	1.04	0.93	1.76
Puntius ranga (Hamilton–Buchanan)	Cyprinidae	100	70	7.73	7.60	7.55
Mystus tengra ((Hamilton–Buchanan)	Cyprinidae	180	125	1.66	2.14	1.88
Rita rita (Hamilton–Buchanan)	Bagridae	260	80	2.95	3.91	3.62
Oreochromis niloticus	Cichlidae	260	190	10.37	9.80	4.48
Notopterus notopterus (Pallas)	Notopteridae	300	150	0.89	1.07	0.48
Chitala chitala (Hamilton–Buchanan)	Notopteridae	800	250	1.12	0.78	0.64
Channa striatus (Bloch)	Chandadae	150	100	3.51	3.20	2.60
Mastacembalis armatus (Lacepede)	Mastacembelidae	540	160	2.08	1.66	1.88
Glossogobius giuris(Hamilton-Buchanan)	Gobiidae	160	95	4.10	4.64	4.97
Chanda nama (Hamilton–Buchanan)	Ambissidae	100	40	5.56	4.48	6.64
Parambassis ranga (Hamilton-Buchanan)	Ambissidae	50	40	4.51	4.17	5.43

*Taxonomic status adapted from Talwar and Jhingran (1991).

of the total catch having body mass range of 500 to 2300 g. Other less abundant catches included *Gadusia chapra*, *Chitala chitala*, *Chanda nama*, *Glossogobius giuris*, *Parambassis ranga*, *Mastacembelus armatus* representing 23-30% of total catch (Table 2).

A persual of the present data showed that *O. mossambicus*

Table 2. Important fish species and their contribution in commercial fishery of the Yamuna River.

Fish group	Fish specie	Total length range [mm]	Body mass range [kg]	Catch contribution [%] range
Indian Major Carps	Labeo rohita, Catla catla, Cirrhinus mrigala	80 - 500	0.5-7.5	1.10 – 2.15
Minor Carps	Labeo calbasu, Cirrhinus reba, Puntius sophore, P. ticto, Puntius ranga	50 - 350	0.2-1.5	15.23 – 18.65
Cat Fishes	Channa striatus, Mystus tengra, Rita rita,, Notopterus notopterus	80 - 260	0.5- 2.3	7.69 - 10.45
Miscellaneous	Mastacembelus armatus,Gadusia chapra, Chitala chitala, Chanda nama, Glossogobius giuris, Parambassis ranga	50 - 800	0.030 – 1.2	23.14 – 30.68
Exotic	O. mossambicus,	180 - 220	0.003 -1.4	24.51 – 26.24
	O. niloticus	190 - 260	0.005 – 1.0	4.48 --10.37
	Cyprinus carpio	230 - 430	0.25 – 8.5	7.82 – 8.32
	C garipinus	250-780	0.35 – 2.5	Stray catch

(Mossambique Tilapia) (Figure 2) formed the most dominant fish species in all the catches from the Yamuna River at all the sampling stations throughout the sampling period (Table 1). The relative abundance (RA) of *O. mossambicus* ranged from 24.5 to 26.24 % from S1 to S3 (Table 1).

Reproductive activity

Gonadal examination of *O. mossambicus* in different catches revealed that immature, maturing, and mature fishes were sampled. Mature females were found at smaller size (130-265 mm TL) while mature males were larger in size (142-280 mm TL). Gonads of 120 examined specimens from different stations of the river showed that mature female represented all reproductive stages (1–6) with varying gonado-somatic index (*GSI* : 0.2 to 6%). In general, the highest *GSI* percentage was recorded during March-April and July–August (Table 3). A consistent pattern of spawning in *O. mossambicus* was also found during July–August when spawning of Indian major carps (*C. catla, L. rohita*, and *C. mrigala*) occured.

Food and feeding

Gut content of tilapia sampled in the present study showed that they consume a variety of food items ranging from macrophytes and algae to plankton and detritus. Trophic spectra of 150 examined specimens of *O. mossambicus* showed that there was similarity in the ingested food at different sites. The analyzed gut contents were *Detritus* (50.62%), *Macrophytes* (21.72%), *Filamentous Algae* (7.83%), cellular algae (14.42%), zooplankton (3.60%), fish (1.73%) and insect parts (0.26%) (Figure 3). Juvenile were found to consume fry.

DISCUSSION

O. mossambicus was introduced into India during 1952 for aquaculture purpose and the utilization of *O. mossambicus* gradually expanded for enhancing reservoir fishery production (Suguan, 1995; Sugunan, 2000). After the expansion of the use of *O. mossambicus* for enhancement of aquaculture production, tilapia now form part of the fish fauna in the Godavari, Krishna, Cauvery, Yamuna and Ganga Rivers (Lakra et al., 2008). Unfortunately, the characteristics (including tolerance to wide ranging ecological conditions, generalist dietary requirements, rapid reproduction with maternal care, and the ability to successfully compete with native fish through aggressive behaviour) that make *O. mossambicus* desirable as an aquaculture species also predispose it for success as an invasive species (Canonico et al., 2005). Invasive populations are now causing environmental and ecological problems in many countries (Canonico et al., 2005) and as such, *O. mossambicus* is listed in the Global Invasive Species Database (2006) as being in the top 100 invasive alien species on the planet.

Results of this study showed abundance of *O. mossambicus* in the fishery and the presence of all reproductive stages (1–6) in the river-caught *O. mossambicus*. This data implied that *O. mossambicus* has established breeding populations in the lower stretch of Yamuna River and the so colonized fishes constituted the bulk of the catches by commercial fisherman. Similar findings were obtained from parts of Yamuna (Singh et al., 2010b) and Jaisamand Lake, Rajasthan (Lakra et al., 2008). Reproductive activity of *O. mossambicus* has been reported to be continuous (non seasonal) in females (De Silva and Chandrasoma, 1980) and the results of the present study provided first evidence of wild spawning of *O. mossambicus* in the Yamuna River forming feral populations. Since Tilapia mossambicus, *O. mossambicus*, is known to exhibit early sexual maturity, maternal care of offsprings, rapid colonization, wide environmental tolerances (Perez et al., 2006; Russell et al., 2012); these attributes have been considered to be important for facilitating successful invasion of this fish in the Yamuna

Table 3. Degree of maturation, gonado-somatic index (GSI) the gonad maturity stage and morphology of ovary in different stages of maturity of *O. mossambicus* of the Yamuna River.

Stage	Degree of maturation	GSI	n	Months of Availability	Ova diameter (mm)	Description
1	Immature or virgin and resting adult	0.2-0.8	12	Throughout the year	0.044 -0.055	Ovaries very small, thin, thread like pale in colour, occupying a small part of the body cavity.
2	Early maturing	0.4-2.2	20	March to September	0.053 -0.085	Ovaries slightly larger and increase in weight and volume with minute opaque whitish eggs occupied about half of the body cavity.
3	Developing	3-5	40	March to October	0.078-0.85	Ovaries occupied about 2/3 of abdominal cavity with large pale yellow eggs.
4	Developed / Prespawning	4-6	20	March to October	0.84- 0.96	Ovary more enlarged occupying almost entire body cavity, with large number of big, turgid, spherical, translucent, deep yellow ripe ova
5	Spawning	3-5.5	18	April to October	1.0- 1.4	Ovary walls thin almost transparent. Riped eggs are visible through the ovarian wall and some riped eggs are present in the oviduct.
6	Spent	2-3	10	April to Late October	0.050 - 0.16	Ovaries are flaccid, shrinked and sac like, reduced in volume. Ovary contains ripped unspawned darkened eggs and a large number of small ova.

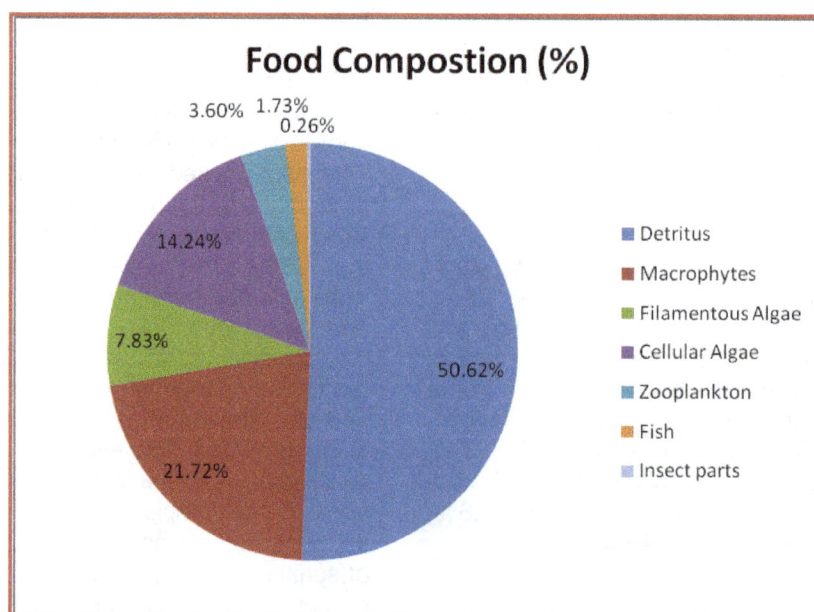

Figure 3. Major food items [%] in the gut of *O. mossambicus* collected (n = 150) from the Yamuna River.

River. The low GSI values observed in the present study are consistent with year-round spawning and are similar to GSI values reported by De silva (1986). The GSI data coupled with the year round presence of mature females, occurring in all months except winter months (December to February), and recruitment of young fish into larger size classes strongly suggest that Mossambiqan Tilapia is spawning year round and is established in river Yamuna.

The gut content analysis showed presence of mainly detritus, plant material, insect parts, algae of similar kind and small fish, which is in agreement with the findings of De Moor et al. (1986), Laundau 1992). Juveniles are carnivorous and eat fry Luna (2012). Gut content analysis showed similar pattern at all the sampling stations. The results indicated that ecological conditions in the Yamuna were homogenizing by the increasing population of O. mossambicus, which could be a great threat to the ecological integrity for this mighty river sustaining rich fish biodiversity.

Conclusion

The invasion of O. mossambicus has increasingly taken-over at all sites of Yamuna River contributing substantially to the fishery of this river, which is considered serious in view of sustainability of indigenous fish diversity. Further investigations should be carried out to determine the extent of spread of O. mossambicus in Yamuna River and to understand its impact on native fish and fisheries. Suitable control and management methods should be found. Such information could contribute to the development of management plans aimed at minimizing possible impacts of this potential invasive species. Moreover, awareness of the implications concerning this invasive species should be generated among scientists, farmers, fishermen, legislators and the general public to provide for the rigorous application of such regulatory measures.

ACKNOWLEDGMENTS

We wish to thank Mr. Mir Shabir Jahar (DBS, Delhi), Mr Khursheed Ahmad Khan (Animal Breeding and Genetics Division, SKUAST-K), Mr Arif Ahmad Shah, Ms Aaliya Mehraj and Ms Mudasir jan (Ichthyology Research Laboratory, University of Kashmir) and Miss Maqsooda Akhtar (Department of Information Technology and Support system, University of Kashmir) for their constructive comments on an earlier draft of this manuscript. Thanks are due to the Research scholars of Ichthyology Laboratory, Bundelkhand University Jhansi, India for their help in field investigations.

REFERENCES

Arthington AH, McKay RJ, Russell DJ Milton DA (1984). Occurrence of the introduced cichlid Oreochromis mossambicus (Peters) in Queensland. Aust. J. Mar. Freshw. Res. 35(2):267-272.

Bilgrami KS (1991). Biological profile of the Ganga: Zooplankton, fish, birds and other minor fauna. In: Krishna Murti CR, Bilgrami KS, Das TM, Mathur RP (eds.) The Ganga: A scientific study .Northern Book Centre, New Delhi. pp. 81-94.

Canonico GC, Arthington A, McCrary JK, Thieme ML (2005) The effects of introduced tilapias on native biodiversity Aquat. Conserv. Mar. Freshw. Ecosyst. 15(5):463-483.

Costal JL, Almeida PR, Moreira FM, Costal ML (1992). On the food of the European eel, Anguilla anguilla (L.) in the upper zone of the Tagus estuary, Portugal. J. Fish. Biol. 41:841-850.

Costa-Pierce BA (2003) Rapid evolution of an established feral tilapia (Oreochromis spp.): the need to incorporate invasion science into regulatory structures. Biol. Invasions 5:71-84

Cucherousset J, Olden JD (2011) Ecological impacts of non-native freshwater fishes. Fisheries 36(5):215–230.

De Moor FC, Wilkinson RC, Herbst HM (1986). Food and feeding habits of Oreochromis mossambicus (Peters) in hypertrophic Hartbeesport Dam, South Africa. S. Afr. J. Zool. 21:170-176

De Silva SS (1986). Reproductive biology of Oreochromis mossambicus populations of man-made lakes in Sri Lanka: a comparative study. Aquac. Res. 17(1):31–47

De Silva SS, Chandrasoma J (1980). Reproductive biology of Oreochromis mossambicus, an introduced species, in an ancient man-made lake in Sri-Lanka. Environ. Biol. Fishes 5(3):253–259

Diana JS (2009) Aquaculture production and biodiversity conservation. Bioscience 59(1):27–38.

Hyslop EJ (1980). Stomach content analyses- A review of methods and their application. J. Fish. Biol. 17:411-429.

Jayaram KC (1981). Fresh water Fishes of India—hand book. Zoological Survey of India, Calcutta. p. 225.

Jayaram KC (1999). The Freshwater Fishes of the Indian Region. Narendra Publishing House, Delhi. pp. 551.

Kapoor D, Dayal R, Ponniah AG (2002). Fish Biodiversity of India. National Bureau of Fish Genetic Resources, Lucknow, India.

Lakra WS, Sarkar UK, Dubey VK, Sani R, Pandey A (2011). River inter linking in india: status, issues, prospects and implications on aquatic ecosystems and freshwater fish diversity. Rev. Fish Biol. Fish. 21:463-479

Lakra WS, Sarkar UK, Kumar RS, Pandey A, Dubey VK, Gusain OM (2010). Fish diversity, habitat ecology and their conservation and management issues of a tropical River in Ganga basin, India. Environmentalist DOI 10.1007/s10669-010-9277-6

Lakra WS, Singh AK, Ayyappan S (eds.) (2008). Fish Introductions in India: Status, Potential and Challenges. Narendra Publishers, New Delhi, India.

Laundau M (1992) Introduction to Aquaculture. John Wiley and Sons, New York. pp. 440.

Luna, Susan M (2012). Oreochromis mossambicus. http://www.fishbase.org/summary/Oreochromis-mossambicus.html (accessed 25, 05, 2012).

Nagelkerke LAJ, Sibbing FA (1996). Reproductive segregation among the Barbus intermedius complex of Lake Tana, Ethiopia. An example of intralacustrine speciation. J. Fish Biol. 49 (6):1244-1266.

Pe´rez JE, Nirchio M, Alfonsi C, Munoz C (2006b). The biology of invasions: the genetic adaptation paradox. Biol. Invasions 8:1115-1121

Philippart JC, Ruwet JC (1982). Ecology and distribution of tilapias. In: Pullin RSV, Lowe-McConnell RH (eds) Biology and culture of tilapias. International Center for Living Aquatic Resource Management, Manila. pp. 15-59.

Pullin RSV (1988). Tilapia genetic resources for aquaculture. In: Pullin RSV (ed) International Center for living aquatic resources management conference proceedings, Manilla, ICLARM. pp. 108.

Rao KL (1975). India's Water Wealth: Its Assessment, Uses and Projections. New Delhi:Orient Longman.

Russell DJ, Thuesen PA, Thomson FE (2012). A review of the biology, ecology, distribution and control of Mozambique tilapia, Oreochromis mossambicus (Peters 1852) (Pisces: Cichlidae) with particular emphasis on invasive Australian populations. Rev. Fish Biol. Fish. (2012) 22:533-554.

Sarkar UK, Lakra WS (2007). Freshwater fish diversity of central India. Edited and published by National Bureau of Fish Genetic Resources, Lucknow. pp. 1-183.

Singh AK, Lakra WS (2011). Risk and benefit assessment of alien fish species of the aquaculture and aquarium trade into India. Rev. Aquac. 3: 3-18 (Wiley-Blackwell).

Singh AK, Lakra WS (2006) Impact of alien fish species in India: emerging scenario. J. Ecophysiol. Occup. Health 6 (3–4): 165–174.

Singh AK, Pathak AK, Lakra WS (2010b). Mapping of invasive exotic fish species of the Yamuna River System in Uttar Pradesh, India. In: Nishida T, Caton AE (eds) GIS / Spatial Analysis in Fishery and Aquatic Sciences, vol. 4, pp. 523–534

Strecker AL, Campbell PM, Olden JD (2011). The aquarium trade as an invasion pathway in the Pacific northwest. Fisheries 36(2):74–85.

Suguan VV (1995). Exotic Fishes and their Role in Reservoir Fisheries in India. FAO Fisheries Technical Paper No. 345.

Sugunan VV (2000). Ecology and fishery management of reservoirs in India. Hydrobiologia 430 (1–3): 121-147.

Talwar PK, Jhingran A (1991). Inland fishes of India and adjacent countries, 2 volumes. Oxford and IBH Publishing Co. Pvt. Ltd., New Delhi, xix / pp. 1158.

Trewavas E (1982). Tilapias: taxonomy and speciation. In: The biology and culture of Tilapias. London.

Vass KK, Das MK, Srivastava PK, Dey S (2009). Assessing the impact of climate change on inland fisheries in River Ganga and its plains in India. Aquat. Ecosyst. Health Manage. 12 (2): 138–151.

18

Plant nutrient release composition in vermicompost as influenced by Eudrilus eugenae using different organic diets

Nweke I. A

Department of Soil Science, Anambra State University, Igbariam Campus Anambra State, Nigeria.

Sub-adult earthworms, Eudrilus, Eugenae were cultured for 100 days in rubber containers made from old tyres to determine the plant nutrient release composition in vermicompost using different organic diets. The diets are Andropogon grass + Pig manure (AGPM) Andropogon grass (AG), Bracharia grass + pig manure (BGPM) and Bracharia grass (BG). Result of the study show that the produced vermicompost in each diet consisted mostly of high quality humus with favorable pH level and high cat ion exchange capacity (CEC) which varied with the diets. Pig manure enhanced the CEC of the vermicompost by 114.82%. Available P, exchangeable Ca and Mg were significantly (p = 0.05) increased in the vermicompost and varied among the various diet treatments. Available P was greatest with Andropogon vermicompost relative to the Bracharia vermicompost. Pig manure enhanced the release of available P of Andropogon vermicompost by over 185% and over 1000% in Bracharia vermicompost relative to the grass vermicompost alone. Exchangeable Ca and Mg of Bracharia vermicompost was enhanced 32 and 74% respectively by Pig manure. Results of the study also show increase in earthworm biomass production and cocoon size in the vermicompost which varied among the diet treatments and non-significant difference among the vermicompost in water holding capacity. Weight loss from composition and decomposition rate was highest in BG relative to BGPM, AG and AGPM. Average decomposition rate of the grass diet was 75.25%. Earthworm (Eudrilus eugenae) through its activities increased the rate of decomposition and degradation of organic wastes and is effective in plant nutrient release if subjected to proper culturing with suitable feed materials.

Key words: Eudrilus eugenae, vermicompost, Brancharia grass, Andropogon grass.

INTRODUCTION

The activities of earthworms in the decomposition, degradation, nutrient recycling and nutrient release of organic matter cannot be over emphasized. Numerous experiments by researchers and scientists using artificial cultures highlighted the tremendous effect of earthworm on plant litter degradation. The amount of soluble mineral matter in compost is increased by the activity of these organisms following their feeding on plant litters and compositing activities, digestion and decomposition pro-

ceeded nutrients were released to the benefit of crops. Worms accelerated nutrient release (Allison 1973). Mba (1984) showed that high earthworm activity resulted in considerable enzymatic activity, microbial biomass as well as high availability of nutrients and decreased cyanide contents in cassava peels. Also Mba (1989), using Eudrilus eugenae on fermented paspalum digitatum reported high activity of the worms and nutrient availability. Edwards et al. (1990) observed that some species of earthworm facilitate the break down and mineralization of surface litter while others incorporate soil organic matter deeper into the soil profile and enhance aeration and water infiltration through burrow formation, hence they reduce and prevent the accumulation of raw organic matter (fallen leaves) remains on the soil surface.

Similarly, Satchell (1967) reported that earthworm not only metabolize carbon but also increase organic matter decomposition by stimulating microbial activity. In another study Satchell (1955) observed three times higher earthworm activities and population in a plot receiving 0.75 ton/ha every four years than that of unmanured. The conjunction work between the earthworm and soil microorganism in decomposition and degradation of organic matter especially resistant plant material is commendable, as earthworms not only decompose material themselves but stimulate other decomposers. The production of cellulases, chitinases and other degrading enzymes from the interaction activities of earthworm and microorganisms help in the breakdown of cellulose of plant tissue, chitin of fungi and perhaps insect cuticle (Griffin, 1972).

The processes of using species of earthworm to compost organic materials are referred to as vermin composting. It is a biotechnological process of waste conversion to produce a better product which is vermicompost. Vermicomposting differs from composting in several ways (Gandhi et al., 1997). It is a mesophilic process, faster than compositing, because the material passes through the earthworm gut where significant transformation takes place and resulting earthworm castings (worm manure) are rich in microbial activity and for strong retention of nutrients, plant growth regulator and other plant growth influencing materials produced by micro-organisms including humates and fortified with pest and root knot nematode repellence attributes as well (Nagavallemma et al., 2004; Asha Aalok et al., 2008; Grappelli et al., 1987; Shiwei and Fu-Zhen, 1991; Arancon et al., 2005b, 2006, 2003; Atiyeh et al., 2002). Containing water soluble nutrient and bacteria, vermicompost is an excellent nutrient rich organic fertilizer and soil conditioner (Appelhof, 1997). Vermicompost contains most nutrients in plant available form such as nitrates, phosphates and exchangeable calcium and soluble potassium (Edwards, 1998; Orozco et al., 1996).

Besides many methods of waste disposal such as burning, incineration, land filling or compost, the most competent and ideal method for the disposal of solid waste that pose a threat to environmental harmony can be achieved through vermiculture (earthworm farming). It is a biotechnological process of converting the organic waste into compost. This biotechnological process could provide an alternative solution to tackle the problem of safe disposal of waste as well as the most needed plant nutrients for sustainable crop production and soil productivity. In fact, earthworms, through a type of biological chemistry, are capable of converting refuse into a useful

product in vermicompost. Thus this study was setup with the aim and objectives to evaluate plant nutrient release composition in vermicompost as influence by Eudrilus eugenae using different organic diets.

MATERIALS AND METHODS

Two species of grass, Bracharia SPP and Andropogon gayanus and Pig manure were used as a source of diet in this study to produce vermicompost using Eudrilus eugenae earthworm. The four different diet treatments were prepared as follows:

Bracharia grass B
Bracharia grass + Pig manure BG
Andropogon A
Andropogon grass + Pig manure AG

1.5 kg of air-dried Bracharia and Andropogon grass were each pre-composted for 2 weeks with or without 0.5 kg of air-dried Pig manure and the mixture (diets) was subsequently air-dried for 2 days. Each treatment was place in rubber containers made from old vehicle tyre. Treatments were replicated 4 times. The worms used in this study were isolated from soil-paspalum grass culture. Eight (8) sub-adult earthworms of Eudrilus eugenae were collectively weighed and inoculated to each of the 16 rubber containers of four treatments. One side of each of the tyre container was covered with fiber sack so that it can have bottom. Each of the treatments was staged together in one frame, while the foot of the frames stood in side a cup containing condemned oil and the entire treatment were covered with a polythene sheet. The oil protect the earthworm from predators and parasites like ants, beetles, centipedes, slugs, leeches and flatworms, that attack earthworms in green house as was reported by Lofty and Edwards (1972).

Treatment types were placed under shade and watered at rate of 130 ml per rubber container. Daily routine involved uncovering (that is, polythene sheet removed) of the entire treatments every morning to monitor earthworms activities and to ensure fresh air circulation in the culture and as well as to minimize carbon di oxide build up. The culture is later covered at mid-day. The cultures were watered on every other day to maintain moisture at the rate of 130ml per rubber container. After 100 days inoculation, the worms were harvested by hand sorting and the adult worms were weighed collectively per rubber container. The resulting vermicompost made up mostly of earthworm casts; and very little grass residues were collected and aliquot was taken for the chemical analysis of the vermicompost and oven dry matter content determination.

Decomposition rate

Another experiment was setup to determine the decomposition rate of Bracharia and Andropogon grass with or without Pig manure using 250 g of grass to 80 g of Pig manure. The grass was stipped in water for 2 days while the Pig manure was moistened overnight. The mixtures were prepared as follows: Andropogon grass + Pig manure (AGPM) Adropogon grass (AG), Bracharia grass + Pig manure (BGPM) and Bracharia grass (BG). Each mixture (treatment) was replicated three times. The mixture were properly mixed and bagged with polythene and tied up properly to prevent air-entry (air tight). The experiment was allowed a period of two (2) weeks after which the rate of decomposition was determined.

Laboratory methods

The resulting vermicomposts were analyzed for pH using a Digital pH meter. The values were determined both in H_2O and 0.1 NKCL

Table 1. Compost production by Eudrilus eugenae and water holding capacity of Vermicompost.

Treatment	Composts Initial dry matter	Vermicompost dry matter g/kg	Vermicompost water holding capacity %
AGPM	41.15	2.73	283.30
(AG)	32.85	2.06	322.47
BGPM	46.47	4.01	335.28
BG	53.66	2.59	243.70
LSD 0.05	0.94	0.11	NS

AGPM = Andropogon gayanus + Pig manure, AG= Andropogon grass, BGPM= Bracharia grass + Pig manure, BG= Bracharia grass, LSD=Least significant difference.

Table 2. Chemical properties of the Vermicompost of different Organic diets

Treatment	% Carbon	CEC Cmolkg^{-1}	pHH$_2$O	Avail. Mgkg^{-1}P	Ca Cmolkg^{-1}	Mg Cmolkg^{-1}	pHn 0.1N kCl
AGPM	28.90	304	6.27	1074.5	83.71	29.28	6.05
AG	31.21	265	6.03	376.52	92.91	30.37	5.84
BGPM	31.86	308	6.39	760.73	107.77	53.64	6.05
BG	35.39	268	6.52	68.32	81.66	30.85	6.12
LSD0.05	0.71	5.07	NS	0.73	0.50	0.15	NS

AGPM= Andropogon gayanus + Pig manure, AG= Andropogon grass, BGPM= Bracharia grass + Pig manure, BG= Bracharia grass. LSD=least significant difference.

with solid to liquid ratio of 1:2.5. Exchangeable cat ions (Ca^{2+} and Mg^{2+}) were extracted with ammonium acetate by Perkin Elmer atomic absorption spectrophotometer (Tel and Rao, 1982) while cat ion exchange capacity (CEC) was obtained by the method of Jackson (1958). Organic carbon (OC) was analyzed by the method of Nelson and Sommers (1982). Available P was determined by Bray II method, Bray and Kurtz (1945).

Data generated from the study was analyzed using analysis of variance (ANOVA) tested on randomized complete block designs (RCBD) according to steel and Torrie (1980).

RESULTS

Daily examination of earthworm activities

Daily examination of the worm culture revealed that worm's activities were prominent at the middle and bottom of the cultures. Castings were noticed in all the cultures at exactly 4 days after inoculation. Degradation of culture materials was most intensive with Andropogon grass + Pig manure culture, followed by Andropogon grass culture in contrast to the Bracharia grass + Pig manure and Bracharia grass treatments were distinct decomposition was observed about 13 days later. This was reflected in the kind of result obtained from the investigation summarized in Tables 1 to 4.

Compost production/Water holding capacity (WHC)

The percentage compost production of the two grass diet differed significantly (P = 0.05). Table 1 showed that the

order of compost production among the organic materials were BG > GBPM > AGPM > AG. The average compost production of the grass was 43.53 g/kg diet. Pig manure enhanced compost production with both grass diets particularly with Bracharia even though the contribution of the manure was negligible.

The vermicompost dry matter content of the treatments in Table 1 were significant (P=0.05), the highest value of 2.06 g/kg was recorded in AG. Average dry matter content of the vermicomposts of the grass was 2.85 g/kg. The water holding capacities of the vermicomposts were similar (Table 1). Pig manure enhanced water holding capacity with both grass diets, particularly with Bracharia (BGPM). Average percentage water holding capacity of the grass was 296.19%.

Chemical composition of vermicompost

The result of chemical analysis of vermicompost of different organic diet summarized in Table 2, showed that the vermicompost differed significantly (P=0.05) in all the chemical parameters assessed except pH. The vermicomposting activities of Eudrilus eugenae increased the nutrient composition of the vermicompost. The two grass diet differed significantly (P=0.05) in their effect on the carbon content of the vermicompost (Table 2). The highest carbon level was observed in unmanured Bracharia grass relative to the manured ones. This indicated that the effect of manure was however not

Table 3: Effect of Vermicompost of different organic diets on Eudrilus Eugenae biomas production, Survival rate and Cocoon size

Treatment	Initial dry weight of worm g/kg diet	Dry earthworm biomass at harvest g/kg diet	Survival rate of worm %	Cocoon size mg/kg diet
AGPM	7.4	20.9	104.2	18.1
AG	7.9	13.8	66.7	17.6
BGPM	7.7	13.0	79.2	14.1
BG	6.9	15.4	95.8	11.9
LSD0.05	0.5	4.0	0.51	0.50

AGPM = Andropogon grass + Pig manure, AG = Andropogon grass. BGPM = Bracharia grass + Pig manure, BG = Bracharia grass. LSD = Least significant difference

significant. With Andropogon and Bracharia grass manured, C level was slightly reduced by vermicomposting.

The CEC values of the vermicompost in Table 2 was significant (P=0.05). The CEC values of AGPM and BGPM vermicomposts were higher than that of the AG and BG vermicomposts. The order of increase in CEC of the vermicompost were BGPM>AGPM>BG>AG. Pig manure enhanced the vermicompost CEC of both grass species by 114.82% relative to the unmanured grass diet. The net charge of vermicompost judged from the pH in water and in 0.1NKCL was negative. The pH was similar among treatments it varies from 6.03 to 6.52 among vermicompost in water and 5.84 to 6.12 in 0.1 NKCL.

The grass species significantly (P=0.05) vary in their influence on the P level of the vermicompost. Available P was greatest with Andropogon relative to the Bracharia. Pig manure contributed significantly to the available P particularly with the Bracharia grass. While Pig manure contribution enhanced the available P of Andropogon vermicompost by over 185%, its contribution to the available P with Bracharia was over 1000% relative to the grass vermicompost.

Exchangeable Ca of the vermicompost vary significantly (P=0.05) among the various treatments. The order of increase in the Exchangeable Ca were BGPM > AG > AGPM > BG. Pig manure enhanced the exchangeable Ca of Bracharia vermicompost by 32%, while the manure contribution to the Exchangeable Ca with Andropogon vermicompost was negligible.

Similar to the Ca level of the vermicompost, the exchangeable Mg vary significantly (P=0.05) among the treatments. Pig manure enhanced the Mg level of Bracharia vermicompost by 74%. The trend of increase of Mg level in the vermicompost, however, were BGPM > BG > AG > AGPM.

Earthworm biomass production

The result of the experiment (Table 3) show that the vermicompost significantly (P=0.05) increased the earthworm (Eudrilus eugenae) weight, survival rate of worm and cocoon size. These parameters differed among the

treatments. The highest weight of 20.9 g/kg diet earthworm biomass was obtained in the AGPM diet.

The percentage increase in weight observed in the treatments relative to initial weight was 182.4% (AGPM), 74.7% (AG), 68.8% (BGPM), and 123.2% (BG). The average production of earthworm biomass was 15.78 g biomass/kg diet. Pig manure enhanced biomass production with both grass diets particularly with Andropogon grass. The survival rate of the worms was influenced by the different diets the highest percentage survival rate of the worms was recorded in AGPM. With regard to the cocoon size, the order of increase were AGPM > AG > BGPM > BG. Cocoon size was significantly (P = 0.05) influenced by the diets (Table 3). With Andropogon diet cocoon size was 28.4% greater than the Bracharia treatment. Pig manure slightly enhanced the cocoon size of the grass species.

Decomposition rate

The result of the study (Table 4) showed that the weight loss from composting and decomposition rate differed significantly (p=0.05) among the treatments.

The weight loss was highest in BG relative to BGPM, AG and AGPM diets. The result of decomposition rate (%) was similar to the weight loss, the highest rate was observed in BG diet. The order of increase in percentage decomposition rate were BG > BGPM > AG > AGPM. Average decomposition rate of the grass diet was 75.25%, Pig manure enhanced decomposition in both grass diets particularly with Bracharia grass.

DISCUSSION

The result of this study indicated that Eudrilus eugenae can be cultured successfully using rubber containers made from old tyre, and that it could produce vermicompost from Andropogon grass, Bracharia grass and Pig manure. Also that Eudrilus eugenae can make use of plant materials as a source of feed. The significant difference in compost production of the two grass diet could be attributed to preference in grass diet by the

Table 4 Decomposition rate of Andropogon grass and Bracharia grass with or without Pig manure.

Treatment	Weight loss from composting g/kg	Percentage (%) rate of decomposition
AGPM	303	69.7
AG	268	73.2
BGPM	217	78.3
BG	203	79.7
LSD0.05	4.10	1.1

AGPM = Andropogon grass + Pig manure, AG = Andropogon grass BGPM = Bracharia grass + Pig manure, BG = Bracharia grass. LSD = Least significant difference

earthworm species in Eudrilus Eugenae. Toutain et al. (1982) noted that earthworm show a preference for particular parts of a plant. The non-significant value observed in the water holding capacity among the grass vermicompost with or without Pig manure cultures might be due to soilless nature of the compost. If the vermicompost where produced in the presence of soil, formation of burrows may have produced macro pores that will increase water infiltration and aeration. Fallani (1978) showed that earthworm activities correlated with higher infiltration rate, while Urbanek and Do lezal (1992) reported that earthworm channels contribute to the water movement and relation under natural condition. The organic carbon content reduction observed among the diets could be due to utilization of carbon by the earthworm as carbon and energy sources during the biosynthesis associated with growth and productivity. The result of the CEC value could be a reflection of high humification activities of compost materials, which involved the breakdown of plant residue by earthworm into smaller particles giving rise to large specific surface area and the formation of good quality colloids. It can also be attributed to increased maturity of the bio product as decomposition proceeded. The amount of available P was found to be increa-=sed in all the diets and the value of exchangeable Ca and Mg was also influenced by the compositing activities of Eudrilus eugenae. The result could be explained on the ability of earthworm to modify characters of organic materials and productivity of the resultant product in vermicompost. Oxygenation and biological activities are highly affected by earthworms through microbial interaction. This obviously must have contributed to the modification of the diets degradation rate, chemical properties and plant nutrient release of the vermicompost. Vermicompost contains most nutrients in plant available form such as nitrates, phosphates and exchangeable calcium and soluble potassium (Edwards, 1998; Orozco et al., 1996). The pH results show that the resultant vermicompost in each diet consisted mostly of high quality humus with favorable pH level. The increase in earthworm biomass production, survival rate of worm and cocoon size could be as a result of increased nutrients in the vermicompost and of nutrients thereof, by earthworm for growth and productivity, especially with regards to Andropogon Pig manure diet. Mba (1988) found out that

hatchability and number of worms per hatch were negatively correlated to weight and cocoon size are varied with the stage (duration days) of vermicompositing at which the cocoons were laid. The result of the decomposition of the two grass species suggests that the addition of animal manure will bring about faster decomposetion of organic materials probably as a result of rich in microbial population.

Results of this study showed that earthworm through the compositing activities increased decomposition, degradation rate and nutrient release of organic diets, Effective in bio waste conversion and if properly cultured can be an effective tool in waste management and bio fertilizer production. The result equally showed that Eudrilus eugenae is effective in plant nutrient release if subjected to proper culturing with suitable feed materials.

REFERENCES

Allison FE (1973). Soil organic matter and its role in crop production. Elsevier Scientific pub. Company Amsterdam, London, New York.

Appelhof M (1997). Worms eat my garbage. 2nd edition, Kalamazoo. MI flower press P3.

Arancon NQ, Yardin EN, Edwards CA, Lee S (2003). The trophic diversity of nematode communities in soils treated with vermicomposts. Pedobiologia, 47:731-735.

Arancon NQ, Edwards CA, Bierman P. Metzger JD, Lucht C (2005b). Effects of vermicomposts produced from cattle manure, food waste and paper waste on the growth and yields of peppers in the field. Pedobiologia 49, 297-306.

Arancon, NQ, Edwards CA, Yardin EN, Oliver T, Byrne RJ, Keeney G (2006). Suppression of two spotted spider mite (Tetranychus Urticae) mealy bugs (Pseudococus SPP) and aphid (Myzus persicae) populations and damage by vermicomposts. Crop production 26, 29-39.

Asha Aalok A, Tripathi K, Somi P (2008). Vermicompositing: A better option for organic solid waste management. J. Hum. Ecol. 24(1): 59-64.

Atiyeh RM, Lee SS, Edwards CA, Arancon NQ, Metzger JD (2002b). The influence of humic acid derived from earthworm-processed organic wastes on plant growth. Biores. Technol 84:7-14

Bray RH, Kurtz IT (1945). Determination of total organic and available forms of Phosphorous in soils. Soil Sci: 59.39-45.

Edwards CA (1998). The use of earthworm in the breakdown and management of organic wastes. In Earthworm Ecology C.A.Edwards (Ed) pp. 327-254, CRC Press Boco Roton, FL.

Edwards WM, Shipitalo MJ, Owens LB, Norton LD (1990). Effect of lumbricus terristaris L. burrows on hydrology of continuous no till corn fields. Geordema 46:73-84.

Fallani E (1978). Die Boden Fauna in Dquerkucturen mit unterschieducher Bodenbedeckung Ergebnisse eines Fellversuchs

Mit einer Erhebung in Nirdian. Ph. D thesis Univ. Giessen.

Gandhi M, Sangwan V, Kapoor KK, Dilbaghi N (1997). Composting of household wastes with and without earthworms. Environ. Ecol. 15(2):432-434.

Grappelli A, Galli E, Tomati U (1987). Earthworm casting effect on Agaricus bisporus fructification. Agrochemical 21:457-462.

Griffin DM (1972). Ecology of soil fungi Chapman and Hall Ltd. London.

Jackson, M.L. (1958). Soil chemical Analysis prentice Hall. Eaglewood cliffs NJ pp. 511-1519.

Mba CC (1984). Decomposition of cassava by Eudrilus eugenae (King Berg). Beitrege Trop. Land witch Verterinamed 22 H.L 41-46.

Mba CC (1988). The effect of diet and incubating media on the production and hatchability of the earthworm Eudrilus eugenae (King Berg) cocoons. Rev. Biol. Trop. 36(1):89-95.

Mba CC (1989). Biomass and vermicompost production by earthworm Eudrilus eugenae (King Berg) Rev. Biol. Trop. 37(1):11-14.

Nagavallemma KP, Wani SP, Stephane Lacroix, Padmaja VV, Vineela C, Babu Rao M, Sahrawat KL (2004). Vermicomposting: Recycling wastes into Valuable organic fertilizer. Global theme on Agro ecosystem Report no.8 Patancheru 502324, Andhra Pradesh, India ICRISAT. p. 20.

Nelson DW, Sommers LE (1982). Total organic carbon and organic matter. In Pege C.A. Miller, R.H.Keeny D.R. (Eds). Method of soil analysis part2, 2nd edn. Agronomy monograph No.9 ASA and SSSA. Madison 1:539-579.

Orozco FH, Cegarra J, Trujillo LM, Roig A (1996). Vermicomposting of coffee pulp using the earthworm Eisenia Foetida: effects on C and N contents and the availability of nutrients. Biology and fertility of Soil, 22: 162-166.

Satchell JE (1955). Earthworm In soil Zoology D.K.McE, Kavan (ed.) Butterworth, London. pp. 180-201.

Satchell JE (1967). Lumbricidae In soil Biology. Burges, A. and Raw, F. (Eds) Academic press, New York pp. 259-322.

Shi-wei Z, Fu-Zhen, H (1991). The Nitrogen uptake efficiency from ^{15}N labeled chemical fertilizer in the presence of earthworm manure (cast) pp 539-542. In Advances in management and conservation of soil fauna. GKVeeresh, D Rajgopal, CA Viraktamath (Eds). Oxford and IBH Publishing Co. New Delhi, Bombay.

Steel GD, Torrie JH (1980). Procedures of Statistics. A biometrical Approach 2nd edn McGraw Hall Book Co Inc. New York p. 633.

Tel DA, Rao R (1982). Automated and Semi-automated method for soil plant analysis. Manual series No711 TA Ibadan.

Toutain F, Villemin G, Albrecht A, (1982). Ultra structural study of biodegradation processes. II Enchytraeids-Leaf litter models. Pedobiologia, 23:145-156.

Urbanek J, Dolezal F (1992). Review of some case studies on the abundance and on the hydraulic efficiency of earthworm channels in Czechoslovak soil, with reference to the subsurface pipe drainage. Soil Biol. Biochem. 24:1563-1572.

19

Macroinvertebrates (oligochaetes) as indicators of pollution: A review

Rafia Rashid and Ashok K. Pandit

Department of Environmental Science, University of Kashmir, Srinagar - 190006, India.

Macroinvertebrates formed an important constituent of an aquatic ecosystem and had functional importance in assessing the trophic status as the abundance of benthic fauna mainly depends on physical and chemical properties of the substratum and thus the benthic communities respond to changes in the quality of water and available habitat. This review discussed the occurrence, composition and distribution of macroinvertebrates of lakes and wetlands, and some environmental factors which regulated their occurrence and distribution. Also, analysis of the benthic community helped in the determination of trophic status of lakes because of their sensitivity to pollution and is, therefore, an important criterion in the ecological classification of lakes.

Key words: Macroinvertebrates, substratum, lake, wetlands, trophic status, habitat.

INTRODUCTION

The benthic macroinvertebrates are associated with bottom or any solid liquid interface, those that are retained by a sieve or mesh with pore size of 0.2 to 0.5 mm which includes a heterogeneous assemblage of organisms belonging to various phyla like Arthropoda, Annelida, Mollusca and others. The benthos occupies an important position in the lake ecosystem, serving as a link between primary producers, decomposers and higher trophic levels (Pandit, 1980). They also play an important role in the decomposer food chain which in turn affects the cycling of minerals (Gardner et al., 1981). Macroinvertebrates are used as indicators of pollution as invertebrate community change in response to changes in physicochemical factors and available habitats (Sharma and Chowdhary, 2011). The importance of macroinvertebrates as bioassessment tools is widely recognised because of their limited mobility, comparatively long life cycles and differential sensitivity to pollution of various types and they reflect the impact of cultural eutrophication on aquatic habitats quite satisfactorily. According to Jumppanen (1976) the first signs of eutrophication and

pollution in a lake are reflected in the benthic flora and fauna as the suspended waste immediately sink to the bottom to decompose and thus cause a change in the benthic organisms. The lakes and wetlands having soft bottom sediments are characterised by annelids either as dominant group or an important contributor to the macrobenthic fauna. Of the fresh water annelids, the oligochaetes display the greatest diversity and have the greatest indicator value.

MACROINVERTEBRATES AS BIOINDICATORS

Liebmann (1942) claims the microscopic benthic organisms being the most useful as true indicators of pollution and investigators like Richardson (1921, 1929), Gaufin and Tarzwell (1952, 1956), Heut (1949), Brinkhurst (1966) and Wilhm and Dorris (1966), among many others have relied almost entirely on them though their suitability and again diminished because of their rapid rate of reproduction and the difficulty of sampling and correctly identifying the species (Liebmann, 1951). Except in case of macroinvertebrates which are generally regarded as most suitable indicators of pollution, the presence of a particular species does indicate the suitability of the environment for its growth and development but the absence of any species does not necessarily indicate the unsuitability of the environment, instead the absence of an entire group of species with same ecological needs indicate adverse environmental conditions (Kaul and Pandit, 1981). According to Jumppanen (1976) also the first signs of eutrophication and pollution in some Finnish lakes are usually seen in the benthic fauna as the suspended wastes immediately sink to the bottom to decompose and thus causing a change in the benthic organisms. Thus, certain species of sponges, for example, respond to various types of poisonous pollutants even in very mild cases, while others (tubificids, sludge worms, maggots and chironomids) can tolerate even the most gross organic pollution and high levels of toxic pollution. Hence, the species analysis of benthic community enables the determination of trophic type of lakes and is, therefore, an important criterion in the ecological classification of lakes (Thut, 1965; Seather and McLean, 1972; Bazzanti, 1975). As the emergence of species like *Tubifex* sp. and *Chironomus* sp. in Nilnag lake indicated the eutrophic status of the lake (Yaqoob et al., 2007). Benthoses of the Shallabugh wetland were represented by Arthopoda (10), Annelida (7) and Mollusca (6). The abundance of some specific pollution indicator species, especially annelids such as *Limnodrilus* sp, *Tubifex tubifex* and *Branchiura sowerbyii*, is depictive of transition in trophic status of the wetland from meso- to eutrophy (Siraj et al., 2010). Dar et al. (2010) reported a few species of annelids like *Tubifex tubifex*, *Limnodrilus* sp. and *Erpobdella octoculata* to be dominant in terms of taxa and abundance. However, Mollusca were poorly represented and Insecta although represented by one taxa namely *Chironomus* sp. was abundant throughout the study period revealing the eutrophic status of Hokera wetland as the organisms recorded mostly occur in eutrophic waters. Lang (1985) studied the eutrophication of lake Geneva and recorded species like *Potamothrix hammoniu*, *P. Heuscheri* and *Tubifex tubifex* to be numerically dominant ones as compared to *P. veidovskyi* (mesotrophic), *Stylodrillus heringianus* (oligotrophic) in the community structure indicating a meso- eutrophic status of lake. Awal and Svozil (2010) identified 481 to 629 organisms in three constructed wetlands in South East metropolitan Melbourne comprising of 16 taxa. There was no significant differences between the wetlands on the basis of one way analysis of variance (ANOVA) for species richness (P> 0.05, F= 0.19) and Shannon-weiner index (P. 0.05, F=2.54) but the data collected was compared with the earlier published data which depicted differences in species richness and diversity. Hence, macroinvertebrates were used as a universal measure of wetland ecosystem integrity and consequently the mana-

gement and conservation of constructed wetlands. Kaul and Pandit (1982) while describing the biotic factors and food chain structure in different wetlands of Kashmir observed the macrozoobenthos to be limited in number of species. They also observed summer predominance of annelids and molluscan predominance in winter. *Tubifex tubifex* and *Glossiphonia weberi* exhibited highest energy content during summer, where as *Chironomus plumosus* and *Viviparus bengalensis* revealed highest values during winter (Gupta and Pant, 1983b). The diversity of benthic macroinvertebrates was much lower in Lake Carl Blackwell. Nineteen genera of benthic macroinvertebrates were found, but nine genera were the maximum being found at one time and station with density ranging from 2310 ind./m^2 in fall to 1625 ind./m^2 in spring of which greatest density (91.5%) was contributed by *Chaoborus* of the assemblage (Howick and Wilhm, 1984). The faunal diversity was minimum at Perumathura where the substratum was highly unstable, but the density was maximum at Murukampuzha where the substratum was relatively stable. There was also the varying pattern of regional and seasonal variations in different groups of benthic organisms (Nair et al., 1984). Singh and Ahmad (1989) compared the benthic fauna of lotic and lentic water bodies and observed oligochaetes/insects to be the chief component of lentic waters while as the polychaetes were the major contributors of lotic system. Oligochaetes were the groups with higher similarity whereas the polychaetes were altogether absent in lentic systems.

FACTORS   AFFECTING   LAKE MACROINVERTEBRATES

The environmental factors that affect the structure of macrozoobenthic community should be considered while scaling the ecological status (Trayanova et al., 2007). Pearson et al. (1986) described long term changes in the benthic community of two areas Loch Linn and Loch Eil, Scotland. They held the view that the changes in population of benthic community over a period of twenty years were related to and dependent upon changing organic inputs which in turn determined the carrying capacity of sedimentary benthos, while as the species composition was dependent upon climatic fluctuations like long term temperature changes. A low species diversity index was observed at thermal effluent site due to deteriorated water quality at that site and hence, the community structure at the effluent site was reported to be under stress (Singh, 1988). The nature of the sediment influenced the population dynamics of the oligochaetes of the lake as dominant oligochaetes of Dal lake, Kashmir includes *Limnodrilus hoffmeister*, *Tubifex tubifex*, *Branchura sowerbyii*, *Aelosoma* sp. and *Nais* sp. which thrive in sediments rich in organic nutrients (Mir and Yousuf, 2003). Among 24 taxa of benthic macroinverebrates in Lake Uluabat, Bursa, Turkey, Insecta and Oligochaeta were the most abundant groups, dominated by species characteristic to nutrient rich waters, including *Pristina aequiseta*, *Nais communis*, *Tubifex tubifex*, *Limnodrilus hoffmeisteri*, *Potamothrix hammoniensis* and *Tanypus punctipennis*. Most of the variance (63.5%) in relationships between species and environmental variables as explained by the first two axes of a canonical correspondence analysis (CCA) and placed most Oligochaeta and Chironomidae near the vectors of high nutrients and chlorophyll-a concentrations, while the sensitive Crustacea and some Oligochaeta (Lumbricidae) species on sectors of the plot with the smallest weight of those variables (Celik et al., 2010). Eighteen taxa of macrozoobenthic organisms, belonging to Annelida, Mollusca and Arthropoda, were recorded during the course of a yearlong study in the Dal lake, Kashmir and marked variations were found in the spatial distribution of various taxa, which was influenced by the texture of the sediment as well as by the macrophytic community structure (Mir and Yousuf, 2002). The macrozoobenthic community was found to be influenced by the type of substrate, the organic matter, the abundance of macrophytes as well as the concentration of calcium (Qadri and Yousuf, 2004). During the survey, 5 macrozoobenthic taxa belonging to Annelida, Mollusca and Arthropoda were recorded among which Annelida formed the most dominant group being represented by two oligochaetes that is, *Tubifex tubifex* and *Branchiura sowerbyii*, which have been designated as an indicators of pollution (Oliver, 1971; Milbrink, 1980; Bazzanti, 1983). The water quality, sediment characteristics and general ecology indicated the association of distribution, diversity and population density with habitat ecology, substratum diversity, altitude and climatic conditions of the concerned area (Roy and Nandi, 2008). Gong et al. (2000) made comparative studies on macrozoobenthos in two shallow mesotrophic lakes (Biandantang and Houhu) and found macrozoobenthos more diverse in Lake Biandantang where macrophytes were abundant than in Lake Houhu where macrophytes were scarce. In shallow lakes, submerged macrophytes are essential for the maintenance of biodiversity of macrozoobenthos because the macrophytes increase habit heterogeneity and availability of suitable food, and may also decrease predation by fish on the macrozoobenthos (Gong et al., 2000). The epiphytic oligochaetes were more diverse and more abundant in the *naiko* than those in littoral Lake Biwa, probably because of higher temperatures, denser aquatic vegetation, and higher primary production (Othaka and Nishino, 2006). Organic matter, ammonium and phosphates were positively correlated with the mean oligochaete abundance, but not with the granulometry. The canonical correspondence analysis (41.2% cumulative variance) indicated that the oligochaetes distributed along both an eutrophication-pollution gradient and a turbidity-conductivity gradient (Armendariz et al., 2011). Increasing temperatures due to climate change were found to influence abundance and timing of species in numerous ways

(Burgmer et al., 2007). Whereas many studies have investigated climate-induced effects on the phenology and abundance of single species, less is known about climate-driven shifts in the diversity and composition of entire community. They analysed time series of entire community of macrozoobenthos in lakes and streams in Northern Europe but, no direct linear effects of temperature and climate indices (North Atlantic Oscillation index) on species composition and diversity was found. However, multivariate statistics showed that trends in average temperature have had profound impacts on species composition in lakes and future climate shifts may thus induce strong variance in community composition (Burgmer et al., 2007). Amakye (2001) while monitoring the seasonal as well as depth wise distribution of macroinvertebrates in the sediments of lake Volta at Yeji area observed the highest density of macroinvertebrates between the shore and depths of 8-10 m and their abundance in July. It was also found that Chironominae were abundant while Orthocladinae and Ephemeroptera were scare in the sediments compared to the formative years of the lake. The observed changes in the composition and diversity of benthos were attributed to increasing anthropogenic influences on the lake which was depicted by the changing chemistry of the lake water. In temporary or permanent wetlands, the total macroinvertebrate biomass and densities were positively related to coarse particulate organic matter abundance (living and nonliving plant matter; CPOM) and negatively related to turbidity. Density of ecologically sensitive EOT (Ephemeroptera, Odonata and Trichoptera) taxa was also positively related to CPOM and negatively related to turbidity. Total taxa richness was negatively related to turbidity, and percent of total macroinvertebrate density consisting of EOT (% EOT) was positively related to CPOM (Stewart and Downing, 2008). Stepwise multiple regression analysis demonstrated that the water depth, conductivity and chlorophyll "*a*" were the key factors affecting macrozoobenthic abundance in the lakes (Yongde and Hongzhu, 2007). The diversity and distribution patterns of certain species were clearly related to water quality (Latha and Thanga (2010). Leech community composition was best described by an ordination incorporating alkalinity, primary productivity and lake area. In general, highest species richness occurred in small eutrophic lakes where as lowest richness was recorded in medium to large lakes with low productivity. Contrary to results for some other taxa, lake pH was not a dominant variable describing only a small amount of variance in the species-environment relationship (Grantham and Hann, 1994). The oligochaete community of the acidified lakes was poorer compared to the neutral ones. Taxa richness, total biomass of the oligochaetes, their relative density and relative biomass in macroinvertebrate communities were lower in the strongly acidified lakes. Changes of major taxa proportions in the total density and biomass of the oligochaetes were recorded with lowering of pH (Ilyashuk, 1999). The oligochaetes were separated into three functional feeding groups, as gatherers (S) that are selectively ingested mainly on the sediment surface and other substrates, gatherers (T) that are selectively ingested mainly in the sediments, and predators. Total density of gatherers (T) as well as their relative density in the oligochaete assemblage and macroinvertebrate community was lower in the acidified lakes (Ilyashuk, 1999). In an eutrophic subarctic lake, the largest populations of animals were found in the deepest part of the lake. However, in the anoxic part of the lake, species were in low number due to the low oxygen levels in water and high organic content of the sediments of the lake (Moore, 1981). Further, the anoxic zone of Nainital lake was found to be devoid of macrobenthos (Gupta and Pant, 1983a). Efitre et al. (2001) quantified the spatial and temporal distribution of macroinvertebrates in Nabugabo lake, Uganda with a focus on habitat associations and they found the total absence of bivalves and crustaceans and less abundance (1.8%) of gastropods. The dominant taxa, however, were ephemeropterans (77.7%), dipterans (11.1%) and smaller contributions were made by annelids (5.4%), odonates (2.8%) and tricopterans (1.3%) to the benthic assemblage. Further, the study revealed that abundance of macroinvertebrates was due to the habitat effects as water lily habitat reflected low level of oxygen near the sediments. Gong and Xie (2001) reported 33 taxa belonging to Mollusca, Oligochaeta and Arthropoda in lake Donghu-China and observed low species diversity in highly eutrophic areas measured in terms of species number, diversity index and k-dominant curves. Abundance of *Limnodrilus hoffmeisteri* was positively correlated to the degree of eutrophication due to its ability to tolerate low dissolved oxygen as the worms exhibit very marked physiological tolerance for oxygen depletion related to excess decomposable organic matter present in the environment, but they do decrease in number when condition are at their worst. Few other organisms can survive under these circumstances, so that worms, which have a very efficient oxygen uptake mechanism, may take up the entire benthic community (Zajic, 1971).

CONCLUSION

From the preceding review, it is evident that macroinvertebrates occupies an important position in the lake ecosystem serving as a link between primary producers and higher trophic levels. They also play an important role in the decomposer food chain which in turn affects the cycling of minerals macroinvertebrate community change in response to changes in physicochemical factors and available habitats and hence, are used as bioassessment tools because of their limited mobility, comparatively long life cycle, differential sensitivity to various types of pollution and reflectance of cultural eutrophication and health status of aquatic habitats, thus can be used as robust bioindicators. In addition, it may

be said that the occurrence, composition and distribution of macroinvertebrates in lakes and wetlands is governed by numerous environmental factors that affect the structure of macrozoobenthic community and their distribution pattern should be considered while evaluating the ecological status.

ACKNOWLEDGEMENTS

The first author acknowledges the immense help received from the scholars whose articles are cited and included in references of this manuscript. The author is also grateful to publishers of all those articles, journals and books from where the literature for this article has been reviewed and discussed.

REFERENCES

Amakye JS (2001). Some observations on macroinvertebrate benthos of Lake Volta at Yeji area (stratum VII) thirty years after impoundment. West Afr. J. Appl. Ecol. 2: 91-102.

Armendariz LC, Capitulo AR, Ambrosio ES (2011). Relationships between the spatial distribution of oligochaetes (Annelida, Clitellata) and environmental variables in a temperate estuary system of South America (Rio de la Plata, Argentina). N. Z. J. Mar. Freshwater Res. 45(2): 263-279.

Awal S, Svozil D (2010). Macro-invertebrate species diversity as a potential universal measure of wetland ecosystem integrity in constructed wetland in South East Melbourne. Aquat. Ecosyst. Health Manage. 13: 472-479.

Bazzanti M (1975). Chironomidi (Diptera) dei Sedimenti del lago di Martignano (Lazio). Bull. Pesca Piscic. Idrobiol. 30:139-142.

Bazzanti M (1983). Composition and diversity of the profundal macrozoobenthic community in the polluted Lake Nemi (Central Italy). Acta Oecol. Appl. 4(3):211-220.

Brinkhurst RO (1966). Detection and assessment of water pollution using oligochaete worms (parts 1 abd 2). Water and Sewage Works, Oct. and Nov. 113: 398-401 and 438-442.

Burgmer T, Hillebrand H, Pfenninger M (2007). Effects of climate-driven temperature changes on the diversity of freshwater macroinvertebrates. Oecologia 151: 93-103.

Celik K, Akbulut N, Akbulut A, Ozatli D (2010). Macrozoobenthos of Lake Uluabat, Turkey, related to some physical and chemical parameters. Panam. J. Aquat. Sci. 5(4):520-529.

Dar IY, Bhat GA, Dar ZA (2010). Ecological distribution of macrozoobenthos in Hokera wetland of J and K, India. J. Toxicol. Environ. Health Sci. 2(5): 63-72.

Efitre J, Chapmah LJ, Makanga B (2001). The inshore benthic macroinvertebrates of Lake Nabugabo, Uganda: Seasonal and spatial patterns. Afr. Zool. 36(2): 205-216.

Gaufin AR, Tarzwell CM (1952). Aquatic invertebrates as indicators of stream pollution. Public Health Rep. 67 (1): 57-64.

Gaufin AR, Tarzwell CM (1956). Aquatic microinvertebrate communities as indicators of organic pollution in Lyttle creek. Sewage Ind. Waste 28: 906-924.

Gong Z, Xie P (2001). Impact of eutrophication on biodiversity of the macrozoobenthos community in a Chinese shallow lake. J. Fresh Water Ecol. 16(2):171-178.

Gong Z, Xie P, Wang S (2000). Macrozoobenthos in two shallow, mesotrophic Chinese lakes with contrasting sources of primary production. J. N. Am. Benthol. Soc. 19 (4):709-724.

Grantham BA, Hann BJ (1994). Leeches (Annelida: Hirudinea) in the experimental lakes area, Northwestern Ontario, Canada: Patterns of species composition in relation to environment. Can. J. Fish. Aquat. Sci. 51 (7): 1600-1907.

Gupta PK, Pant MC (1983b). Seasonal variation in the energy content of benthic macroinvertebrates of Lake Nainital U.P, India. Hydrobiologia 99: 19-22.

Gupta PK, Pant MC (1983a). Macrobenthos of Lake Nainital (U.P, India) with particular reference to pollution. Water Air Soil Pollut. 19: 397-405.

Howick GL, Wilhm J (1984). Zooplankton and benthic macroinvertebrates in Lake Carl Blackwell. Proc. Okla. Acad. Sci. 64:63-65.

Heut M (1949). La pollution des eaux. L Analyse biologique des aeux polluees. Bull. Centr. Belg. Etude Doc. Eaux, 5 and 6: 1-31.

Ilyashuk BP (1999). Littoral oligochaete (Annelida: Oligochaeta) communities in neutral and acidic lakes in the Republic of Karelia, Russia. Boreal Environ. Res. 4: 277–284.

Jumppanen K (1976). Effects of waste waters on a lake ecosystem. Ann. Zool. Fennici. 13: 85-138.

Kaul V, Pandit AK (1981). Benthic communities as indicators of pollution with reference to wetland ecosystems of Kashmir. In: Proceedings of the WHO Workshop on Biological Indicators and Indices of Environmental Pollution. (A. R. Zafar, K.R. Khan, M.A. Khan and G. Seenayya, eds.). Cent. Bd. Prev. Cont. Water Poll./ Osmania University, Hyderabad, India. pp. 33-52.

Kaul V, Pandit AK (1982). Biotic factors and food chain structure in some typical wetlands of Kashmir. Pollut. Res. 1(1-2): 49-54.

Lang C (1985). Eutrophication of Lake Geneva indicated by the oligochaete communities of the profundal. Hydrobiologia 126: 237-243.

Latha C, Thanga VSG (2010). Macroinvertebrate diversity of Veli and Kadinamkulam lakes, South Kerala, India. J. Environ. Boil. 31: 543-547.

Liebmann H (1942). Die Bedeutung der mikroskopfschen untersuchung fur die biologische Wasseranalyse. Vom Wasser 15: 181-188.

Liebmann H (1951). Handbuch der Frishwasser und Abwasserbiologie. Munich Oldenbourg.

Milbrink G (1980). Oligochaete communities in population biology: The European situation with special reference to lakes in Scandinavia. In: Aquatic Oligochaeta Biology. (R.D. Brinkhurst and D.G. Cook, eds.). Plenum Press, N.Y and London. pp. 433-455.

Mir M F, Yousuf AR (2002). Distributional pattern of macrozoobenthic fauna of Dal Lake, Kashmir. pp. 32-33. In: National Seminar on Recent Research Trends in Life science (Kachroo, P. ed.). University of Kashmir, Srinagar.

Mir MF, Yousuf AR (2003). Oligochaete community of Dal lake, Kashmir. Orient. Sci. 8: 83-87.

Moore JW (1981). Factors influencing the species composition, distribution and abundance of benthic Invertebrates in the profundal zone of an eutrophic northern lake. Hydrobiologia 83: 505-510.

Nair NB, Dharmaraj K, Azis PKA, Arunachalam M, Kumar KK (1984). A study on the ecology of soft bottom benthic fauna in Kadinamkulam backwater, south west coast of India. Proc. Indian Natl. Sci. Acad. 5:473-482.

Oliver JH (1971). A study of biological communities in the Scioto river as indices of water quality. Ohio. Biol. Sury. and. Water Research Cance. Ohio State Univ, Columbus, Ohio. p.181

Othaka A, Nishino M (2006). Studies on the aquatic oligochaete fauna in Lake Biwa, central Japan. IV. Faunal characteristics in the attached lakes (Naiko). Limnology 7:129–1422.

Pandit AK (1980). Biotic factor and food chain structure in some typical wetlands of Kashmir. Ph.D. thesis, University of Kashmir, Srinagar-190006, J&K, India.

Pearson TH, Duncan G, Nuttal J (1986). Long term changes in benthic communities of Loch Linnhe and Loch Eil, Scotland. Hydrobiologia 142: 113-119.

Qadri H, Yousuf AR (2004). Ecology of macrozoobenthos in Nigeen lake. J. Res. Dev. 4:59-65.

Richardson RE (1921). Changes in the bottom and shore fauna of the middle Illinois river as a result of increase southward of sewage pollution. Bull. III. Nat. Hist. Surv. 14: 33-75.

Richardson RE (1929). The bottom fauna of the middle Illinois river. Bull. III. Nat. Hist. Surv. 17: 387-475.

Roy M, Nandi NC (2008). Macrozoobenthos of some lacustrine wetlands of West Bengal, India. The 12th World Lake Conference, 506-513.

Seather BA, McLean J (1972). A survey of the bottom fauna in wood Kalamalka and Shaha lakes in the Okanagan valley, British Columbia. Fish Res. Bd. Canada, Tech. Rpt. 342: 1-20.

Sharma KK, Chowdhary S (2011). Macroinvertebrate assemblages as biological indicators of pollution in a Central Himalayan River, Tawi (J&K). Int. J. Biodivers. Conserv. 3(5):167-174.

Singh AK, Ahmad SH (1989). A comparative study of the benthic fauna of lentic and lotic water bodies in around Patna Bihar, India. J. Environ. Biol. 10(3): 283-291.

Singh S (1988). Evaluation of macrobenthic community through species index in Kanti Oxbow lake, Muzzaffarpur, Bihar. J. Int. Fish. Soc. India 20(2): 30-34.

Siraj S, Yousuf AR, Bhat FA, Parveen M (2010). The ecology of macrozoobenthos in Shallabugh wetland of Kashmir Himalaya, India. J. Ecol. Nat. Environ. 2(5): 84-91.

Stewart TW, Downing JA (2008). Macroinvertebrate communities and environmental conditions in recently constructed wetlands. Wetlands 28(1): 141-150.

Thut R (1965). A study of the profundal bottom fauna of the Lake Washington, Seatle. M.S. Thesis, p.79.

Trayanova A, Moncheva S, Donchova V (2007). Macrozoobenthic communities as a tool for assessment of the ecological status of Varna lagoon. TWB,Transitt. Warters. Bull. 3: 33 36.

Wilhm JL, Dorris TC (1966). Species diversity of benthic macroinvertebrates in a stream receiving domestic and oil refinery effluents. Am. Midl. Nat.76: 427-449.

Yaqoob KU, Wani SA, Pandit AK (2007). A comparative study of macrobenthic community in Dal and Nilnag lakes of Kashmir Himalaya. J. Himalayan Ecol. Sustain. Dev. 2: 55-60.

Yongde C, Hongzhu W (2007). Ecology of macrozoobenthic communities in two Plateau lakes of Southwest China. Chin. J. Oceanol. Limnol. 26: 345-352.

Zajic JE (1971). Water Pollution, Disposal and Reuse. Vol.1. Marcel Dekker Inc. New York. pp 1-389.

20

Spatial relations of migratory birds and water quality management of Ramsagar reservoir, Datia, Madhya Pradesh, India

R.K. Garg[1] , R.J. Rao[2] and D.N. Saksena[2]

[1]Centre of Excellence in Biotechnology, M.P. Council of Science and Technology (MPCST), Vigyan Bhawan, Nehru Nagar, Bhopal-462003 (M.P.), India.
[2]School of Studies in Zoology, Jiwaji University, Gwalior-474011 (M.P.), India.

Ramsagar reservoir is constructed over Nichroli nallah, in the Basin of Sindh River in district Datia, Madhya Pradesh, India. The present study has been carried out from April, 2003 to March, 2005. Eighteen species of birds belonging to 5 different orders and 11 families were recorded. Out of these, family Ardeidae with 4 species was dominant followed by Charadridae, Anatidae, Rallidae, Palacrocoracidae, Sturnidae, Muscicapidae, Alcedinidae, Dacelonidae, Cerylidae and Meropidae. Coot (*Fulica atra,* Linnaeus) have been the most common and abundant species of family Rallidae in the reservoir. Reservoir is getting shallower each year due to silt deposition and accumulation of decomposed vegetation resulting winter birds visiting the site having decreased over the years. The shallow areas of the reservoir are facing the danger of eutrophication, which in turn may cause anoxic conditions thereby destroying the habitats for migratory birds forever.

Key words: Birds migration, diversity, aquatic environment, pollution, aquatic habitat, water quality management.

INTRODUCTION

Avifauna is an important constituent as well as an important link in the food chain of any ecosystem. Birds have been considered as useful biological indicators because they are ecologically versatile and inhabit all kinds of habitats (Jarvinen and Vaisanen, 1979; Sivaperuman and Jayson, 2006). The aquatic avifauna is quite susceptible to the changes in wetlands; similarly, they are more conspicuous in an ecosystem and hence can be easily observed for monitoring the changes taking place in them (Morrison, 1986). Some birds are migratory and are responsible for fluctuations in the population of birds that occur during different seasons of the year. This helps us to know whether the area is ecologically healthy or getting polluted, as total absence of birds from area may

be considered as pollution indication or human disturbance such as excessive hunting or human pressure (Borale et al., 1994). Submerged vegetation can attract more number of migratory birds in freshwater bodies (Sahu and Rout, 2005).

Various reports are available on the avifauna of Madhya Pradesh such as Sharma and Singh (1986) which studied wetlands birds in National Chambal Sanctuary. These studies have clearly showed seasonal fluctuations in avifauna along with their preferred localities. Sharma et al. (1995a) surveyed National Chambal Sanctuary in summer season and showed that no migratory birds were observed in the Sanctuary during this period. They also suggested that the migratory birds prefer muddy or stony

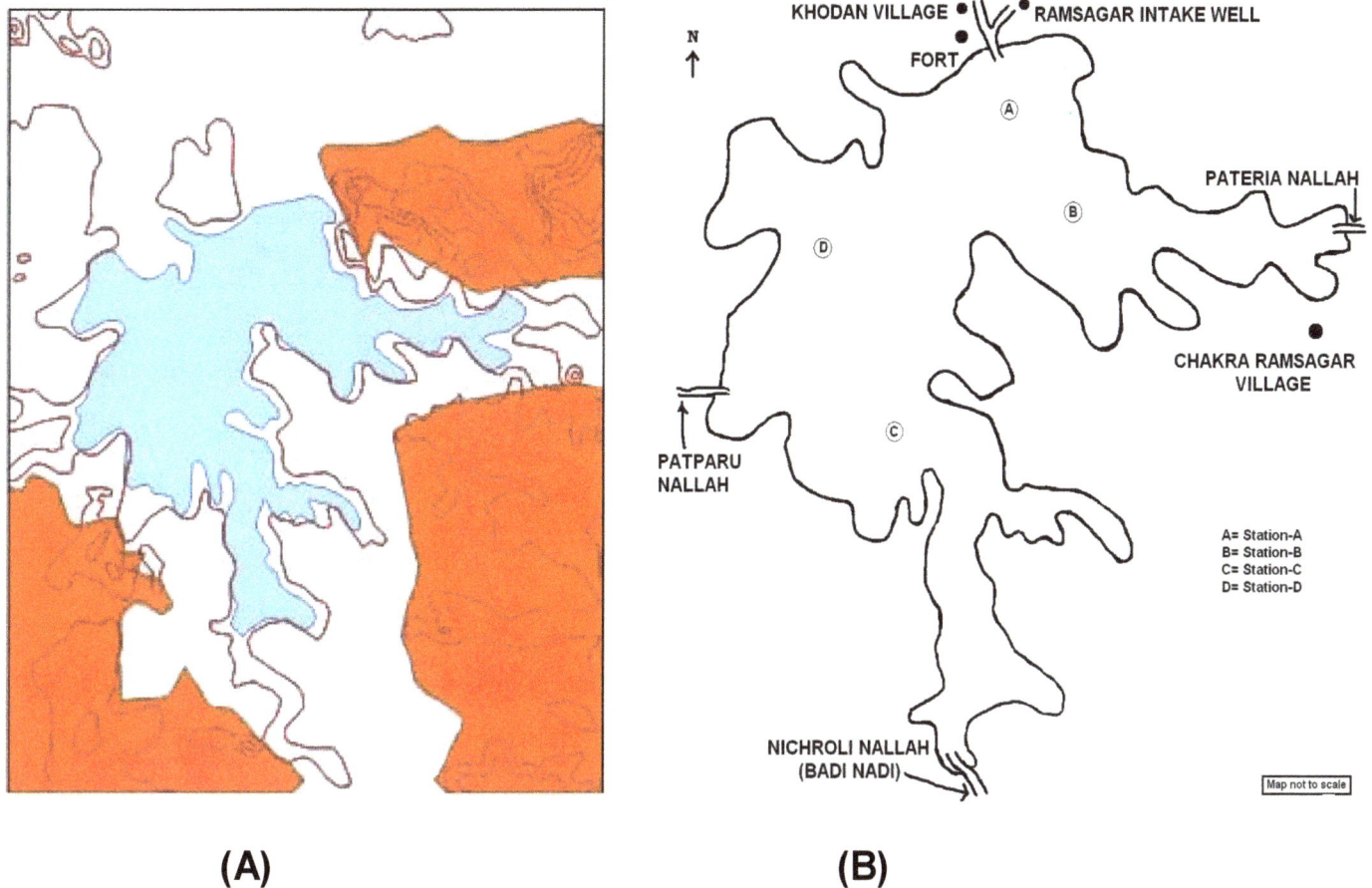

(A)   **(B)**

Figure 1. (A) Water spread area of Ramsagar reservoir (B) Study sites where birds were recorded.

habitat with well grown *Typha* grass. Faunal diversity of river Mahanadi, Madhya Pradesh includes 55 species of birds on the banks of this river.

In the present study, we attempted to investigate the status of migratory, local migratory and residential birds in the catchments area of Ramsagar reservoir, Datia Madhya Pradesh (India). The evaluation of avifauna of this reservoir may suggest habitat protection and may call for some measure of conservation.

MATERIALS AND METHODS

Description of study site

Ramsagar, a small man-made reservoir with 140.097 ha water spread area, was built over a Nichroli Nallah in the basin of Sindh river. The reservoir is located approximately 8 km northwest of Datia city in Madhya Pradesh and approximately 80 km south of Gwalior. Geographically, it lies between 25° 40' N latitude and 78° 23' E longitude and at an altitude of 229 m from mean sea level. Reservoir is used for different purposes like drinking water supply, irrigation, fisheries and thus is true a multipurpose tank. Four study sites namely: Station-A, B, C and D were established for counting of birds covering whole area of reservoir (Figure 1 A and B).

Detailed methodology

Birds were observed during winter season at most active period in the day that is, morning (6.00 am to 10.00 am) and late afternoon (4.30 pm to 7.00 pm) during April, 2003 to March, 2005. A direct visual count with binoculars was done and wherever possible an actual count was taken. Where there were a large number of birds, an estimate (up to the nearest 100 individuals) was made. Along with the water birds, other terrestrial birds sighted in that locality were also noted. Birds sighted during the study period were categorized according to their status as residents (R) (birds that have been known to breed in the study area and were encountered during every visit) and local migrants (LM) (birds which were encountered many times during the study period and breeding in neighboring areas). Some birds sighted occasionally during specific season and which were not residents of the area have been considered as migrant birds (M). The birds were identified following Grimmett et al. (1999), Ali and Ripley (2001).

RESULTS AND DISCUSSION

The family, taxonomic name, common name and status of aquatic and terrestrial avifauna in Ramsagar reservoir area are given in Table 1. In the present investigation, eighteen species of birds belonging to 5 orders and 11 families were recorded. Family Ardeidae with 4 species

Table 1. Taxonomic position of avifauna of Ramsagar reservoir, Datia, Madhya Pradesh.

Order	Family	Genus and Species	Common name	Status
Anseriformes	Anatidae	*Netta rufina* (Pallas)	Redcrested Pochard	M
		Nettapus coromandelianus (Gmelin)	Cotton Teal	R
Coraciiformes	Alcedinidae	*Alcedo atthis* (Linnaeus)	Small Blue Kingfisher	RM
	Dacelonidae	*Halcyon smyrnensis* (Linnaeus)	Whitebreasted Kingfisher	R
	Cerylidae	*Ceryle rudis* (Linnaeus)	Pied Kingfisher	R
	Meropidae	*Merops superciliosus* (Linnaeus)	Bluecheeked Bee-eater	R
Gruiformes	Rallidae	*Fulica atra* (Linnaeus)	Coot	RM
		Amaurornis phoenicurus (Pennant)	Whitebreasted Waterhen	R
	Phalacrocoracidae	*Phalacrocorax carbo* (Linnaeus)	Large Cormorant	RM
		Phalacrocorax niger (Vieillot)	Little Cormorant	RM
Ciconiiformes	Ardeidae	*Ardea cinerea* (Linnaeus)	Grey Heron	RM
		Ardeola grayii (Sykes)	Pond Heron	R
		Ardea alba (Linnaeus)	Large Egret	RM
		Egretta garzetta (Linnaeus)	Little Egret	R
	Charadridae	*Vanellus indicus* (Boddaert)	Redwattled Lapwing	R
		Himantopus himantopus (Linnaeus)	Blackwinged Stilt	R
Passeriformes	Sturnidae	*Sturnus contra* (Linnaeus)	Pied Myna	R
	Muscicapidae	*Saxicoloides fulicata* (Linnaeus)	Indian Robin	R
05	11	18		

Abbreviations: R = Resident, RM = Resident migrant, M = Migrant.

was relatively dominant (22.22%) followed by Charadridae (11.11%), Anatidae (11.11%), Rallidae (11.11%), Palacrocoracidae (11.11%), Sturnidae (5.56%), Muscicapidae (5.56%), Alcedinidae (5.56%), Dacelonidae (5.56%), Cerylidae (5.56%) and Meropidae (5.56%) (Figure 2). The most common and abundant species of Ardeidae were the Little Egret (*Egretta garzetta*) and Pond Heron (*Ardeola grayii*) followed by Grey Heron and Large Egret as these birds are heterogeneous in their feeding habits. Large Cormorant (10), Little Cormorant (40), Grey Heron (02), Pond Heron (05), Large Egret (02), Little Egret (10), Coot (100), Redwattled Lapwing (05), Pied Kingfisher (01), Small Blue Kingfisher (02), Pied Myna (02), Indian Robin (02) Blue Cheeked Bee-eater (07) and Blackwinged Stilt (02) were seen in the reservoir during 2003-2004 (Table 2). During 2004-2005 Large Cormorant (25), Little Cormorant (15), Grey Heron (04), Pond Heron (20), Large Egret (18), Little Egret (30), Redcrested Pochard (11), Cotton Teal (25), Coot (30), Redwattled Lapwing (04), Blackwinged Stilt (02), Pied Kingfisher (02), Small Blue Kingfisher (02) Whitebreasted Kingfisher (01) and Whitebreasted Waterhen (12) were seen in the reservoir (Table 2). The available fauna, *viz.,* crabs, snails, calms, worms, insect's

larvae and pupae in the surrounding area of water body constitute their feed.

Resident migrant birds (that is, birds that breed in one part of the area in one season and move to other parts within the state or country in a different season) such as Large Cormorant, Little Cormorant, Grey Heron, Large Egret and Coot were winter migrants in this region. Out of these Coots were seen in large numbers in winter season during both the years of study which reflected that coots prefers winter season for their food, breeding etc. It was also observed that the maximum bird species were recorded during spring season, early monsoon and late winter while comparatively less number of species were observed during late summer, late rainy and early winter season. In few cases hunting of migratory birds was also reported. Local people are neither aware of the importance of migratory birds nor the legislation. Awareness among the local population is needed so that hunting of birds should be prevented. From this study it is very clear that the Ramsagar reservoir is an ideal habitat for migratory and local migratory birds, especially the winter visitors. Fish and macrophytic resources of the Ramsagar reservoir are important sources of food for wetland birds.

The major threats to wetlands are expressed as percen-

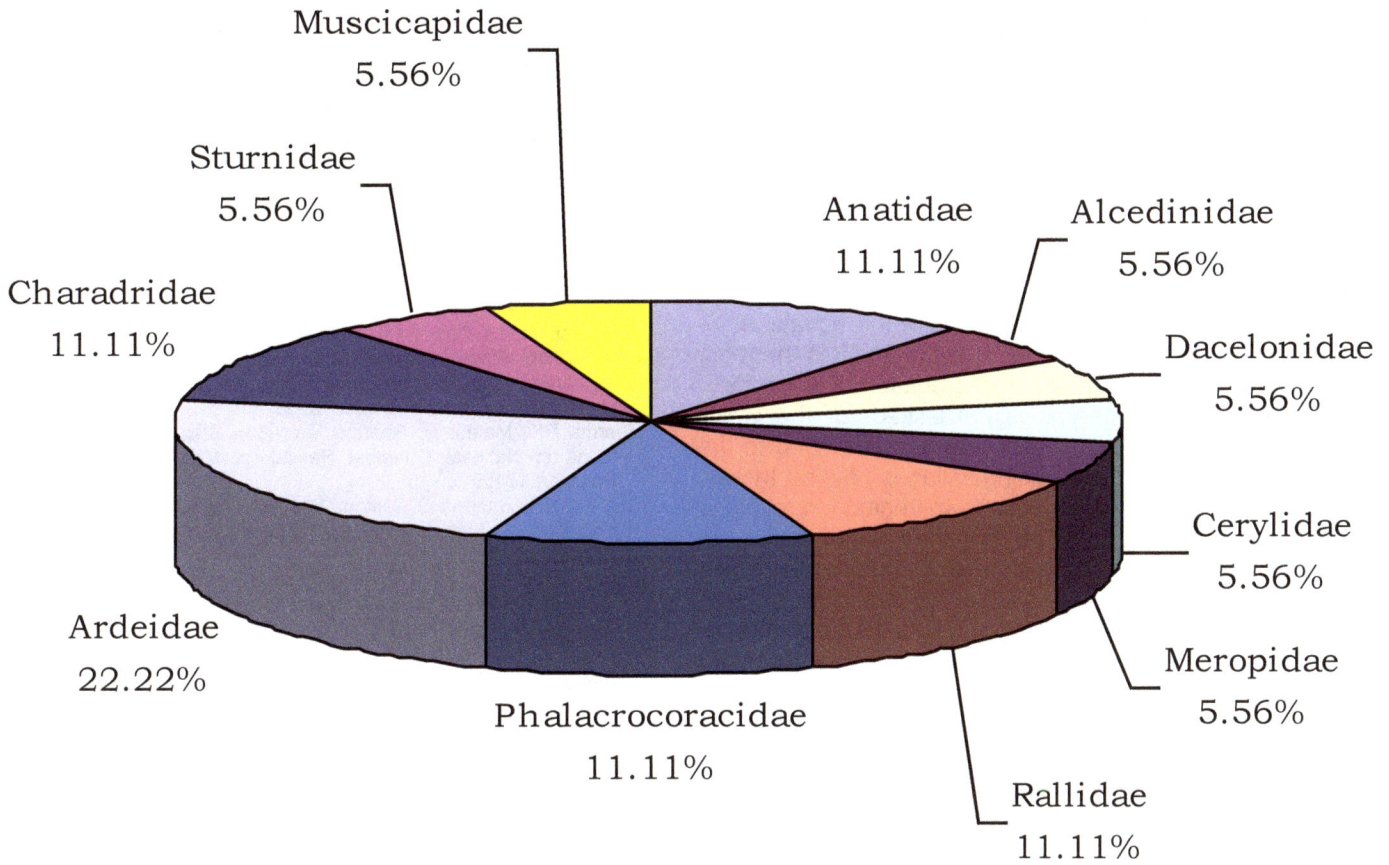

Figure 2. Percentage of contribution of different families of bird species in Ramsagar reservoir.

Table 2. List of avifauna of Ramsagar reservoir and their composition for two years.

Common name	Taxonomic name	Status	Number of individuals during 2003-2004	Number of individuals during 2004-2005
Blackwinged Stilt	*Himantopus himantopus* (Linnaeus)	R	02	01
Blue cheeked Bee-eater	*Merops superciliosus* (Linnaeus)	R	07	04
Coot	*Fulica atra* (Linnaeus)	RM	100	11
Cotton Teal	*Nettapus coromandelianus* (Gmelin)	R	-	25
Grey Heron	*Ardea cinerea* (Linnaeus)	RM	02	04
Indian Robin	*Saxicoloides fulicata* (Linnaeus)	R	02	02
Large Cormorant	*Phalacrocorax carbo* (Linnaeus)	RM	10	25
Large Egret	*Ardea alba* (Linnaeus)	RM	02	18
Little Cormorant	*Phalacrocorax niger* (Vieillot)	RM	40	15
Little Egret	*Egretta garzetta* (Linnaeus)	R	10	30
Pied Kingfisher	*Ceryle rudis* (Linnaeus)	R	01	30
Pied Myna	*Sturnus contra* (Linnaeus)	R	02	02
Pond Heron	*Ardeola grayii* (Sykes)	R	05	20
Redcrested Pochard	*Netta rufina* (Pallas)	M	-	12
Redwattled Lapwing	*Vanellus indicus* (Boddaert)	R	05	25
Small Blue Kingfisher	*Alcedo atthis* (Linnaeus)	RM	02	04
Whitebreasted Kingfisher	*Halcyon smyrnensis* (Linnaeus)	R	-	01
Whitebreasted Waterhen	*Amaurornis phoenicurus* (Pennant)	R	-	12

Abbreviations: R = Resident, RM = Resident migrant, M = Migrant

percentage of sites by the World Conservation Monitoring Centre (2002). The activities that contribute towards the loss of the resources are: hunting and allied activities, human settlement, drainage of agriculture, disturbance due to recreation, reclamation for urban and industrial development, pollution, catchment degradation, alteration of water, soil erosion and siltation. Wetlands in India are facing one or multiple of above mentioned factors. The lack of proper management and ignorance of the importance of healthy wetlands became evident. However, most accepted view is that the crisis of the aquatic environment is basically an economic issue and it is the most decisive factor playing a role in degradation of wetlands.

The Ramsagar reservoir is harnessed for the water supply to the Datia city. It also provides livelihood to the human population of adjoining villages. Major threats include diversion of water, overgrown vegetation resulting in increased grazing pressure and agriculture practices in the catchments area. The numbers of winter birds visiting the site have decreased over the years. It has been observed that the reservoir is getting shallower each year due to silt deposition and accumulation of decomposed vegetation. The shallow areas of the reservoir are facing the danger of eutrophication, which in turn may cause anoxic conditions thereby destroying the habitats for migratory birds forever.

ACKNOWLEDGEMENTS

The authors are thankful to University Grants Commission, New Delhi for financial support under SAP-DRS-I (No.F-03.07.2002 SAP-II) to School of Studies in Zoology, Jiwaji University, Gwalior (M.P.). We also extend our gratitude to the Head, School of Studies in Zoology for providing all necessary facilities for conducting this research work.

REFERENCES

Ali S, Ripley SD (2001). Hand Book of Birds of India and Pakistan. Oxford University Press, New Delhi, India.

Borale RP, Patil JV, Vyawahare PM (1994). Study of population of local migratory (Aquatic) birds observed in and around Dhule, Maharastra. Pavo., 32:81-86.

Grimmett R, Inskipp C, Inskipp T (1999). Birds of the Indian sub-continent. Oxford University Press, New Delhi, India.

Morrison ML (1986). Bird population as indicator of environmental changes In: Current Ornithology, Vol.-3. (Ed.: Johnston). Plenum Publishing Corporation, London.

Sahu HK, Raut SD (2005). Checklist of water Birds in Mayurbhanj District, Orissa. Zoos Print J. 20: 1992-1993.

Sharma RK, Singh LAK (1986). Wetland birds in National Chambal sanctuary. Crocodile research centre Wildlife Institute of India, Hyderabad, April, 1986. Mimeo: p. 36.

Sharma RK, Mathur R, Sharma S (1995). Status and distribution of fauna in National Chambal Sanctuary, Madhya Pradesh. Indian Forester, 121: 912-916.

Sharma RK, Sharma S, Mathur R (1995a). Faunistic survey of river Mahanadi vis-à-vis environment condition in Madhya Pradesh. Tiger Paper, 22:21-26.

Sivaperuman C, Jayson EA (2006). Status and conservation of bird fauna in the Vembanad-Lole Ramsar site, Kerala, India. In: Proc. Nat. Conf. on Wetland Biodiversity. February, 2-3, 2006: 31-37.

WCMC (2002). World Conservation Monitoring Centre, Global Biodiversity: Status of the Earth's Living Resources. Chapman and Hall, London.

21

Isolation of heterotrophic thiosulfate-oxidizing bacteria and their role in a conserved tidal flat in the Ariake Sea, Japan

Irfan Mustafa[1,2], Hiroto Ohta[1], Takuro Niidome[1] and Shigeru Morimura[1]

[1]Graduate School of Science and Technology, Kumamoto University, 860-0862, Japan.
[2]Biology Department, Faculty of Sciences, Brawijaya University, 65145, Indonesia.

Intolerable sulfide emission was spotted at several areas in tidal flats of the Ariake Sea, Japan. Sulfide is naturally produced in tidal flats and rapidly oxidized by sulfur-oxidizing bacteria (SOB). This makes them important players in controlling released sulfide by sulfate-reducing bacteria. A part of SOB can grow heterotrophically and we isolated them from a conserved muddy tidal flat in the Midorikawa Estuary, Kumamoto. The obtained heterotrophs oxidized sulfur compounds in presence of organic carbon. Various metabolic pathways were detected among them during oxidation of thiosulfate and an Isolate showed capability of sulfide oxidation. Phylogenetically, they were close-related to the genera *Paracoccus*, *Bacillus*, *Dyella*, and *Pseudomonas*. This suggested that the isolated SOB were affiliated to diverse classes and functioned diversely in oxidative side of the sulfur cycle. Additionally, population number of heterotrophic SOB was detected in high abundance, suggesting that they played a significant role in the sulfur cycle of the Midorikawa Tidal Flat.

Key words: Heterotrophic sulfur-oxidizing bacteria, isolation, tidal flat, thiosulfate, sulfide.

INTRODUCTION

Ariake sea is a semi-closed sea that has limited contact to the open ocean for exchange of materials and thereby is susceptible to environmental changes. In ecological term, the sea is dependent on the surrounding tidal flats to control nutrition that flow to the semi-closed sea. However, environmental deterioration of the tidal flats has been observed in recent decades (Du et al., 2008). Substantial decreases in populations of existing benthic animals due to sulfide emission have been reported (Moqsud et al., 2006). Carrion of death animals augmented the amount of existing organic matter, which caused extensive anoxia in the sediments. This worsened the condition as poisonous sulfide was more produced to massive quantities (Du et al., 2008).

Sulfide is normally emitted from the anoxic part of tidal flats near the surface layer of muddy sediments. The compound is excreted by sulfate-reducing bacteria (SRB) as a metabolic waste product from the anaerobic respiration of organic compounds with sulfate that is supplied by seawater during high tide (Muyzer and Stams, 2008). Large amount of sulfide is produced in the sediments, but only insignificant amounts are released to the atmosphere

(Bodenbender et al., 1999). Small amount of the sulfide reacts with existing metals and settle into the sediments, while a much larger proportion of that is removed by biological oxidation (Jorgensen, 1977). Sulfide is oxidized to elemental sulfur or sulfate by various sulfur-oxidizing bacteria (SOB) that exist in tidal flats. The bacteria are the key organisms which are responsible for the oxidation of inorganic sulfur compounds (Jorgensen and Nelson, 2004; Lenk et al., 2011).

The sulfur cycle of tidal flats in the Ariake Sea had been analyzed through interdisciplinary scientific approaches (Du et al., 2008; Azad et al., 2005). Early information about the bacteria that occupied conserved tidal flats was reported by Liem et al. (2014). They surveyed the diversity and relative abundance of bacterial populations based on 16S rRNA gene. Both SRB- and SOB-related clones are found as major groups indicating active sulfur cycle in the tidal flats. Such culture-independent approach is necessary for providing information on composition of potential sulfur bacteria. However, the role of SOB in oxidation of reduced sulfur compounds in the tidal flat remains unknown.

In the present study, we isolated heterotrophic SOB from a conserved Midorikawa Tidal Flat and observed their attributes in the oxidation of thiosulfate and sulfide. Thiosulfate was used for isolation in this study regarding its stability and abundance in tidal flats. The role of the heterotrophic SOB in the sulfur cycle was discussed.

MATERIALS AND METHODS

Sampling site description

The Midorikawa Tidal Flat is a part of the Midorikawa Estuary, which is located in the western part of Kumamoto Prefecture, Kyushu Island, Japan. At the estuary, fresh water from the Midorikawa River meets seawater from the semi-closed Ariake Sea. The flat becomes submerged by seawater twice a day with a tidal range of 3–4 m (Azad et al., 2005). During low tide, the muddy flat becomes exposed and a number of foraging migratory birds and many burrowing animals, such as mudskippers and crabs, can be observed.

Sampling and isolation procedures

The top 2-cm of sediment was sampled into a sterile 50 mL polypropylene Falcon conical tube (BD Bioscience, Durham, NC, USA) during low tide in June 2013. The sampling tube was kept on ice during trip to the laboratory and stored at 4°C before inoculation. Thiosulfate mineral (TM) liquid medium was used as enrichment medium to grow sulfur-oxidizing bacteria. TM medium was prepared as follows (per liter): 1 g NH_4Cl, 0.1 g $CaCl_2 \cdot 2H_2O$, 10 g NaCl, 0.5 g KH_2PO_4, 2 g K_2HPO_4, 0.8 g $MgSO_4 \cdot 7H_2O$, 3 g $Na_2S_2O_3 \cdot 5H_2O$ (Nacalai Tesque, Kyoto, Japan) in distilled water and 2 mL of 0.5% phenol red as pH indicator. The first three compounds were combined with 1 mL trace metal solution (Robertson and Kuenen, 1983) and autoclaved in 121°C for 20 min and the subsequent components were added after filter sterilization through a cellulose acetate membrane filter (0.2 μm). The final pH of the medium was adjusted to 7.3. A sediment sample of 3 g was transferred to 100 mL

TM medium in a 300-mL conical flask. The inoculated flask was plugged with sterile cotton and incubated aerobically at 30°C on a rotary shaker at 120 rpm. Incubation was stopped when the pH of the liquid culture began to drop below 7.0. The liquid culture was then subjected to serial dilution prior to spreading onto TM medium plates containing 1.5% agar and 0.02% yeast extract. The inoculated plates were incubated at 30°C for 7 days under aerobic condition. Morphologically distinct colonies were picked up and streaked onto TM medium plates and they were incubated under the same conditions.

Characterization on oxidizing thiosulfate

Obtained isolates were inoculated into 40 mL of modified TM mediums in 100 mL conical flasks and they were incubated aerobically at 30°C on a rotary shaker at 130 rpm for 10 days. Three modified TM mediums were prepared in combinations of the presence of thiosulfate and pyruvate as an organic carbon source. The first medium namely autotrophic medium was prepared in addition of 0.1% $NaHCO_3$ and absence of pyruvate. The second was mixotrophic medium containing both thiosulfate and 0.1% Na-pyruvate. The third medium was heterotrophic medium and it was prepared in presence of 0.1% Na-pyruvate and omission of thiosulfate. All of the modified mediums contained 0.02% yeast extract. The experiments were replicated two times.

Characterization on oxidizing sulfide

Glass tube was filled with two-layer agar medium and each of which occupied 10 ml. The lower layer contained 1% agar and 400 μl of 10% Na_2S while the upper one contained TM medium with 0.5% agar, 0.1% $NaHCO_3$ and 0.02% yeast extract. Thiosulfate was omitted from the medium leaving sulfide as a sole energy source. Na_2S solution was added in the tube immediately before the lower agar was poured. When the lower agar had turned to solid, 0.5 mL cell suspension in 1% salt solution was transferred on it. Then it was covered by the upper layer agar medium. The inoculated tubes were incubated aerobically at 30°C. The experiments were replicated two times and, as controls, an inoculated medium without Na_2S and a sterile medium with Na_2S were prepared.

PCR amplification of partial 16S rRNA gene

Single colony of purified isolate was picked up from the agar medium using sterile toothpick and transferred directly to the PCR reaction mixture as a template. The universal bacterial primer set of 27F (5'-AGAGTTTGATCCTGGCTCAG-3') and 518R (5'-GTATTACCGCGGCTGCTGG-3') and AmpliTaq Gold (Applied Biosystems, Carlsbad, CA) were used to amplify the partial 16S rRNA gene. PCR was performed using a T-Gradient Biometra Thermocycler (Biometra, Goettingen, Germany) under the following conditions: initial denaturation at 95°C for 5 min, followed by 25 cycles of denaturation at 95°C for 1 min, annealing at 50°C for 1 min, and extension at 72°C for 2 min. To confirm that amplicons were of correct size, the PCR reactions were subjected to electrophoresis on 1.5% (w/v) agarose gels in 1X TAE buffer, gels were stained with ethidium bromide, and reaction products were visualized by illumination with ultraviolet light at 302 nm. PCR products confirmed to be of the correct size were purified using the UltraClean PCR Clean-Up Kit (MoBio, Carlsbad, CA).

Sequence analysis

The purified PCR products of the partial 16S rRNA genes were sent

Table 1. Growth of five isolates under various conditions to determine ability in oxidizing thiosulfate.

Isolate code	Lithoautotrophic medium ($Na_2S_2O_3$ + 0.1% $NaHCO_3$)		Mixotrophic medium ($Na_2S_2O_3$ + 0.1% Na-pyruvate)		Heterotrophic medium (0.1% Na-pyruvate)	
	pH changes	Turbidity	pH changes	Turbidity	pH changes	Turbidity
MKH02	[a]-1.2	[b]-	-3.6	++	+1.0	++
MKH04	-0.3	+	-0.7	++	+0.1	+/-
MKH16	+0.4	-	+1.4	++	-0.1	-
MKH20	-0.2	-	-0.6	++	+1.2	++
MKH41	+0.8	-	+1.5	++	+0.8	++

[a]+ and – marks, respectively, indicate increase and decrease of the medium's pH after incubation;[b]-, no growth; +/-, poor growth; +, turbidity clearly apparent; ++, stronger turbidity.

to the TaKaRa company to be sequenced using the BigDye Terminator v3.1 Cycle Sequencing Kit on an ABI 3730xl DNA Analyzer (Applied Biosystems, Foster City, CA). The obtained sequences were uploaded and compared to the GenBank database using the web-based BLAST analysis tools at the National Center for Biotechnology Information (www.ncbi.nlm.nih.gov/BLAST/) to analyze their phylogenetic affiliations and identify their closest relatives.

Estimation of heterotrophic thiosulfate-oxidizers abundance in the environment

One-gram sediment sample was subjected to serial dilution in glass tubes containing 9 ml of 1% salt solution. The series included eight 10-fold dilution steps. From each of those, 1 ml suspension was transferred into 40 ml TM liquid medium in 100 ml conical flask containing 0.1% Na-pyruvate as carbon source and 0.02% yeast extract. The medium was prepared in duplicate and thus incubated aerobically at 30°C. The activities of thiosulfate oxidizers were detected as discoloration of the pH indicator.

RESULTS

Isolation of thiosulfate oxidizers

After 7-days incubation, bacterial colonies grew on the all of the agar plates up to 10^5-fold dilution. Prolonged incubation period allowed the colonies to thicken and expand slightly but did not change their numbers. Well-separated colonies were only observed on agar plates with 10^3- to 10^5-fold dilutions. From those plates, colonies were streaked onto fresh agar plates with the same composition. Twenty-five colonies survived on the agar plates. Most of them changed the medium color to yellow while others turn it to pink. As representatives, three acid producers (MKH02, MKH04, and MKH20) and two alkaline producers (MKH16 and MKH41) were chosen for characterization.

Characteristics of the isolates in thiosulfate oxidation

The five isolates distinctly responded to the three conditions of the mediums (Table 1). Isolates of MKH02 and MKH20

could grow in both mixotrophic and heterotrophic medium, but not in autotrophic medium. Another isolate, MKH04, was capable of inorganic carbon fixation and thus it could grow under all conditions as a facultative lithoautotroph. Those mentioned isolates lowered the pH of the mixotrophic medium. In contrast, the remaining two isolates MKH16 and MKH41 increased the pH of the medium during mixotrophic growth and they could not grow in medium lacking organic carbon. Moreover, both of them changed the pH of the mixotrophic medium in a very short time which was less than 12 h while other isolates took 3-8 days to establish growth and decrease the pH of the medium.

Characteristics of the isolates in oxidizing sulfide

Sulfide supplied in the lower agar had diffused to all part of the upper agar within 2 days. The diffusion was recognized by discoloration of the pH indicator to pink. Solution of Na_2S gives alkaline pH since the sulfide ion reacts with H^+ and leaves OH^-. White thin mat was then observed a few millimeters below the surface of upper agar of all isolates. After 10 days of incubation, the mat in the tube inoculated with MKH41 isolate was thicker than that in other tubes. We decided to subculture the mats into fresh agar medium with the same condition. Prolonged incubation of the subcultured isolates faded the mats but again thickened the mat of MKH41 isolate (Figure 1). In controls without addition of Na_2S showed no existence of mats in all isolates. Thin mats appeared in controls with uninoculated medium but they faded away after prolonged incubation. Additionally, discoloration of upper agar to yellow was not observed in all tubes.

Abundance of heterotrophic thiosulfate oxidizers in the environment

In previous analysis, all of the five isolates were capable of growth under simultaneous existence of both thiosulfate and pyruvate and changed the color of the medium during their growth. According to the mentioned characteristics,

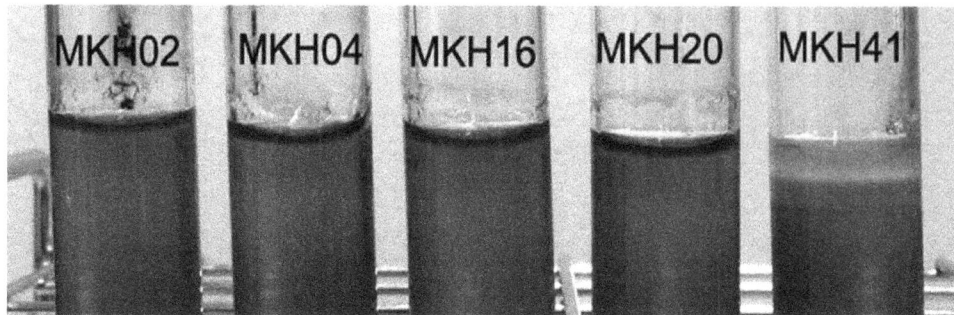

Figure 1. Growth of the heterotrophic SOB in Na₂S agar medium. The white band indicated the bacterial mat of MKH41 isolate that was capable of sulfide oxidation.

Table 2. Identification of heterotrophic thiosulfate-oxidizing bacteria isolated from the Midorikawa tidal flat.

Isolate code	Nearest species/GenBank (Accession No.)	Sequence identity (%)	Class of the nearest species	Origin of the nearest species
MKH02	*Paracoccus homiensis* (NR043733)	99	*Alphaproteobacteria*	Sea-sand sample from South Korea
MKH04	*Paracoccus limosus* (NR109093)	98	*Alphaproteobacteria*	Activated sludge from sewage treatment plant in South Korea
MKH16	*Bacillus jeotgali* (NR025060)	99	*Bacilli*	Korean traditional fermented seafood
MKH20	*Dyella ginsengisoli* (NR041370)	98	*Gammaproteobacteria*	Soil from ginseng field in South Korea
MKH41	*Pseudomonas xanthomarina* (NR041044)	99	*Gammaproteobacteria*	Coastal invertebrate *Halocynthia aurantium* from the Sea of Japan near Russia

we could estimate population number of heterotrophic thiosulfate oxidizers within the tidal flat sediment. The discolorations and turbidities were observed in all of the liquid mediums with dilution up to 10⁷-folds after two-week incubation. In all positive dilutions, changes of pH were early observed within 1-2 days of incubation and they all turned to alkaline. Prolonged incubation allowed them to lower the pH leaving the highest two dilutions in alkaline.

Homology analysis

The homology search of 16S rRNA gene showed that the isolates were distributed into three different classes, the *Alphaproteobacteria*, the *Gammaproteobacteria* and the *Bacilli*. Most of the isolates were phylogenetically related to marine microorganism strains already isolated from the regions surrounding Japan. Results of the homology search are presented in Table 2.

DISCUSSION

Characteristics of the isolates in oxidizing thiosulfate and sulfide

Various metabolic types were demonstrated by the isolates

regarding thiosulfate oxidation. Isolates of MKH02 and MKH20 grew equally in both heterotrophic and mixotrophic mediums. In those cases, pyruvate served as a sole carbon and energy source. Although the isolates grew under pyruvate, they were capable of oxidizing thiosulfate when it was present. Decrease of the pH of the medium was caused by released of sulfuric acid, the oxidation product of thiosulfate (Friedrich et al., 2001).

In contrast, MKH04 isolate showed growth when it was supplied with thiosulfate. Significant growth was observed in both mixotrophic and autotrophic medium. It indicated that the energy for growth was mainly obtained from oxidation of thiosulfate to sulfate. The isolate could fix inorganic carbon to support its assimilative metabolism when no organic carbon was supplied. In the present study, it was the only one that was capable of autotrophic growth and the growth was demonstrated after 8-day incubation. Coexistence of pyruvate in the medium supported the lithotrophic growth of MKH04 in oxidizing thiosulfate. It was recognized by pH reduction of the medium which was more intense in mixotrophic than in autotrophic growth. Known facultative lithoautotrophs, *Thiobacillus* sp. and *Burkholderia* sp., are reported to produce more biomass and sulfate during mixotrophic than autotrophic growth (Gottschal and Kuenen, 1980; Anandham, 2009).

In the sulfate producers, thiosulfate oxidation is commonly catalyzed by Sox enzyme system and the enzyme is known to be harbored in *Alphaproteobacteria* and *Gammaproteobacteria* (Friedrich et al., 2005). We checked the presence of *soxB* enzyme in the above-mentioned isolates, MKH02, MKH04 and MKH20, by using PCR-based detection. Each of them showed 260 bp amplification of partial *soxB* gene (data not shown) using a serial of degenerate primer sets as described previously (Petri et al., 2001). The positive amplification could suggest the presence of Sox enzyme system in the isolates. Hence the acid producers in this study were likely to use the Sox enzyme in oxidizing thiosulfate.

Another isolate, MKH16, showed no growth on both heterotrophic and lithoautotrophic condition. In contrast, the isolate grew very well under mixotrophic medium. Concomitant existence of organic carbon and thiosulfate appeared to be necessary for its growth. This strain showed a distinguished metabolic pathway in oxidizing thiosulfate by releasing alkaline compound. In alkaline producers, tetrathionate as well as OH$^-$ are released as the main products of thiosulfate oxidation catalyzed by tetrathionate synthase (Muyzer et al., 2013; Sorokin, 2003).

Rises of pH during the growth of MKH41 isolate was showed in all of the three conditions. However, only in autotrophic condition the turbidity was not detected. It was obvious that the growth of the isolate relied on the availability of organic carbon although thiosulfate oxidation was noticed in autotrophic medium. The pathway used in the oxidation of thiosulfate was likely similar with the MKH16 isolate that produced tetrathionate as the final oxidation product. In tetrathionate producers, oxidation of thiosulfate is not an energy-consuming reaction, but instead it releases two electrons for every tetrathionate formed (Podgorsek and Imhoff, 1999). Therefore, a slight increase of pH in autotrophic medium might be a result of thiosulfate oxidation by the inoculated cells but they were unable to grow due to inability of inorganic carbon fixation.

In sulfide oxidation test, MKH41 was the only isolate that survived the successive subculture suggesting that the isolate could oxidize sulfide aerobically. However, the isolate was known to be a heterotroph and no organic carbon was supplied to the medium. The growth might be supported by trace organic carbon existing in the gelling agent. Sulfide oxidation by known tetrathionate producers has been reported and the oxidation product was also end to tetrathionate (Sorokin, 2003).

In the present study, the white mat established under the medium surface might contain elemental-sulfur precipitate which was a product of chemical reaction between tetrathionate and the existing sulfide. Indirect sulfide oxidation by a heterotroph, *Catenococcus thiocyclus*, is observed during growth under an acetate-limited continuous culture. Elemental sulfur and tetrathionate are yielded in the culture (Podgorsek and Imhoff, 1999).

Homology analysis

In the present study, we obtained two isolates grouped in class *Gammaproteobacteria*, the MKH20 and MKH41. In the Midorikawa Tidal Flat, according to culture-independent study performed by Liem et al. (2014), *Gammaproteobacteria* are discovered to be the major group composing more than 20% of the total clones. The class is known to comprise a wide array of autotrophic and heterotrophic SOB (Teske et al., 2000; Meyer et al., 2007) and commonly occupies as the main group in brackish water sediment and coastal marine sediment (Wilms et al., 2006; Asami et al., 2005). The MKH20 and MKH41 were taxonomically related to *Dyella* and *Pseudomonas*, respectively. Several strains of *Dyella* have been isolated as thiosulfate oxidizer from coastal sediment in India (Krishnani et al., 2010) and as PAH degrader in Taiwanese mangrove sediment (Chang et al., 2008). Strains related to *Pseudomonas* reportedly oxidize thiosulfate to tetrathionate under heterotrophic conditions. The genus commonly occupies the interface sulfide-oxygen layer (Sorokin et al., 1999) and a strain was reported to be found in coastal invertebrate at Sea of Japan (Romanenko et al., 2005).

Two other isolates, MKH02 and MKH04 were closely related to genus *Paracoccus*. Such alphaproteobacterial genus is widely found in marine and estuarine sediment and known for its lithotrophy under thiosulfate (Teske et al., 2000; Roh et al., 2009). Alphaproteobacterial group is also known to contribute to the majority of bacterial population after the *Gammaproteobacteria* in Midorikawa Tidal Flat (Liem et al., 2014). As the closest genus for MKH16 isolate, some *Bacillus* strains, have been isolated from saline coastal soil and seawater above hydrothermal vents in South Korea and the North Atlantic, respectively (Siddikee et al., 2010; Rajasabapathy et al., 2014). They exhibit heterotrophic thiosulfate oxidation.

Role of the heterotrophic SOB in the sulfur cycle

In the sulfur cycle of Midorikawa Tidal Flat, MKH41 might oxidize both sulfide and thiosulfate to tetrathionate. The oxidation of thiosulfate in this experiment occurred in a very short time and that possibly occur in the sediment as well. In alkaline producers, oxidation rate of thiosulfate occurs in much faster time than that of sulfide (Sorokin, 2003). Tetrathionate which is released to the environment reacts immediately with the existing sulfide. Such chemical reaction yields thiosulfate and elemental sulfur (S^0). Then the thiosulfate is oxidized back to tetrathionate, hence generating a cycle that consumes sulfide and produces elemental sulfur. This indirect sulfide oxidation may occur in the tidal flat as long as tetrathionate is produced. The tetrathionate producers may compete with SRB that are capable of reduction and disproportionation of thiosulfate to sulfide (Sorokin et al., 1999).

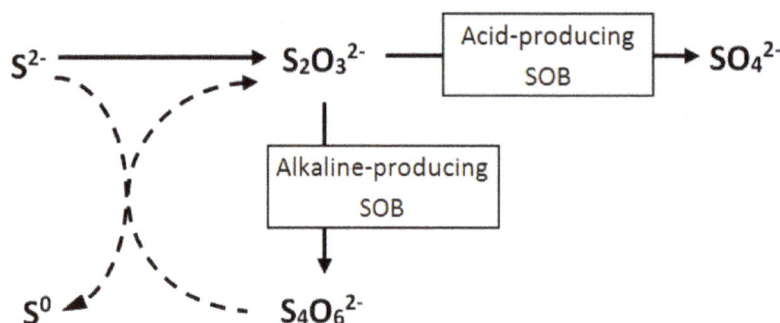

Figure 2. Estimated role of the isolated heterotrophic SOB in the sulfur cycle. Solid arrows indicated the biotic reaction while dashed arrows indicated abiotic reaction. Tangent lines showed that the spontaneous reactions are occurred interdependently.

On the other hand, thiosulfate is oxidizable by the acid-producing SOB. Thiosulfate which is the main product (60%) of sulfide oxidation in tidal flats is partly oxidized to sulfate (Jorgensen, 1990). Heterotrophic sulfate producers obtained from this study was likely to take part in catalyzing such reaction. The possible position of the heterotrophic SOB in the sulfur cycle is summarized in Figure 2. Additionally, populations of heterotrophic SOB in Midorikawa Tidal Flat were estimated in considerable number, up to 10^7 cells/gram of sediment, suggesting their significant role in the sulfur cycle. The high number of the SOB was possibly supported by the tidal flat condition which is rich in organic materials in addition of unlimited supply of oxygen.

Conclusions

Various heterotrophic SOB belonged to *Alphaproteobacteria*, *Gammaproteobacteria* and *Firmicutes* were isolated from Midorikawa Tidal Flat. Regarding the final product of thiosulfate oxidation, they were divided into sulfate producers and tetrathionate producers. The heterotrophs oxidized reduced sulfur compounds in presence of organic matter and hence integrated the sulfur cycle with the carbon cycle. Considering their high abundance, their roles could be significant in the conserved tidal flat.

Conflict of interests

The authors did not declare any conflict of interest.

REFERENCES

Anandham R, Indira Gandhi P, Kwon SW, Sa TM, Kim YK, Jee HJ (2009). Mixotrophic metabolism in *Burkholderia kururiensis* subsp. *thiooxydans* subsp. nov., a facultative chemolithoautotrophic thiosulfate oxidizing bacterium isolated from rhizosphere soil and proposal for classification of the type strain of *Burkholderia kururi*. Arch. Microbiol. 191:885-894.

Asami H, Aida M, Watanabe K (2005). Accelerated sulfur cycle in coastal marine sediment beneath areas of intensive shellfish aquaculture. Appl. Environ. Microbiol. 71:2925-2933.

Azad MAK, Ohira S, Oda M, Toda K (2005). On-site measurements of hydrogen sulfide and sulfur dioxide emissions from tidal flat sediments of Ariake Sea, Japan. Atmos. Environ. 39:6077-6087.

Bodenbender J, Wassmann R, Papen H, Rennenberg H (1999). Temporal and spatial variation of sulfur-gas-transfer between coastal marine sediments and the atmosphere. Atmos. Environ. 33:3487-3502.

Chang BV, Chang IT, Yuan SY (2008). Biodegradation of phenanthrene and pyrene from mangrove sediment in subtropical Taiwan. J. Environ. Sci. Health 43:233-238.

Du YJ, Liu SY, Hayashi S (2008). Experimental study on the deterioration and natural remediation of the Ariake Sea tidal mud caused by the sea laver treatment acid practice and the upward seepage of pore water liquid. Environ. Geol. 55:889-900.

Friedrich CG, Bardischewsky F, Rother D, Quentmeier A, Fischer J (2005). Prokaryotic sulfur oxidation. Curr. Opin. Microbiol. 8:253–259.

Friedrich CG, Rother D, Bardischewsky F, Quentmeier A, Fischer J (2001). Oxidation of reduced inorganic sulfur compounds by bacteria: Emergence of a common mechanism? Appl. Environ. Microbiol. 67:2873-2882.

Gottschal JC, Kuenen GJ (1980). Mixotrophic growth of *Thiobacillus* A2 on acetate and thiosulfate as growth limiting substrates in the chemostat. Arch. Microbiol. 126:33-42.

Jorgensen BB (1977). The sulfur cycle of a coastal marine sediment (Limfjorden, Denmark). Limnol. Oceanogr. 22:814-832.

Jorgensen BB (1990). A thiosulfate shunt in the sulfur cycle of marine sediments. Science 249:152-154.

Jorgensen BB, Nelson DC (2004). Sulfide oxidation in marine sediments: geochemistry meets microbiology, 63-81, in: Amend JP, Edwards KJ, Lyons TW. (Eds.), Sulfur biogeochemistry-past and present, The Geological Society of America-special paper 379, Colorado.

Krishnani KK, Gopikrishna G, Pillai SM, Gupta BP (2010). Abundance of sulphur-oxidizing bacteria in coastal aquaculture using *soxB* gene analyses. Aquac. Res. 41:1290-1301.

Lenk S, Arnds J, Zerjatke K, Musat N, Amann R, Mussmann M (2011). Novel groups of *Gammaproteobacteria* catalyse sulfur oxidation and carbon fixation in a coastal, intertidal sediment. Environ. Microbiol. 13:758-774.

Liem TT, Nakano M, Ohta H, Niidome T, Masuda T, Takikawa K, Morimura S (2014). Microbial community composition of two environmentally conserved estuaries in the Midorikawa River and Shirakawa River. Int. J. Biol. Sci. Eng. 05:28-34.

Meyer B, Imhoff JF, Kuever J (2007). Molecular analysis of the distribution and phylogeny of the *soxB* gene among sulfur-oxidizing bacteria - Evolution of the Sox sulfur oxidation enzyme system. Environ. Microbiol. 9:2957-2977.

Moqsud MA, Hayashi S, Du YJ, Suetsugu D (2006). Temporal variation of sulphide content, pH, salinity and oxidation-reduction potential of tidal flat mud in the Ariake Sea, Japan. Environ. Inform. Arch. 4:225-232.

Muyzer G, Kuenen JG, Robertson LA (2013). Colorless sulfur bacteria, In: Rosenberg E, DeLong EF, Lory S, Stackebrandt E, Thompson F. (Eds.), The Prokaryotes-prokaryotic physiology and biochemistry, Springer-Verlag, Heidelberg. pp. 555-588.

Muyzer G, Stams AJM (2008). The ecology and biotechnology of sulphate-reducing bacteria. Nat. Rev. Microbiol. 6:441-454.

Petri R, Podgorsek L, Imhoff JF (2001). Phylogeny and distribution of the soxB gene among thiosulfate-oxidizing bacteria. FEMS Microbiol. Lett. 197:171-178.

Podgorsek L, Imhoff JF (1996). Tetrathionate production by sulfur oxidizing bacteria and the role of tetrathionate in the sulfur cycle of Baltic Sea sediments. Aquat. Microb. Ecol. 17:255-265.

Rajasabapathy R, Mohandass C, Colaco A, Dastager SG, Santos RS, Meena RM (2014). Culturable bacterial phylogeny from a shallow water hydrothermal vent of Espalamaca (Faial , Azores) reveals a variety of novel taxa. Curr. Sci. 106(1):58-69.

Robertson LA, Kuenen JG (1983). *Thiosphaera pantotropha* gen. nov. sp. nov., a facultatively anaerobic, facultatively autotrophic sulphur bacterium. J. Gen. Microbiol. 129:2847-2855.

Roh SW, Nam Y Do, Chang HW, Kim KH, Kim MS, Shin KS, Yoon JH, Oh HM, Bae JW (2009). *Paracoccus aestuarii* sp. nov., isolated from tidal flat sediment. Int. J. Syst. Evol. Microbiol. 59:790-794.

Romanenko LA, Uchino M, Falsen E, Lysenko AM, Zhukova NV, Mikhailov VV (2005). *Pseudomonas xanthomarina* sp. nov., a novel bacterium isolated from marine ascidian. J. Gen. Appl. Microbiol. 51:65-71.

Siddikee MA, Chauhan PS, Anandham R, Han GH, Sa T (2010). Isolation, characterization, and use for plant growth promotion under salt stress, of ACC deaminase-producing halotolerant bacteria derived from coastal soil. J. Microbiol. Biotechnol. 20:1577–1584.

Sorokin DY (2003). Oxidation of inorganic sulfur compounds by obligately organotrophic bacteria. Microbiology 72:641-653.

Sorokin DY, Teske A, Robertson LA, Kuenen JG (1999). Anaerobic oxidation of thiosulfate to tetrathionate by obligately heterotrophic bacteria, belonging to the *Pseudomonas stutzeri* group. FEMS Microbiol. Ecol. 30:113-123.

Teske A, Brinkhoff T, Muyzer G, Moser DP, Rethmeier J, Jannasch HW (2000). Diversity of thiosulfate-oxidizing bacteria from marine sediments and hydrothermal vents. Appl. Environ. Microbiol. 66:3125-3133.

Wilms R, Sass H, Kopke B, Koster J, Cypionka H, Engelen B (2006). Specific bacterial, archaeal, and eukaryotic communities in tidal-flat sediments along a vertical profile of several meters. Appl. Environ. Microbiol. 72:2756-2764.

22

Threat reduction assessment approach to evaluate impacts of landscape level conservation in Nepal

Ram Prasad Lamsal[1] , Bikash Adhikari[1], Sanjay Nath Khanal[1] and Khet Raj Dahal[2]

[1]Department of Environmental Science and Engineering, Kathmandu University, Nepal.
[2]Department of Civil Engineering, Kantipur Engineering College, Tribhuvan University, Nepal.

Major challenges to the landscape level conservation intervention are to monitor and evaluate the conservation impacts in an accurate and cost-effective manner. Threat reduction assessment (TRA) has been proposed as a method to measure conservation success and as a proxy measurement of conservation impacts and monitoring threats. We conducted TRAs to evaluate the effectiveness of Nepal's Terai Arc Landscape (TAL) program in mitigating threats to forests of seven corridor and bottleneck sites. We modified Margoluis and Salafsky (2001) framework and scoring approach and calculated TRA index. Threats were standardized to allow comparisons across the sites and effectiveness of management modes in reducing threats between the community-based management (CBM) and conventional government managed system (GMS). TRA index of CBM was significantly higher from those of GMS as evident by various parametric and non-parametric tests including principal component analysis. However, the TRA approach is not immune to bias as it depends on subjective analysis, but it could be a simple and cost-effective conservation monitoring tool to be easily implemented by local communities and stakeholders.

Key words: Terai arc landscape (TAL), threat reduction assessment (TRA), community based management (CBM), government managed system (GMS).

INTRODUCTION

Nepal is exceptionally rich in biodiversity; however, it has experienced enormous challenges in biodiversity conservation particularly in the Terai region (Wagley and Ojha, 2002). Over time, a high proportion of the Terai forests have been modified by cutting, cultivation, burning, grazing and other anthropogenic actions (Chakraborty, 1999; FAO, 2009) and many of these forests have been significantly reduced in quality and quantity over time. The main threats to the Terai's biodiversity are forest encroachment and land use conversion, illegal logging, forest fire, wildlife poaching, uncontrolled grazing, comercial mining and invasive species (World Wildlife Fund, (WWF), 2004; National Planning Commission (NPC), 2010; Sapkota, 2009).

Nepal has experienced a series of policies and strategies for the management of forests and conservation of biodiversity (Multi-stakeholder Forestry Program (MSFP), 2013; NPC, 2013). Recently, the landscape-based conservation approach has been adopted as an opportunity to scale up conservation initiatives (WWF, 2004); and Terai arc landscape (TAL) programme, as the recent example, a very ambitious and long-term programme initiated to secure biodiversity conservation and sustainable development (NPC, 2010).

The TAL is part of an overall conservation strategy aimed at protecting the biodiversity both inside and outside protected areas. The various management interventions undertaken by the TAL program contribute to the emergence of a new agenda to improve the management and protection of species and ecosystems as well as people's livelihood (Baland and Platteau, 1996; Treves et al., 2005; Barbier and Burgess, 2001). Thus, search for common and efficient methodology or strategy for program improvement and change assessment is one of the priority concerns. Understanding of pressures and threats may form basis to design pragmatic regimes for the protection of biodiversity, assessment of performance and identify the changes (Haines-Young and Potschin, 2009).

Despite the challenge, complexity and time taking to determine the changes in conservation status of biodiversity, "biodiversity monitoring" and "biodiversity threat assessments" are the two main commonly used approaches currently in use to measure biodiversity impacts (GEF, 1998, 2008). To address the challenges faced in implementing biological indicator approaches to measuring conservation impacts and using results for decision making (Noss, 1999), scientists have responded to the need for practical and meaningful measures of conservation impacts by developing the TRA method (Margoluis and Salafsky, 1999; Lindner, 2012).

The TRA method is a low-cost and practical alternative to high cost and time-intensive approach (Lindner, 2012). This is a measurement tool that provides useful information at an acceptable cost and complements biological indicator approaches to measure conservation success. The TRA approach to measure conservation success is based on three key assumptions (Margoluis and Salafsky, 1998): a) All biodiversity destructions are human-induced; b) All threats to biodiversity at a given site can be identified and c) Changes in all threats can be measured or estimated.

The TRA method identifies threats, ranks them based on the criteria and assesses the progress in reducing them (Rome, 1999). The threats reduction can be evaluated using qualitative or quantitative measures and can serve a monitoring tool and alternative method of measuring conservation impacts (Margoluis and Salafsky, 1998; Rome, 1999). The TRA begins by following the procedural approach developed by IUCN (1998), Mugisha and Jacobsen (2003), Okot (2011), Margoluis and Salafsky (1999) which involves:

a) Defining the project area and listing all direct threats present at the site;
b) Ranking each threat based on 3 criteria: area, intensity and urgency (area refers to the percentage of the habitats in the site that the threat affects, intensity refers

to the impacts of the threat within the site and urgency refers to the immediacy of the threat). Out of total threats, the highest ranked threat for each criterion receives the highest score, and the lowest ranked threat receives the lowest score;
c) Adding up the scores across all three criteria for total ranking;
d) Determining the degree to which each threat has been met;
e) Calculating the raw score for each threat and multiplying the total ranking by the percentage calculated to get the raw score for each threat; and
f) Calculating the final threat reduction index score by adding up the raw scores for all threats, dividing by the sum of the total rankings, and multiplying by 100 to get the TRA index.

Landscape level conservation with CBM has been lauded as a better approach to manage different resource regime than conventional, top-down GMS. However, the CBM has been appreciated for its success to achieve conservation and livelihood goals (Roche, 2007; Aryal et al., 2012) and empirical data are already generated in providing its effectiveness. However, in Nepal, both the GMS and the CBM approach have been operating concurrently for a decade. This study evaluates and compares the ability of landscape level conservation to mitigate threats, at the two different management regimes of CBM and GMS, as a proxy measure of conservation success.

Objectives and hypothesis

This study firstly identifies pressures and threats to biodiversity in TAL and develop TRA index; secondly determines and compares the effectiveness of conservation interventions between CBM and GMS; and thirdly identifies the suitability of TRA method in monitoring and performance assessment at landscape conservation. Moreover, the study was designed to test two main hypotheses, which include: a) areas where CBMs are being implemented have reduced threats as compared to area of GMSs; and b) TRA method is appropriate for monitoring and measuring the performance and impacts.

METHODOLOGY

Field sites

TAL is a transboundary landscape between Nepal and India consisting of a total area of 23,199 km² in Nepal with forest area of 14000 km². Four corridors (Mohana-Laljhadi, Basanta, Khata and Barandavar) and three bottleneck areas (Mahadevpuri, Lamahi and Dovan) of TAL were selected for study. The seven intervention sites had a total of 341 community forests, 114 government and 56 civil society institutions, totaling 511, which were considered as the population (N). Field study was conducted in 2012 and 2013 by

Table 1. Population and sample of respondent institutions.

Sites	CFUGs		Government staffs		Civil Society groups		Total	
	N	n	N	n	N	n	N	n
Basanta	105	30	32	28	13	9	150	66
Khata	49	15	9	8	4	4	62	27
Mahadevpuri	30	8	9	8	6	4	45	20
Lamahi	55	13	23	16	11	12	89	39
Dovan	35	9	7	7	5	4	47	21
Mohana Laljhadhi	52	11	22	15	`8	7	82	36
Barandabhar	15	4	12	7	9	6	36	16
Total	341	90	114	89	56	46	511	225

N = population size; n = sample size; one for Lamahi is added from district headquarters.

selecting 225 representatives, one per institution, (n), with sampling error of 5% using Cochran's sample size formula for categorical data collection. The sample size of each site was determined as proportionate to the population size of the site. Site sample sizes were determined by using Equation 1:

$$n_h = \left(\frac{Nh}{N} \right) \times n \qquad (1)$$

Where n_h is the sample size for site h, N_h is the population size for site h, N is total population size, and n is total sample size.

The participants were divided into three groups: Community forest user groups (CFUGs), n = 90; Government staff, n = 89; and Civil society groups, n = 46) (Table 1). Civil society respondents were identified as forestry sector stakeholders comprising federations of community based forest management groups, NGOs, INGOs, political parties, user groups of other natural resource management and development groups, private sector, professional organizations, donors and indigenous leaders. All three groups belonged to the forestry sector working with rural communities.

Methods

Series of interviews and discussions elicited an array of perspectives and a large amount of information. Four sets of questions were given to the participants to understand threats as per their experiences and perceptions. Firstly, participants were given a list of possible risks to the forest and biodiversity and asked to respond by indicating their level of agreement or disagreement on a 5-point Likert scale starting from '1 = strongly disagree' to '5 = strongly agree'.

Secondly, they had to answer how worrisome they estimated each threat using the same Likert scale to their respective site based on the five principal risks for which they thought improved preventive and remedial measures are required. Thirdly, open questionnaire survey was supplemented by discussions and field visits about the risks perceived by respondent such as potentially damaging to forests and biodiversity.

Participants were asked to consider threats to habitat integrity, quality and ecosystem functioning while natural phenomena such as earthquakes were not considered threats. Participants ranked the threats based on the relative importance and their experiences. Ranking scales of 1 (minimum) to 5 (maximum) were used throughout the exercise and all threats were ranked along one continuum. Total sum score was computed after all the threats were ranked with score. The respondents were individually asked to award mark,

based on their evaluation of the extent to which management efforts had mitigated the threats. The scores for each threat were discussed to reach a consensus about a realistic score for the success of the management approach. After the scoring and ranking exercise, total ranking scores were multiplied by the percentage of the threat met to get a raw score for each threat. The TRA index was computed as (Equation 2) (Margoluis and Salafsky, 1999):

$$TRA\ index = \frac{Sum\ of\ raw\ score}{Sum\ of\ possible\ ranking} \times 100 \qquad (2)$$

Due to the proximity and topographical similarity between management modes CBM and GMS, it was possible to observe large differences in threat variables due to the social and management factors of the management categories of the forest area studied. Finally, the result obtained was presented and responses were received from field level government staff (N=37) regarding the assessment of TRA approach using the standard 5-point Likert scale: Strongly disagree = 1; disagree = 2; neutral = 3; agree = 4; and strongly agree = 5.

Variables

The independent variables, the presumed causes, in this study were the characteristics of respondents and types of forest management modes in relation to threat mitigation as listed in Table 2.

The dependent variables, the presumed effect of interest were the five priority threats which were assessed by using quantitative information as listed in Table 3 on both CBM and GMS.

RESULTS

Demographic characteristics

The sample largely mirrors the population and the respondents were well represented across the sites based on their size. Accordingly, site wise, highest number of 66 respondents, (29.33%) was from Basanta corridor, while lowest number of 20 respondents, (8.9%) was from Mahadevpuri bottleneck. Among the respondent categories, 90 respondents (40%) were community representatives, 89 respondents (39.6%) were government staffs and 46 respondents (20.4%) were from civil society.

Table 2. Independent variables.

Name	Type*	Explanation	Unit	Sources
Site name	N	Name of sites (1 to 7)	Number	Office record
Forest name	N	Name of forests	Number	Office record
Respondent groups	N	1= Community; 2= Government and 3= Civil society group	Number	Survey Design
Management modes	C	1= CBM (Community based management); 2= GMS (Government managed system)	Number	Office record

Table 3. Dependent variables.

Name	Variables	Type*	Unit	Sources
Different	Listing of threat variables	O	Likert scale	Survey design
CTRI	Threat reduction in CBM	C	Percent	
GTRI	Threat reduction in GMS	C	Percent	
CTR1	Encroachment and land use conversion in CBM	C	Percent	
CTR2	Poaching and trade in CBM	C	Percent	
CTR3	Forest fire in CBM	C	Percent	
CTR4	Commercial mining in CBM	C	Percent	Office records and field verification with
CTR5	Invasive species and grazing in CBM	C	Percent	map and questionnaire
GTR1	Encroachment and land use conversion in GMS	C	Percent	
GTR2	Poaching and trade in GMS	C	Percent	
GTR3	Forest fire in GMS	C	Percent	
GTR3	Commercial mining in GMS	C	Percent	
GTR5	Invasive species and grazing in GMS	C	Percent	

*N = Nominal; C = continuous, O = ordinal.

Age is an important factor that influences the working ability of the respondents. Results of analyses of data collected for this study reveal that the major age group of the respondents was of the 31 - 40 years age group (44.4%) followed by the 41 - 50 age group (28%), the 20 - 30 age group (18.1%) and the 51 - 60 years old group (9.3%).

Education, as a major component of empowering people and means of enhancing human capital varied among the respondents. In terms of the educational attainments, 36% of respondents had a capacity of simply to read and write; 38.2% of respondents attained school; 23.1% had a college degree and 2.7% had higher educations. Gender of respondents is considered as one of the variables influencing the perception on local forest resources, and in this study approximately 61% respondents were male followed by 39% of female respondents.

Patterning was also apparent in terms of respondents' socio-economic status. In terms of economic status, respondents indicated that they represented from high level (20%), medium level (56%) and lower level (20%). Social inclusion analyses showed that Brahmin and Chettri together added up 44% of the total participants followed by 28.4% indigenous group, 17.8% Madhesi and 9.8% Dalit community (Figure 1).

Threats in TAL

The threats were ranked based on value derived from Friedman test as a measure of non-parametric alternative to the one-way ANOVA with repeated measures to test for differences between groups when the dependent variable being measured is ordinal. The test statistics was found significant with χ^2_{23} = 1418.03 and p = 0.000. Out of a total of 24 threats, five primary and common threats to the biodiversity across the TAL area were identified as (a) encroachment and land use conversion, b) poaching and trade (timber, NTFP and wildlife), (c) forest fire, d) commercial mining and e) invasive species and grazing (Table 4).

Table 5 shows the Chi-square test result based on proportion of respondents identifying and agreeing on existing or potential severity of threats on their locations. In general, higher number of threats were found statistically significant (p<0.05) with the some site-wise differences in: a) all five primary threats in Dovan bottlenecks were not statistically significant (p>0.05); b) threats of invasive species and grazing in Khata(p=0.097) and poaching and trade in Mahadevpuri (p=0.247); encroachment (p= 0.056) and poaching and trade (p=0.113) in Barandavar were not significant. This reveals that the threats to biodiversity at a given site can be different depending on

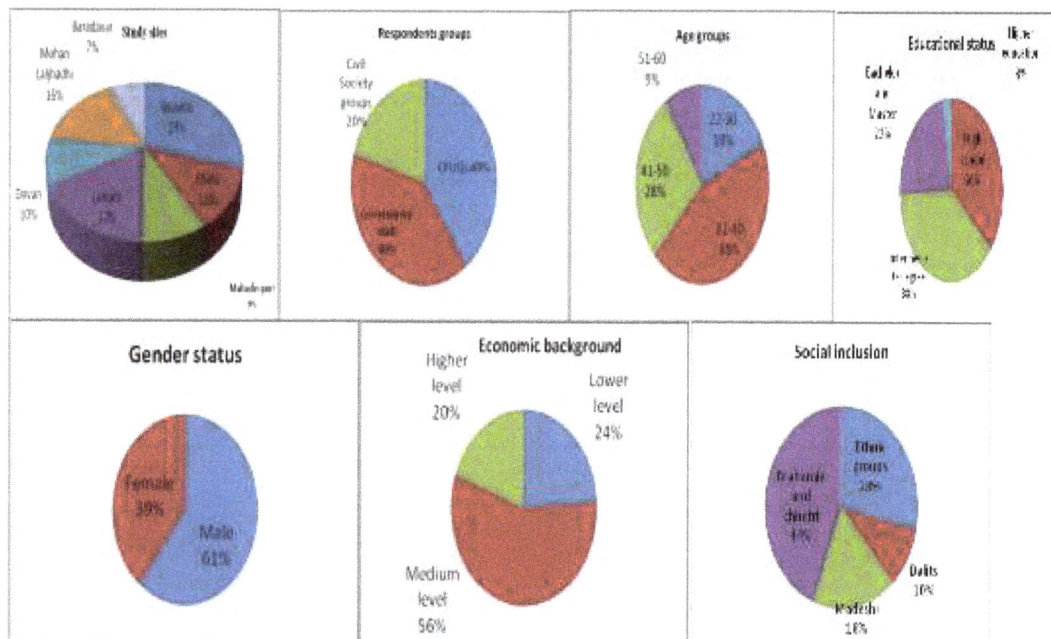

Figure 1. Demographic characteristics of the respondents (Source: field survey 2012 and 2013).

Table 4. Mean rank of threats based on Friedman test.

S/N	Threats	Mean Rank	S/N	Threats	Mean Rank
1	Encroachment and land use conversion	22.57	13	Land degradation and river cutting	10.78
2	Poaching and trade)	22.52	14	Charcoal burning	11.07
3	Forest fire	18.82	15	Poor management	12.41
4	Commercial mining	18.96	16	Lack of manpower and budget	11.34
5	Invasive species and grazing	18.95	17	Poor institutional capabilities	12.55
6	Unclear boundaries	11.58	18	Community rights denied	12.03
7	Highways and development projects	13.32	19	Bad community and staff relations	11.47
8	Human wildlife conflicts	11.52	20	Lack of awareness	12.49
9	Increased human population	13.16	21	Policy conflicts	11.18
10	Political interference	13.38	22	Illiteracy	12.44
11	Armed conflicts and insurgency	11.58	23	Poor law and order	10.51
12	Fuel-wood sell	13.24	24	Corruption and poor governance	12.11

Source: Field survey, 2012.

nature and magnitude of direct threats and indirect threats. Therefore, assessing how much the threat had changed at landscape level since project implementation also required support of experienced respondents on identification, quantification and interpretation of site level data which has been often challenging.

Reduction of primary threats

Twenty four threats were identified at the entire seven study sites. The most frequently reported common threats in all sites of both CBM and GMS were forest encroachment

and land use conversion followed by poaching; trade of timber, NTFP and wildlife; forest fire; commercial mining and non-human factors such as invasive species and livestock grazing.

Encroachment was a main reason of land use change in recent years that occurred in all study areas. However, the trend has been slowed or halted due to the landscape conservation intervention such as security of land tenure and access to resources for local people through CBM, strengthening protected area system and expansion of buffer zone. As shown in Table 7, this was the largest threats in terms of area, intensity, urgency and greatly reduced in CBM against GMS. The paired t test revealed

Table 5. χ^2 test result on site specific risk of primary threats.

Sites	Encroachment and land conversion			Poaching and trade			Forest fire			Commercial mining			Invasive species and grazing		
	χ^2	n	P	χ^2	n	p	χ^2	n	p	χ^2	n	p	χ^2	n	p
Basanta	31.55	44	0.000	22.06	40	0.000	17.58	38	0.000	15.25	37	0.000	21.16	40	0.000
Khata	9.56	16	0.008	16.22	18	0.000	6.89	14	0.000	6.89	14	0.032	4.667	14	0.097
Mahadevpuri	12.40	14	0.02	2.80	10	0.247	6.70	12	0.035	9.80	15	0.007	16.30	15	0.000
Lamahi	25.95	29	0.000	15.42	25	0.000	34.39	31	0.000	15.42	25	0.000	22.88	28	0.000
Dovan	1.60	8	0.45	5.20	10	0.074	4.90	9	0.086	0.10	7	0.951	0.10	7	0.951
Laljhadi	35.09	27	0.000	21.27	23	0.000	27.46	25	0.000	12.18	18	0.002	24.18	24	0.000
Barandavar	5.765	10	0.056	4.353	9	0.113	1.53	8	0.000	7.882	11	0.019	18.47	14	0.000

that the threat of encroachment has been found lower at CBM ($\bar{x}$ =37.26 ± 1.29) than GMS ($\bar{x}$ =25.33 ± 1.54) with difference of $\bar{x}$ =11.92 ± 1.88 (t_{224}=6.324; p =0.000) but it was still common in both.

CBM has reduced poaching including illegal logging and deforestation by creating local village level institutions. Local people conduct regular patrolling against illegal activities inside forest. The over extraction of flora and poaching of fauna diversity have been reduced (CBM, $\bar{x}$ =37.97 ± 1.05 against the GMS, $\bar{x}$ =18.04 ± 0.68) resulting in difference of $\bar{x}$ =19.92 ± 1.37 and t_{224}=14.55; p =0.000). Interventions were created to combat the threats posed by poaching. This initiative was comprised of processes which address the complex and sensitive issues at local, national levels and was implemented in cooperation with the major stakeholders.

The traditional approach of focusing on legislation alone was not sufficient; and involving local communities were crucial to manage forest fires. Access to forest ownership have encouraged local participation and community based practices resulting in reduction in damaging and unwanted forest fires that led to more effective fire prevention and suppression. Legal obligations in fire management by government agencies have not been successful while local communities themselves were unable to manage intense and large fires. Nevertheless, awareness programs and community based forest fire management activities have been assisted by this program to manage forest fires. Result shows that the reduction of threats on fire was significant in CBM ($\bar{x}$ =37.00 ± 1.04) when compared with GMS ($\bar{x}$ =18.11 ± 0.68) with the difference of 18.89% and was statistically significant ($\bar{x}$ =18.89 ± 1.33 with t_{224} = 14.13; p = 0.000).

Although collection, processing, transportation and trade of boulder, stone and sand have become a serious issue in biodiversity conservation, it has been reduced in CBM ($\bar{x}$ =41.05 ± 1.05) and in GMS ($\bar{x}$ =16.51 ± 0.73) (t_{224}=17.77; p=0.000). Active community participation have gradually managed open grazing and invasive species particularly *Mikania micrantha* which have been widespread

from east to west in Terai forests of Nepal which were significantly reduced in CBM ($\bar{x}$ =41.32 ± 1.04) as compared to GMS ($\bar{x}$ =17.75 ± 0.76) (t_{224}=17.16; p=0.000) (Table 6).

Threat reduction index

Threat reduction analysis conducted showed that at all levels of area, intensity and urgency, forest encroachment and land use conversion represents the largest threat with a total average rank value of 12.3, followed by poaching of timber and wildlife (rank value 9.49), forest fire (rank value 8.49), commercial mining (rank value 7.75), and invasive species and grazing (rank value 3.83). The extent of reducing threats differed between CBM and GMS. CBM illustrates reduction of threat with a range of 37.00 to 41.32%, whereas GMS shows the range between 13.51 to 25.3% depending on specific threats.

Raw factor (percent threat reduction/100) and raw score (raw factor/total rank value) were used to estimate TRI. The result showed CBM with a total TRI of 38.47 with 10.32% in encroachment and land use conversion, 8.36% in poaching and trade, 6.94 in forest fire, 7.23 in commercial mining and 5.63 in invasive species and grazing. However, the GMS only showed a total TRI of only 19.31 with 6.96% in forest encroachment and land use conversion, 3.96% in poaching and trade, 3.36 in forest fire, 2.80 in commercial mining and 2.33 in invasive species and grazing (Table 7).

The TRI at CBM showed that there was significantly higher threat reduction than conventional GMS (mean difference of 19.16 ± 1.238, t_{224}=15.74; p = 0.000). With reference to the overall performance of CBM and GMS, the ANOVA test revealed the difference at p=000 (CTRI, $F_{6,218}$ = 41.596; and GTRI, $F_{6,218}$ = 59.195)

Principal component analysis (PCA) on major threats

The results of the KMO measure of sampling adequacy

Table 6. t-Test on comparing threats between CBM and GMS.

Comparisons	Mean difference	SE	t value	Df	Sig (2 tailed)
CTR1 - GTR1	11.92	1.88	6.34	224	.000
CTR2 - GTR2	19.92	1.37	14.55	224	.000
CTR3 - GTR3	18.89	1.34	14.14	224	.000
CTR4 - GTR4	24.54	1.38	17.77	224	.000
CTR5 - GTR5	27.79	1.62	17.16	224	.000

Source: field survey, 2012.

Table 7. Threat reduction index.

Threats	Average value of threats*			RV	CBM				GMS			
	Area	Intensity	Urgency		PTR	RF	RS	TRI	PTR	RF	RS	TRI
Encroachment and land use conversion	4.35	3.99	4.21	12.55	37.26	0.37	4.64	10.32	25.33	0.25	3.13	6.96
Poaching and trade (timber. NTFP and wildlife)	3.45	3.02	3.43	9.9	37.97	0.38	3.76	8.36	18.04	0.18	1.78	3.96
Forest fire	2.9	3.07	2.47	8.44	37.00	0.37	3.12	6.94	18.11	0.18	1.51	3.36
Commercial mining	2.46	2.57	2.9	7.93	41.05	0.41	3.25	7.23	16.51	0.16	1.26	2.80
Invasive species and grazing	1.84	2.35	1.99	6.18	41.32	0.41	2.53	5.63	17.75	0.17	1.05	2.33
Total	15	15	15	45			17.31	38.47				19.40

*Measured in scale (1 to 5): Vey low, low and medium; RV = rank value = area + intensity + urgency ; PTR= percent threat reduction; RF = raw factor = PTR/100; RS= raw score = RF/total rank value; TRI = threat reduction index= RS/corresponding individual RV.

Table 8. Rotated component matrix.

	Components	
	1	2
Eigen value	4.27	3.14
Variance explained	42.7	31.4
GTR3	0.969	
GTR5	0.924	
GTR2	0.910	
GTR4	0.861	
GTR1	0.604	
CTR2		0.880
CTR3		0.873
CTR5		0.841
CTR1		0.829
CTR4		0.778

Extraction Method: principal component analysis; rotation method: Varimax with Kaiser normalization; a. rotation converged in 3 iterations.

values of 1.0 or higher. These two dimensions, explained 74% of the variance. The two underlying dimensions were labeled as follows: 1. Threats on GMS; and 2. Threats on CBM. In addition, reliability was performed on each of the two factors, based on the assessment items retained in each dimension.

Factor one, which is identified as GMS threats explained 42.70% of the variance with an Eigen value of 4.27 and a reliability coefficient of 0.83. Factor two, which is labeled as threats on CBM, explained 31.3% of the variance with an Eigen value of 3.13 and a reliability coefficient of 0.78. In the rotated factors, GTR1 to GTR5 all have high positive loadings on the first factor (and low loadings on the second), whereas CTR1 to CTR5 all have high positive loadings on the second factor (and low loadings on the first).

Factor loading from GMS ranged between 0.969 and 0.604. Forest fire (0.969), invasive species and grazing (0.924), poaching and trade (0.910), commercial mining (0.861) and encroachment (0.604) were of great importance in the settlement of factor 1 of GMS. Similarly, factor loading from CBM ranged between 0.880 and 0.778. Poaching and trade (0.880), forest fire (0.873), invasive species and grazing (0.841), encroachment (0.829) and commercial mining (0.778) outstandingly contributed to the formation of factor 2 in CBM.

Analysis of additional threats

Nineteen additional threats were identified as the threats

revealed 0.791 and Bartlett's Test of Sphericity revealed a significance at a level of 0.000 (χ^2 =2049.96, df=45). Thus, the variables must be related to each other for the factor analysis to be appropriate. In order to examine underlying dimensions of the threat reduction, a factor analysis with a varimax rotation was performed. The results are presented in Table 8 with the factor at the level of 0.50 (or higher). Two factors emerged with Eigen

Table 9. Comparing means of threats using McNemar test (df =1).

Additional threats	NF		CF		McNemar χ^2_1	p
	Yes	No	Yes	No		
Armed conflicts and insurgency	158	67	131	94	20.7	0.000
Bad community and staff relations	73	152	55	170	27.40	0.000
Charcoal burning	67	158	33	192	36	0.000
Poor law and order	128	97	130	95	4.0	0.046
Corruptions and poor governance	96	129	110	115	4.55	0.033
Fuelwood sale	137	88	101	124	11.01	0.000
Community rights restricted	74	151	96	129	25.671	0.000
Development projects	155	70	171	54	31.36	0.000
Human wildlife conflicts	159	66	161	64	37.16	0.000
Illiteracy	152	73	154	71	27.04	0.000
Increased population	145	80	122	123	8.73	0.003
Lack of awareness	144	81	126	99	9.78	0.002
Lack of manpower and budget	128	97	114	111	1.37	0.242
Land degradation and river cutting	152	73	133	92	17.47	0.000
Policy conflicts	152	73	119	106	11,02	0.001
Political interferences	159	66	134	91	23.12	0.000
Poor management	102	123	113	112	0.42	0.520
Unclear boundaries	163	62	141	84	30.74	0.000
Poor institutional capabilities	127	98	101	124	0.045	0.830

to sustainable management of resource. Comparison between CBM and GMS indicates significant differences in mitigation of additional threats. The specific threats identified and mitigated at different areas, however, offer a deeper understanding of conservation effectiveness. Closed questions with 3 options - yes, no, do not know were analyzed applying McNemar Chi Square test where "do not know "was taken closer to "no" and recoded as same variable and yes as the other. A p value of < 0.05 was taken as significant. The responses were compared between CBM and GMS and statistically significant threats as indicated by McNemar test (Table 9).

Statistically significant threats with p<0.05 included: armed conflicts and insurgency; b) bad community and staff relations; c) community rights restricted c) development projects; d) human wildlife conflicts; e) illiteracy; f) increased population; g) lack of awareness; h) land degradation and river cutting; i) policy conflicts; j) political interferences and k) unclear boundaries. Similarly, significant threats at marginal level were: a) poor law and order; b) corruptions and poor governance. However, statistically not significant threats at p>0.05 were: a) lack of manpower and budget (p=0.242); b) poor management (p=0.52) and c) poor institutional capabilities (0.83).

Assessment of TRA method

Reliability analysis was undertaken in order to understand whether the questions in this questionnaire all reliably measure the same latent variable (perception towards TRA), a Cronbach's alpha was run on a sample size of 37 respondents and the value 0.801 which indicated a high level of internal consistency within the given scale was found. One sample median test showed the mixed results of the 10 response questions on assessment of TRA. The test with reference to value 2.5 and 50% cut point revealed a significant difference toward positive conclusion on its simplicity to use, easy to understand, useful, cost effecttiveness and replicable with p = 0.000 and not positive conclusion on its accuracy (p = 0.324); training requirement (p=0.099); and comparatively better (p = 0.099) (Table 10).

Conclusion

In general, TRA acts as useful tool for monitoring and evaluating conservation interventions, with specific weakness as it indirectly measures threats in biodiversity conservation. Despite the merits, biases could have occurred in the process of selecting the sites and respondents to participate in the survey and discussion. The results could be subjective and the scores for management performance may not be directly linked to specific intervention on biodiversity conservation.

The assessment highlighted that the potential for involving communities in monitoring trends in biodiversity should be integrated with biodiversity conservation. The results provided a current snapshot of the variety and severity of threats throughout the TAL conservation system. It involved key stakeholders in identifying threats

Table 10. One sample median test on effectiveness of TRA method.

	OP of category			+/ -		OP of category			+/ -
	<2.5	> 2.5	p			<2.5	> 2.5	p	
Simple to use	0	1	.000	+	No training required	0.65	0.35	.099	-
Easy to understand	0	1	.000	+	Creates baseline	0.08	0. 92	.000	+
Useful	0	1	.000	+	Replicable	0	1	.000	+
Cost-effective	0	1	.000	+	Apt for all scales	0.11	0.89	.000	+
Accurate	0.59	0.41	.324	-	Comparatively better	0.35	0.65	.000	+

OP= Observed proportion; test proportion=50%; p = 0.000 for all; + = positive and - = negative weight.

and prioritizing problems from a multidisciplinary perspective and found that TRA approach could be used in TAL as a tool of monitoring and assessing impacts of conservation based on its scope and limitations.

In conclusion, the study findings indicated that the overall current management approaches under TAL fall short of addressing threats. Nevertheless, a trend in the data suggested that threats have been better and significantly mitigated at CBM as compared to GMS, indicating the CBM as a potentially more successful approach to conservation than the traditional top-down approach. It can therefore be concluded that CBM has performed better, as an approach to landscape conservation than the traditional top-down GMS. However, both approaches have not addressed all the threats which is expected.

Conflict of interests

The authors did not declare any conflict of interest.

REFERENCES

Aryal A, Brunton D, Shrestha TK, Koirala RK, Lord J, Thapa YB, Adhikari B, Ji W, Raubenheimer D (2012). Biological diversity and management regimes of the northern Barandabhar Forest Corridor: an essential habitat for ecological connectivity in Nepal. Trop. Conserv. Sci. 5(1):38-49.
Baland J, Platteau J (1996). Population Pressure and Management of Natural Resources: Income-sharing and Labor Absorption in Small scale Fisheries. FAO Economic and Social Development Paper. No. 139. Rome, Italy: Food and Agricultural Organization of the United Nations.
Barbier BE, Burgess JC (2001). The economics of tropical deforestation. J. Econ. Surv. 15(3):413-431.
Chakraborty RN (1999). Stability and outcomes of common property institutions in forestry: evidence from the Terai region of Nepal. Ecol. Econ. 36 (2001):341-353
FAO (2009). Nepal Forestry Outlook Study. Ministry of Forests and Soil Conservation Singha Durbar, Kathmandu, Nepal. Working Paper No. APFSOS II/WP/2009/05
GEF (1998). Guidelines for Monitoring and Evaluation for Biodiversity Projects. Global Environment Division, GEF/World Bank.
GEF (2008). GEF Impact Evaluation for GEF Protected Area Projects in East Africa. Impact Evaluation Information Document No. 12 September 2008. [Online] https://www.thegef.org/gef/sites/thegef.org/files/documents/Impact_Eval_Infodoc12 (Accessed 12 February 2013).

Haines-Young R, Potschin M (2009). Methodologies for defining and assessing ecosystem services. A research study. Centre for Environmental Management University of Nottingham, Nottingham.
IUCN (1998). Evaluating Effectiveness: A Framework for Assessing the Management of Protected Areas. Adrian Phillips, Series Editor, Best Practice Protected Area Guidelines Series No. 6, IUCN.
Lindner R (2012). Evaluating evaluation: Exploring evaluation methods to assist WWF-UK programme management. A thesis submitted in partial fulfillment of the requirements for the degree of Master of Science and the Diploma of Imperial College.
Margoulis R, Salafsky N (1999.). A Guide to threat reduction for conservation, Biodiversity Support Program, Washington, DC. [Online] www.worldwildlife.org/bsp/.../threat/tra.pdf [Accessed: 3 March 2012]
Margoulis R, Salafsky N (1998). Measures of success: Designing, managing, and monitoring conservation and development projects. Washington, D.C.: Island Press.
Margoulis R, Salafsky N (2001). Is Our Project Succeeding? A Guide to Threat Reduction Assessment for Conservation. Washington, DC, USA: Biodiversity Support Program.
MSFP (2013). National Forestry Sector Strategy Development (concept note), Multi-stakeholder Forestry Program, Nepal
Mugisha A, Jacobsen SK (2003). Threat reduction assessment of conventional and community-based conservation approaches to managing protected areas in Uganda. Department of Wildlife Ecology and Conservation, Program for Studies in Tropical Conservation, University of Florida.
Noss R (1999). Assessing and monitoring forest biodiversity: A suggested framework and indicators. For. Ecol. Manage. 115:131-146
NPC (2010). Three Years Periodic Plan, Nepal
NPC (2013). Three Year Periodic Plan for 2013-2015, National Planning Commission, Government of Nepal
Okot JE (2011). An Ecological Assessment of Biodiversity Pressure in Mgahinga Gorilla National Park South Western Uganda. An unpublished MSc Thesis, University of Klagenfurt, Austria.
Roche R (2007). Livelihoods Approaches as a Conservation tools. University of Rhode Island.
Rome A (1999). Ecotourism Impact Monitoring: A Review of Methodologies and Recommendations for Developing Monitoring Programs in Latin America Ecotourism Technical Report Series 1, Alex C. Walker Foundation, The Nature Conservancy.
Sapkota IP (2009). Species diversity, regeneration and early growth of sal regeneration in Nepal: Responses to Inherent disturbance regimes, Doctoral Thesis, University of Agricultural Science, Alnarp.
Treves LN, Holland MB, Brandon K (2005). The Role of Protected Areas in Conserving Biodiversity and Sustaining Local Livelihoods. Annu. Rev. Environ. Resour. 30:219-52
Wagley M, Ojha H (2002) Analyzing Participatory Trends in Nepal's Community Forestry. Policy Trend Report 2002:122-142
WWF (2004). Terai Arc Landscape Strategy Plan (2004-2104), Ministry of Forests and Soil Conservation, Nepal.

23

Biofuel potential and land availability: The case of Rufiji District, Tanzania

Simon L. A. Mwansasu[1,2] and Lars Ove Westerberg[2,3]

[1]Institute of Resource Assessment, University of Dar es Salaam, P O Box 35097, Dar es Salaam, Tanzania.
[2]Department of Physical Geography and Quaternary Geology, Stockholm University, SE-106 91 Stockholm, Sweden.
[3]Bolin Centre for Climate Research, Stockholm University, S-10691 Stockholm, Sweden.

Africa's attractiveness to potential biofuel investors is based on the assumption that there is plenty of unused land available for investment in different countries of the continent. However, their postulations are not based on any concrete studies on land available at country, regional or local level. This study investigates land availability for potential biofuel investment at the local level, using Rufiji district in Tanzania as a case study. We have analyzed different land cover/land use types and separated them into areas of potential biofuel investment and areas where biofuel investment is not possible by a process of elimination. The results suggest that land available is inadequate to meet the needs of biofuel investors. The land assumed to be unused or underutilized by biofuel investors is either part of the fallow system or used to harvest natural resources and for other traditional uses. Expropriating the assumed idle land will have impact on the livelihoods of the local communities.

Key words: Biofuel investment, land available, Rufiji District.

INTRODUCTION

The alleged existence of abundant underutilized land in Africa has attracted biofuel investors from wealthy countries to the continent (Cotula et al., 2009; Madoffe et al., 2009). The assertion is part of a long held dogma, where African lands are perceived to be unoccupied and therefore in need of investments (Neville and Dauvergne, 2012). However, there is a huge difference between those assertions and the appraisal of land available for biofuel production according to the International Energy Agency (Haugen, 2010). The discrepancy between the assertions of the potential biofuel investors and the assessment by the International Energy Agency can be

attributed to little research on land availability in Africa, and emphasizes the need for more research and more high quality data (Cotula et al., 2009; Ahlberg, 2011). Nevertheless, there are recent studies estimating land availability for biofuel production at the global level using both coarse resolution remote sensing data (Cai et al., 2011) and high resolution remote sensingdata (Fritz et al., 2013). Using high resolution remote sensing data, Fritz et al. (2013) substantially lowered the amount of estimated land available for biofuel production. Yet, the remote sensing studies have neither considered land availability at country, regional or local level, nor have they considered other activities that might be competing for land apart from biofuel production.

Sulle and Nelson (2009:7) define biofuels as "liquid, solid or gaseous fuels that are predominantly or exclusively produced from biomass". In general, biofuels, such asbiodiesel, ethanol and biogas are derived from crops, plant residues or garbage. The acquisition of land for biofuel and biodiesel production has increased worldwide, in particular in Sub-Saharan Africa (Havnevik, 2009), where the acquisition has been received with mixed feelings. Some construe the biofuel sector as important in revolutionizing agriculture and alleviating poverty. Others are afraid that the biofuel sector will inevitably lead to harmful land use changes once the land is converted to estate agriculture (Martin et al., 2009; Schoneveld et al., 2011).

Biofuel production is also evolving as a critical policy matter in agricultural development and natural resources management (Sosovele, 2010). In most African countries, policy institutions are passive in decision making (Romijn and Caniëls, 2011), and at the advent of biofuel investment in Africa, most countries did not have policies in place to monitor and control biofuel investments. As a consequence, national and local government agencies were trapped in a confusing role between defending interests of the local people and those of the biofuel investors (Cotula et al., 2009). Moreover, in dealing with biofuel investors, local communities are at a disadvantage in protecting their interests, because they do not comprehend the full effects of biofuel investments (Beyene et al., 2013).

There is a belief that Tanzania will reap the benefits of biofuel investments in terms of capital, expertise and knowledge transfer (Kweka, 2012). In addition, it is hoped that biofuels will lessen the economic burden of importing petroleum, thus improving environmental conservation and livelihoods (Martin et al., 2009). The optimism in the benefits of biofuel investments has resulted in Tanzania becoming a major destination for potential biofuel investors for the supposed existence of enormous unexploited lands (Habib-Mintz, 2010). However, Sulle and Nelson (2009) contend that land pursued for biofuel investment might be physically unoccupied but not unused. The land might be in fallow, or it may be common land used for example, charcoal production, and fuel wood and timber collection. If such land is lost to biofuel investment, not only will the livelihoods of the locals be affected, but this will also lead to shortened fallows that in turn will adversely affect soil fertility (Daley and Scott, 2011).

Despite all the optimism and potential of the biofuel sector, Tanzania lacks a coherent biofuel policy base (Sosovele, 2010). The existing policy does not address a wide range of energy options and has shaky institutional and legal frameworks. Under such circumstances, developing the biofuel industry will be a difficult task, some stakeholders in the biofuel sector have advised the government to halt the biofuel investments until appropriate policies are in place (Sosovele, 2010).

Biofuel potential in Rufiji district

Rufiji district covers a total area of 13,339 km^2 according to official figures (URT, 2013). The population density is among the lowest of any district in Tanzania according to the 2012 census, with 16 inhabitants per km^2, against the national average of 51 inhabitants per km^2. These figures might present a picture of huge tracts being available for biofuel investment in Rufiji District.

The choice of Tanzania by a Swedish company, SEKAB (now taken over by Eco Energy, to be referred to as SEKAB/Eco Energy), was based on the presumed availability of apt land for large scale biofuel investment (Havnevik, 2009). Authors have quoted various figures regarding what SEKAB/Eco Energy intended to acquire in Rufiji district, ranging from 250,000 (Neville and Dauvergne, 2012) to 500,000 ha (Cotula et al., 2009). Another company, Africa Green Oil (AGO), was negotiating with six villages in Rufiji District for 30,000 ha of land. In the course of their negotiations, they settled on 5000 ha, of which in the end only 2800 ha were actually available (Neville and Dauvergne, 2012). In Nyamatanga, one of the villages where AGO acquired land, the local population have not only lost agricultural land, but also income generated from the selling of products they were collecting from the acquired land (Daley and Scott, 2011). The direct engagement of biofuel companies with the villages without any government oversight has left the local people in a precarious position as far as their interests are concerned (Beyene et al., 2013).

The AGO narrative (seeking 30,000 ha of land but finding that only 2800 ha were actually available) demonstrates that there is a huge gap between the biofuel investor's wishes and the actual land available for biofuel investments. According to Mwakaje (2012), Rufiji district offers one of the best case studies for biofuel investments because it has attracted a considerable number of potential biofuel investors. This paper aims to investigate the hypothesis that there is abundant, idle or unused land that can be used for large scale biofuel production at the local level in developing countries like Tanzania. The study will therefore contribute to developing methods of assessing land availability for biofuel investments at the country, regional or local level supplementing those done at the global level.

METHODOLOGY

Study area

Rufiji District is located in the Coast (Pwani) Region (7°30'S to 8°40'S and 37°50' to 39°40'E) in Eastern Tanzania and is dominated by the Rufiji River that runs almost in the middle of the district embracing the flood plain on both sides and an extensive mangrove delta at the river mouth (Figure 1)

Rufiji district is one of the six districts in the Coast Region of

Tanzania. About 75% of the region's economy comes from the agricultural sector, mostly managed by smallholder farmers who do not practice improved farming. As a result, yield per acre is relatively low. Rufiji district has 482,466 ha of arable land (20.7%) out of which only 90,000 is under active crop production (URT, 2007). FAO (2010, p. 17) defines arable land as "land under temporary crops (double-cropped areas are counted only once), temporary meadows for mowing or pasture, land under market and kitchen gardens and land temporarily fallow (less than five years). The abandoned land resulting from shifting cultivation is not included in this category". This is an important distinction, as potentially arable land, such as land under fallow for prolonged periods, is not included in the definition.

Livelihood schemes in Rufiji demonstrate a strong interconnection of activities between the floodplain, the forested areas in the north and south and the lakes located close to the flood plain (Hamerlynck et al., 2010). In principle, there are three agricultural systems: the flood plain agriculture, practiced by the majority; the delta agriculture and the hill agriculture. The latter is characterized by low fertility and low yields (Havnevik, 1983). In all three systems, shifting cultivation is practiced. In the delta, where mangroves are cleared for agriculture, sedges replace crops during fallow phases (Semesi, 1989). In forests, north and south of the floodplain, cultivated fields are left fallow for a period of two to three years (Durand, 2003), and also in the flood plains, where cultivation is presently expanding, shifting cultivation is practiced (Hamerlynck et al., 2010).

Data sources

The study is based mainly on a literature review and on secondary data, mostly obtained from authorities and NGOs and from government offices in the Rufiji district council. Semi structured interviews were also conducted with relevant government officials.

Land use/land cover

Land use/land cover digital maps and boundaries of protected areas (game and forest reserves) were obtained from the database at the Tanzania Natural Resources Information Centre (TANRIC) at the Institute of Resource Assessment, University of Dar es Salaam. Land use/land cover types are based on Landsat TM images of 1994/95 (Hunting Technical Services, 1997). Of 64 land use/land cover digital sheets at the scale of 1:250,000 covering Tanzania, Rufiji District is covered by four sheets. We have modified the original classification of land use/land cover types based on extensive field experience from working in Rufiji District. For example, classes such as dense bushland, open bushland, bushland with emergent trees, have been merged into a single class called bushland. Likewise, closed woodland and open woodland have been merged into a single class called woodland.

Boundaries of protected areas

The best available map delineations of protected area boundaries have been used. The protected areas in Rufiji District consist of one game reserve and nineteen central government forest reserves including the Rufiji Delta (Appendix 1). A list of these forest reserves (Appendix 2), provided by the Rufiji District authorities contain discrepancies in size as compared to the size generated in GIS (Appendix 1), despite the fact that on the maps, they appear similarly in shape. In some cases, the area of certain forest reserves is not indicated at all in the official list. In addition to central government forest reserves, the list from Rufiji District officials contains local government forest reserves (owned by the

district council) and community based village forest reserves (owned by village governments). It also includes a number of proposed community village forest reserves whose sizes are not indicated. The total area from the district list for all types of forest reserves (community, local government and central government) is 2278.2 km^2, while the area under protection, as calculated in GIS, reaches 5227.1 km^2. Though the mangroves are a forest reserve, they have been considered separately. Unlike the rest of the forest reserves, mangrove forest reserves have no definite boundary, but are defined by the intertidal range. Thus, the boundary delineation is based on extent of mangroves as mapped from the images. The Selous Game Reserve, one of the largest faunal reserves in the world with an area estimated to be 54,600 km^2, cuts across several regions and districts, with 6.5% of its area in Rufiji District alone.

The digital district boundary used is the same as the one that appears on various documents. The area of Rufiji District from this digital source is 12,998.5 km^2, which is 97.4% of the figure quoted in official documents. This discrepancy in area is common in many administrative units (region, district) between the official figures and digital sources even from those obtained from the Survey and Mapping Division – the ultimate mapping authority in Tanzania.

GIS manipulation

We have applied a Geographical Information System (GIS) to produce maps and to generate data. The process of obtaining the area that might be considered for biofuel investment was done by elimination or subtraction (Figure 1). First, relevant digital land use/land cover sheets coverage was merged. Then, the land use/land cover map of Rufiji district was clipped (extracted). This was followed by superimposing boundaries of the protected areas (the game reserve, forest reserves, and the extent of mangroves) on the district land use/cover map. The areas covered by protected areas were then subtracted, leaving possible areas to be considered for biofuel investment.

The figure obtained from the GIS manipulation was used to deduct arable land (URT, 2007) from various land use/land cover-types to obtain the possible biofuel investment areas. The main limitation of this study was the inability to map or segregate arable land from different land use/land cover types (non GIS in Figure 2).

RESULTS AND DISCUSSION

Various land use/cover type

Land use/land cover in Rufiji District is dominated by woodland, wooded grassland and the floodplain (Table 1). Cultivation is represented by two land use/land cover types, mixed cropping and scattered cultivation. The sum of the two cultivation land use/land cover types is relatively low. Given that the size of the farms on an average is approximately 1.2 ha per household (Turpie, 2000), it is likely that cultivated land is underrated, as such small areas cannot easily be detected with the 30 x 30 m resolution of the Landsat images. However, also without taking cultivated land into consideration, the results suggest that a huge portion (40.2%) of the district is covered by protected areas (game and forest reserves), a portion of Rufiji district that cannot be considered for biofuel investment.

The results can be analyzed under two scenarios (Table 2). The first scenario assumes that protected

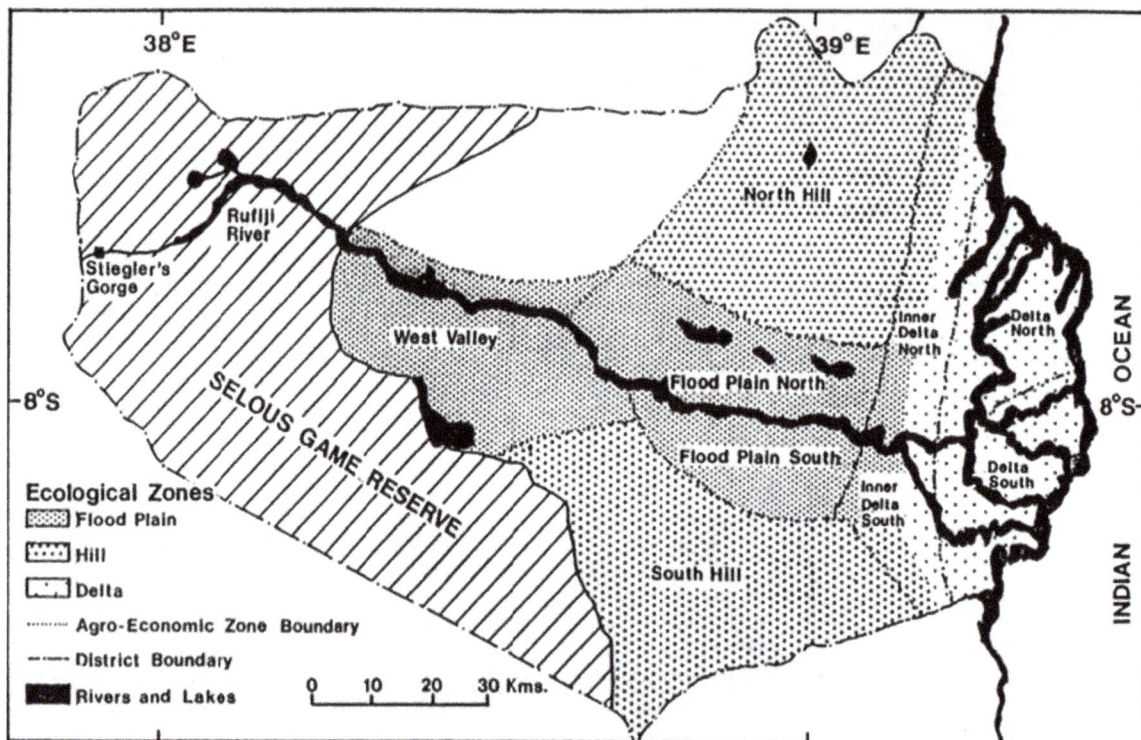

Figure 1. Rufiji District Agro-Ecological Ecological Zones (AEZ). Source: Havnevik (1981).

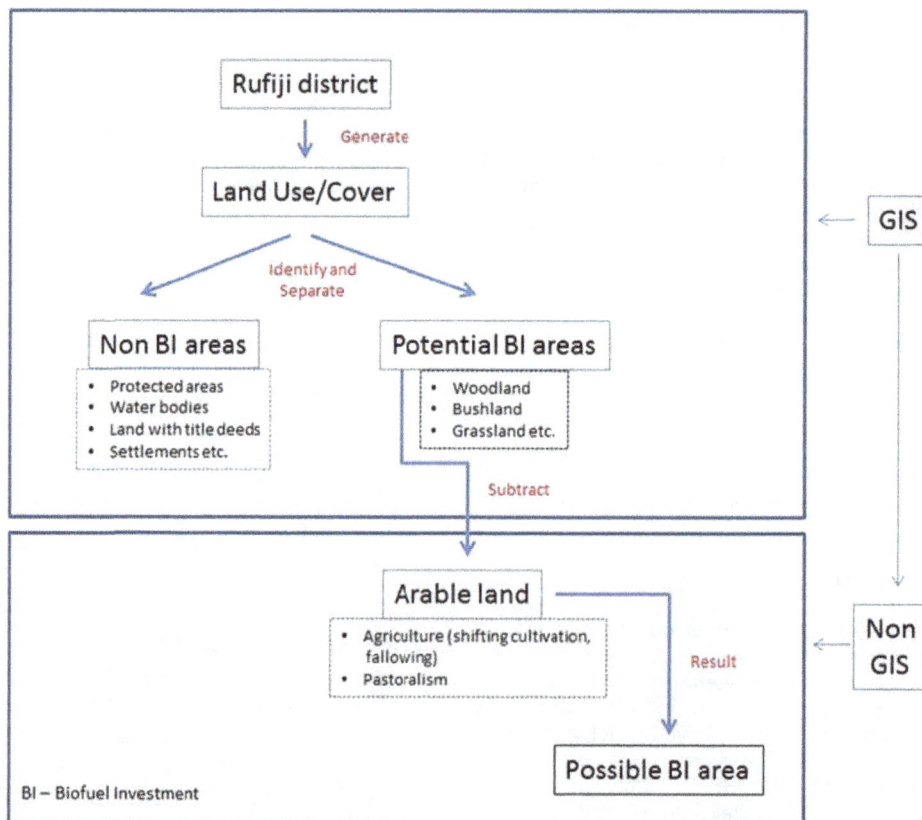

Figure 2. Flowchart of GIS manipulation.

Table 1. Distribution of land use/cover types in Rufiji District.

Land use/cover	Total area		Protected areas		Non-protected areas	
	Area (x 100 km^2)	%	Area (x 100 km^2)	%	Area (x 100 km^2)	%
Mangroves	4.8	3.7	4.8	3.7		
Natural/Riverine Forest	2.4	1.9	0.9	0.7	1.6	1.2
Forest Plantation	0.0	0..			0.0	0
Woodland	54.5	41.9	23.7	18.3	30.7	23.7
Bush land	7.4	5.7	1.2	0.9	6.2	4.8
Scattered Cultivation	8.7	6.7	3.0	2.3	5.7	4.4
Wooded Grassland	21.6	16.6	15.1	11.6	6.4	4.9
Flood Plain	19.5	15	1.1	0.8	18.4	14.2
Mixed Cropping	6.2	4.8	0.3	0.2	5.9	4.6
Bare Soil/Sand Dunes	0.8	0.6	0.2	0.1	0.6	0.5
Permanent Swamp	2.3	1.7	1.4	1.1	0.8	0.6
Lakes/Major River	1.8	1.4	0.5	0.4	1.3	1
Settlements/Urban Areas	0.1	0			0.1	0
	130.0	100	52.2	40.1	77.7	59.9

Source: University of Dar es Salaam - Land Use / Cover based on Landsat TM of 1994/95.

Table 2. Comparison of biofuel investment scenarios.

Scenario 1		Scenario 2	
District total area (x 1,000 km^2)	13.0	District Total Area (x 1.000 km^2)	13.0
Protected areas	5.2	Protected areas	5.2
Arable land	4.8	Arable land under crop production	0.9
SEKAB/Eco energy investment request	2.5	SEKAB/Eco Energy Investment request	2.5
	12.6		8.6
Balance after SEKAB/Eco Energy investment	0.4	Balance after Eco-Energy investment	4.4

areas and presently cultivated land will not be considered for biofuel investment, while the second scenario assumes that only arable land under crop production will be considered for biofuel investment. The most conservative figure among the many figures is quoted by different authors for biofuel investment in Rufiji District, as suggested by SEKAB/Eco Energy, 2500 km^2. In the first scenario, only some 450 km^2 will remain for other land needs. In the second scenario, if the wishes of SEKAB/Eco Energy were to be granted, some 4400 km^2 would be available. However, there are other important issues to consider. First, the area under forest reserves is a very conservative estimate by any means. Only central government forest reserves have been considered, while some of the reserves, owned by district and village councils, whose figures are in some cases not available (Table 2), were neglected. Second, although only 900 km^2 of 4824 km^2 is estimated to be under crop production according to the Coast Region Social-economic profile (URT,2007), the area used for agriculture may be considerably higher, as the estimation of areas of arable land under crop production is very difficult in places where shifting cultivation and

land fallowing is the norm. Third, only one potential investor (SEKAB/Eco Energy) has been considered, leaving out others like AGO. And finally, land availability has been gauged against the most conservative figure among those quoted for SEKAB/Eco Energy.

Africa Green Oil's proposed investment

The proposed investment proposal of Africa Green Oil (AGO) sheds some light on the flawed perception of biofuel investors about vast lands being available for biofuel investment. The initial request was 30,000 ha in six villages - Mangwi, Nyamatanga, Nyanjati, Ruaruke A, Ruaruke B and Rungungu (Figure 3). The total area of the six villages obtained from a scanned map of village survey in the north eastern part of Rufiji district by the Regional Secretariat Surveyor is 35,003 ha. This means that AGO was requesting 85.7% of land in those six villages. This suggests that AGO had only vague ideas about the total area of the six villages before making the claim for 30,000 ha. Some preliminary investigation of land use

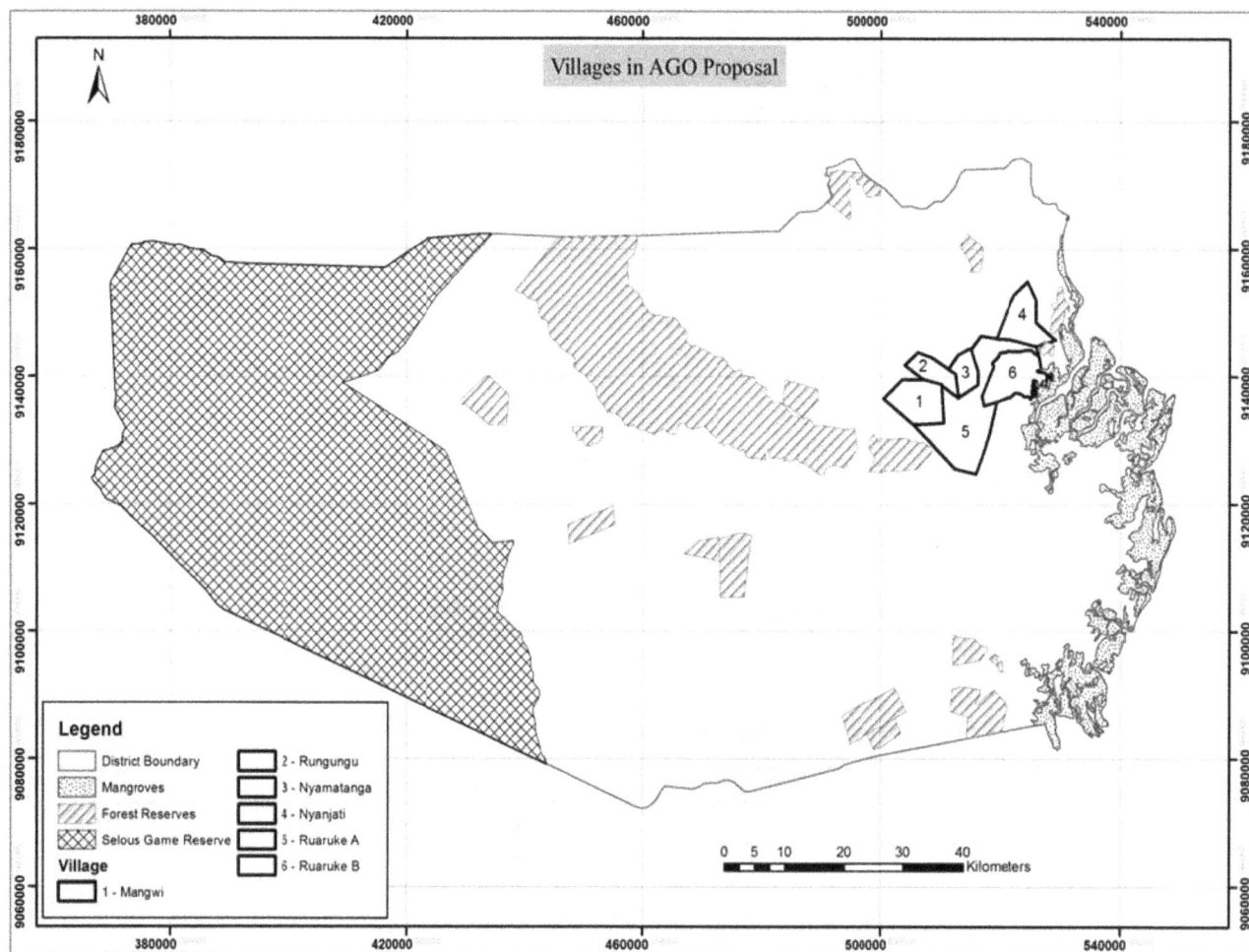

Figure 3. Rufiji district- Location of villages proposed for investment by AGO. Source: Regional Secretariat Surveyor – Coast Region. Registered Plan No. 45274 (30/01/2007).

and availability in the six villages, could have prompted AGO to further investigate the possibility of biofuel investment before committing resources and then realizing the futility of their expectations. The procedure outlined in Figure 2, with the necessary modification could constitute a starting point for accessing land availability for biofuel investment at the local level.

Possible consequences of biofuel investment in Rufiji District

Expansion of agricultural areas for biofuel production should not deprive the local communities of their land (Haugen, 2010). In Rufiji district, livelihoods are often complemented with the use of natural resources obtained from rivers, lakes and forests (UNDP, 2012). Land acquisition by biofuel investors like AGO has resulted in the local population losing income generated from the selling of products they were collecting from land (Daley and Scott, 2011). After losing their land, the displaced communities will be compelled to seek alternative areas for

settlements, farming and grazing (Madoffe et al., 2009). Seeking alternative areas after being displaced can be best illustrated by the Ujamaa villagization program that was implemented in Tanzania in the 1970s. It was aimed at settling people in designated villages, but some of the people refused to be settled in assigned villages and eventually settled in the inner delta (Figure 1), a transi-tion zone between the mangroves and the floodplain, where they cleared mangroves to establish new farms to support their livelihoods (Ochieng, 2000). After all, seeking refuge in the forests, including the mangroves of Rufiji Delta, in times of crisis is not a new phenomenon in Rufiji district. During the Maji Maji rebellion against the colonial German government, the Rufiji villagers made use of the forests as safe havens for the duration of the war. After the war, they continued to live in the forests and river islands of the delta to avoid forced labor, colonial government levy and controls imposed on their use of natural resources (Sunseri, 2003). Displacing people by biofuel investments could possibly result in the same situation exacerbating mangrove degradation.

Conclusion

This study has demonstrated the possibility of assessing land availability for potential biofuel investment at the local level. However, the assessment must take into consideration the relevant biofuel investment policies. The case of Rufiji district has revealed that the existence of huge amounts of unused land or under-utilized is an incorrect perception. This suggests that biofuel investment in Rufiji district is only possible if the land currently used (or fallowed) by the people for their livelihoods is assumed to be unused. The unused land may be physically unoccupied but used for shifting cultivation or extraction of natural resources like harvesting of forest and non-forest products. Taking such land by whatever means will amount to land grabbing with the implied consequences for the livelihoods of people who have been using, are still using and will be using the land for their livelihoods. The procedure applied to assess land availability for biofuel investments in Rufiji district could be used with the necessary adjustment or modifications in other areas at the local level.

Conflict of Interests

The author(s) have not declared any conflict of interests.

ACKNOWLEDGEMENT

This research was made possible as a result of Sida/SAREC funded collaboration between the Geography Departments of Stockholm University and Institute of Resource Assessment, University of Dar es Salaam, under the Integrated Natural Resources Management Project.

REFERENCES

Ahlberg L (2011). Study Estimates Land Available for Biofuel Crops. http://cee.illinois.edu/cai_biofuel_land. Accessed 26/08/2013

Beyene A, Mung'ong'o C,Atteridge A, Larsen R (2013). Biofuel Production and its Impacts on Local Livelihoods in Tanzania - A Mapping of Stakeholder Concerns and Some Implications for Governance. Stockholm Environment Institute, Working Paper 2013-03

Cai X, ZhangX, Wang D(2011). Land availability for biofuel production. Environ. Sci. Technol. 45(1):334-339.

Cotula L, Vermeulen S, Leonard R, Keeley J (2009). Land Grab or development opportunity? Agricultural investment and international land deals in Africa. London/Rome, IIED/FAO/IFAD.

Daley E, Scott S (2011). Understanding Land Investment Deals in Africa - Tanzania Case Study Report. Oakland Institute Tanzania Case Study Report. Mokoro Ltd.

Durand J (2003). Implementation of the Rufiji Forest Action Plan. With Special Emphasis on Community Based Natural Resources Management and a Case study of Ngumburuni Forest. Rufiji Environment Management Project. Technical Report No. 45.

FAO (2010). Data Structure, Concepts and Definitions common to FAOSTAT and Country STAT Framework. Food and Agriculture Organization of the United Nations.

Fritz S, See L, van der Velde M et al. (2013). Downgrading Recent Estimates of Land Available for Biofuel Production. Environ. Sci. Technol. 47(3):1688-1694

Habib-Mintz H (2010). Biofuel investment in Tanzania: Omissions in implementation. Energy Policy 38:3985–3997

Hamerlynck O, Duvail S, Hoag H, Yanda P, Jean-Luc P (2010). The Large-Scale Irrigation Potential of the Lower Rufiji Floodplain: Reality or Persistent Myth? Shared Waters, Shared Opportunities: Hydropolitics in East Africa Calas, B. & Mumma Martinon C.A. (Eds.) 2010

Haugen HM (2010). Biofuel potential and FAO's estimates of available land: The case of Tanzania. J. Ecol. Nat. Environ. 2(3):030-037.

Havnevik K (1983). Analysis of Rural Production and Incomes, Rufiji District Tanzania. Institute of Resource Assessment (IRA) paper no. 3. DERAP Publications Bergen.

Havnevik K (2009). Outsourcing of African lands for energy and food – challenges for smallholders. Paper (first draft) presented at IPD's African Task Force, Pretoria, South Africa.

Hunting Technical Services (1997). National Reconnaissance Level Land Use and Natural Resources Mapping Project. Forest Resources Management Project. Ministry of Natural Resource and Tourism. Final Report, Volume I and II.

Kweka O (2012). On Whose Interest is the State Intervention in Biofuel Investment in Tanzania? Cross-cultural Communication 8(1):80-85

Madoffe S, Maliondo S, Maganga F, Mtalo E, Midtgaard F, Bryceson I (2009). Biofuels and neo-colonialism. 06 June 2009. Pambazuka News, Available at http://www.pambazuka.org/en/category/features/56727 Accessed 13/06/2012

Martin M, Mwakaje A, Eklund M (2009). Biofuel development initiatives in Tanzania: development activities, scales of production and conditions for implementation and utilization. J. Clean. Prod. 17 (2009) S69-S76.

Mwakaje AG (2012). Can Biofuel Plantations Stimulate Rural Development in Tanzania? Insights from Rufiji District. Energy Sustain. Dev. 16(3):320-327.

Neville KJ, Dauvergne P (2012). Biofuels and the politics of mapmaking, Political Geography

Ochieng C (2002) Research Master Plan for the Rufiji Floodplain and Delta 2003-2013. Environmental Management and Biodiversity Conservation of Forests Woodlands, and Wetlands of the Rufiji Delta and Floodplain. Rufiji Environment Management Project

Romijn H, Caniëls M (2011). The Jatropha Biofuels Sector in Tanzania 2005-9: fEvolution Towards Sustainability. Research Policy 40(4):618-636.

Schoneveld G, German L, Nutakor E (2011). Land-based Investments for Rural Development? A Grounded Analysis of the Local Impacts of Biofuel Feedstock Plantations in Ghana. Ecol. Soc. 16(4):10

Semesi A (1989). The mangrove resources of the Rufiji delta, Tanzania. Paper presented at a workshop on Marine Sciences in East Africa.4-16 November, 1989. Institute of Marine Sciences, University of Oar es Salaam.

Sosovele H. (2010). Policy Challenges Related to Biofuel Development in Tanzania. Africa Spectrum 45(1):117-129.

Sulle E, Nelson F (2009). Biofuels, land access and rural livelihoods in Tanzania. IIED, London. ISBN 978-1-84369-749-7.

Sunseri T (2003). Reinterpreting a Colonial Rebellion: Forestry and Social Control in German East Africa, 1874- 1915. Environmental History, Vol. 8, No. 3 (Jul., 2003), pp. 430-451.

United Nations Development Programme (2012). Rufiji Environment Management Project. Equator Initiative Case Study Series. New York, NY.

United Republic of Tanzania (URT) (2013). 2012 population and housing census – population distribution by administrative units

United Republic of Tanzania (URT) (2007). Coast Region Socio-economic profile. Second edition. United Republic of Tanzania.

Appendix 1. Protected areas in Rufiji District as generated from GIS.

Protected area	Name	Area (x 100 km^2)	Total (x 100 km^2)
Game Reserve	Selous	35.5	35.5
Forest Reserves	Marenda	0.0	
	Mtita	0.3	
	Kingoma	0.1	
	Ruhoi	7.9	
	Mchungu	0.1	
	Kikale	0.0	
	Mtanza	0.4	
	Ngulakula	0.2	
	Kipo	0.1	
	Nyumburuni	0.5	
	Iyondo	0.2	
	Katundu	0.5	
	Utete	0.1	
	Mohoro	0.2	
	Mohoro River	0.0	
	Tambulu	0.5	
	Namakutwa	0.5	
	Nyamyete	0.1	12.0
Mangroves	Rufiji Delta	4.8	4.8
			52.3

Appendix 2. Forest Reserves from Rufiji District Council.

Forest reserve	Authority	Reference	Year established	Size (x 100 km^2)
Nyamakutwa-Namuete FR	Central Government	Jb.2320	1930	0.4
Muhoro FR	Central Government	Jb.615	1930	0.2
Muhoro River	Central Government	Jb.602	1930	0.0
Ngumburuni FR	Central Government		1930	0.3
Kingoma FR	Central Government		1930	
Mtita FR	Central Government	Jb.1026/RE/R/7/1	1930	0.3
Mangroves	Central Government	Jb. 634	1930	6.8
Utete FR	Central Government	Jb.625	1930	0.1
Utete warm spring FR	Central Government		1930	0.1
Tamburu FR	Central Government	Jb. 1620	1930	0.6
Kipo FR	Central Government	Jb. 1084	1930	0.2
Kikale FR	Central Government	Jb 1983	1930	0.1
Mpanga FR	Central Government	Jb.1959b	1930	0.5
Mtanza FR	Central Government	Jb.	1930	0.5
Rupiage FER	Central Government		1930	0.4
Katundu FR	Central Government	Jb 1086	1930	0.6
Mbumi FR	Central Government		1930	0.1
Mchungu FR	Central Government	Jb.1082	1930	0.1
Ngulakula FR	Central Government		1930	0.2
Nandundu FR	Central Government	Jb.RE/R/2/1	1930	0.0
Marenda FR	Central Government		1930	0.0
Kiwengoma FR	Central Government	Jb. 2310	1930	0.4
Kirengoma FR	Central Government	Jb. RE/R/6/1	1930	0.0
Kumbi FR	Central Government	Jb. E/R/2/1	1930	0.0

Appendix 2. Contd

Nerumba FR	Central Government	Jb.E/R/2/1	1930	0.0
Ruhoi LAFR	Rufiji district Council	Jb.508	1965	6.9
Kichi LAFR	Rufiji district Council		2000	1.5
Mtanzamsona VLFR	Village Council		2009	0.9
Tawi VLFR	Village Council	Jb.2351	2007	0.3
Nyamwage VLFR	Village Council	Jb.1200	2007	0.1
Nambunju VLFR	Village Council	Jb.2353	1998	0.2
Mbwara VLFR	Village Council	Jb.2354	2007	0.2
Mkoko VLFR	Village Council		2011	0.1
Utunge VLFR	Village Council		2010	0.4
Yelya VLFR	Village Council	Jb.1300	2007	0.1
Nzenge VLFR (prop)	Village Council		2011	0.1
Nyamitandai VLFR (prop)	Village Council		2011	0.2
Mbingo VLFR (prop)	Village Council		2009	
Urembo VLFR (prop)	Village Council		2009	
Jogoobahari VLFR (prop)	Village Council		2009	
Mkupuka VLFR (prop)	Village Council		2011	
Muyuyu VLFR (prop)	Village Council		2011	
Mangwi VLFR (prop)	Village Council		2011	
Ruaruke VLFR (prop)	Village Council		2009	
Minganje VLFR (prop)	Village Council		2009	
Nyambawala VLFR (prop)	Village Council		2009	
Mtunda VLFR (prop)	Village Council		2009	
Nyambawala B VLFR (prop)	Village Council		2009	
				22.8

*Source: Tarimo, Gaudence (District Forest Officer) and Mongo, Kennedy (District Fisheries Officer). Rufiji District Council (2011).

24

Assessing the diversity and intensity of pesticide use in communal area cotton production in Zimbabwe

W. Mubvekeri[1], J. Bare[2], Caston Makaka[2] and F. Jimu[1]

[1]Cotton Research Institute, Kadoma, Department of Research and Specialist Services Zimbabwe.
[2]Department of Biological Sciences, Faculty of Science and Technology Midlands State University Gweru, Zimbabwe.

A survey was conducted in Checheche, Nemangwe, Sanyati and Tafuna areas of Zimbabwe to assess the level of insecticide use and use of protective clothing in smallholder cotton production areas where the Cotton Research Institute conducted cotton experiments. Compliance with the closed season legislation, the Plant Pest and Diseases Act, Chapter 19, Section 8 of 1988 was checked because of its role in seasonal pest survival. Generally, pest management was found to be anchored on the use of insecticides with 71.9% of the farmers having positive indications regarding dependence on insecticides for pest control. Fifty nine percent of the farmers did not use scouting as a method to determine the need to spray insecticides. The closed season that helps break life cycle of insects was predominantly not observed. Integrated pest management approaches need to be promoted for the sack of the environment and the future of humanity.

Key words: Closed season, insecticide, integrated pest management.

INTRODUCTION

Cotton plays a significant role in the economy of Zimbabwe as it is the second largest export crop after tobacco (Esterhuizen, 2009). In 2008 the crop earned the country about US $150 million (Esterhuizen, 2009). However, in spite of its contribution to national economies, cotton is regarded as the most environmentally "toxic" crop on the planet (Cummins, 2003). Cotton covers 2.5% of the world's cultivated land yet it accounts for 24% of the world's insecticide use making it the most insecticide intensive crop globally (Laura, 2010). Chemical insecticides are used extensively in cotton production to control insect pests, with the primary target being bollworms (Vitale et al., 2007). Bollworm pressure has a positive impact on insecticide use (Qaim et al., 2003; Cotton Handbook Zimbabwe, 1998). Studies have shown that in Zimbabwe chemical pesticides alone can account for 70% of the variable costs in cotton production (Mudimu et al., 1995). Chemical insecticides when used carelessly can harm not only the environment, but also valuable pest predators, and the health of growers. The purpose of this study was to assess and establish the range and quantities of pesticides that are used by cotton growers in Zimbabwe. The results of the study would provide baseline information for further survey at a national scale. The objectives of the study were: To determine level of pesticides use in smallholder cotton production sector of

Zimbabwe and determine the extent of use of protective clothing

MATERIALS AND METHODS

Study areas

The study was carried out in the cotton growing areas of Checheche, Nemangwe, Sanyati, and Tafuna.

Checheche

Checheche is located in the South Eastern Lowveld, approximately 80 km north northeast of Chiredzi town along the highway to BirchnoughBridge. It is in Natural Region V. The study area is located approximately 20°49' S and 32° 15' E. The altitude of the area ranges from 395 m in the south to 404.4 m in the north.

Nemangwe

The area is located approximately 30 km west of Gokwe Growth Point. The study area lies within the 18° 11' 00 S to 18°° 12' 17 S and 28° 50' 25 E to 28° 51' 10 E coordinates in Natural Region IV. The altitude is about 1175 m. Soils are of loamy sand texture. The area is dominated by Mopani (*Colophospermunmopani*) woodlands. It is a smallholder communal area. The main landmark is Half-way Business Centre (H. B. C).

Sanyati

Sanyati area is located in Natural Region III on 17° 54' 47 S and 29°15' 15 E. The general altitude for the area is 832 m. Soils are of loamy sand texture. It is a smallholder communal area as well to the west of the area in Munyati river that flows northward.

The survey

The survey involved questionnaire interviews to collect baseline data on level of synthetic pesticide usage, and challenges regarding cotton stalk destruction in the areas under study.

Method of sampling

Sampling of respondents for questionnaire interviewees

Personal interviews were carried out using a designed questionnaire. Interviews were conducted in the villages of cotton growers who hosted Cotton Research Institute experiments in the same areas selected above. At each site, villages in which farmers hosting Cotton Research Institute experiments were located, identified and 50% of the villages were picked using simple random sampling. Heads of the selected villages were requested to provide names of all cotton growers in their villages. Fifty percent of the cotton growers in each village were randomly picked using simple random sampling method, and interviewed using prepared questionnaires. The interviewing team underwent a process of rehearsals to prepare them for the interviews. In Checheche, 13 farmers from Matikwa village in ward 26 of Chief Garawa were interviewed. In Nemangwe, eight farmers from Ndziko village in ward 12 of Chief Nemangwe were interviewed. In Sanyati, 11 out 12 farmers from Madhuveko village in ward 12 of Chief Wozhele were also interviewed. The twelfth farmer had gone to attend a funeral of a relative.

The questionnaire

The questionnaire was in four parts.

Part one

The first part sought to gather the farmer's location, average cotton hectares, and years of experience as a cotton grower.

Part two: pesticides

The second part was concerned with pesticides. Section one gathered information about a variety of insecticides and average quantities of each that a cotton grower applied seasonally. Seven insecticides were used on lepidopteran pests.

Section two sought to collect information on the standard of grower protection against insecticides. Chemicals sprays contaminate the environment and human beings and more-so those who conduct the spraying which poorly protected. Insecticides are mostly toxic chemicals. The assumption is that people who care less about their own personal safety against poisons would care lesser against poisoning the environment as well.

Part three: Slashing and destruction of cotton stalks to determine level of farmer compliance.

This part of the questionnaire collected information regarding slashing and destruction of cotton stalks. The information would help understand the causes behind ratoon cotton production and whether the cotton growers appreciated the ecological value of residue destructtion as a cultural, non-chemical pest management tool. The information would also help to determine the level of compliance with regulations guiding cotton stalk destruction. The team leader of the plant inspectorate was also asked to tell the constraints regarding enforcement of the closed season regulation through personal one on one communication.

Statistical analysis

Statistical Package for Social Scientists (SPSS) was used for data entry and analysis of frequencies. Data from questionnaires were analysed for frequencies. Cross tabulation was done using SPSS to determine relationships between variables.

RESULTS

Factors affecting grower compliance with plant pests and diseases act; chapter 19:08 of the republic of Zimbabwe

Cotton growers gave a variety of reasons why the closed season legislation was being ignored by some of the growers. Farmers who rent rather than own land were also cited though rarely as the ones who leave standing cotton over the off season. The ratoon cropping, and laziness were cited as the major reasons for not complying with the closed season legislation. The price and unavailability of planting seed, and labour constraint were also common responses (Figure 1).

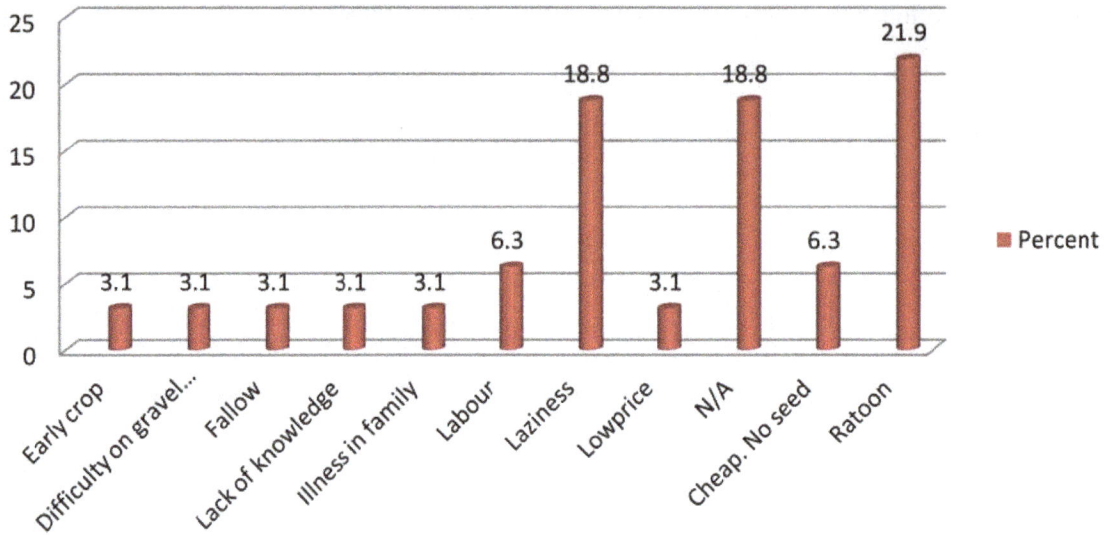

Figure 1. Diversity of reasons the farmers gave for not destroying cotton stalks.

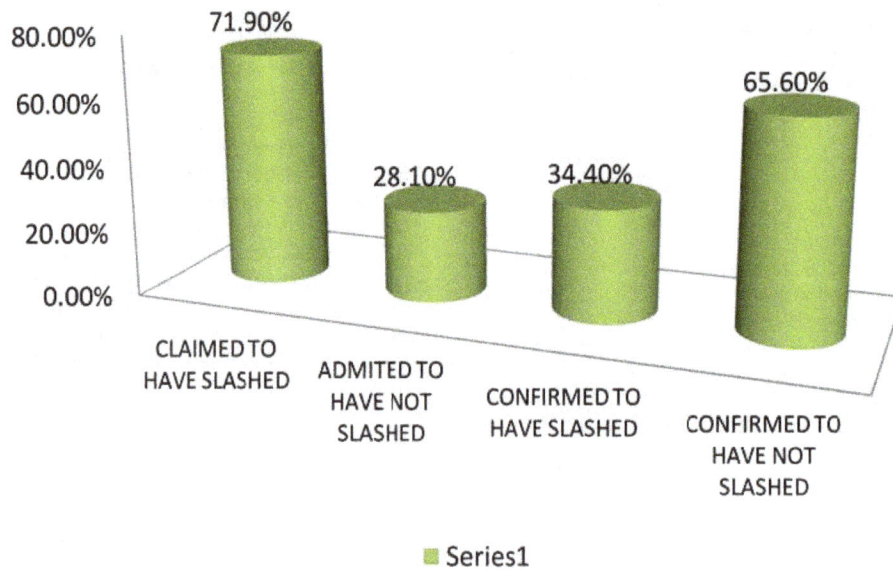

Figure 2. Graphical presentation showinglevelof compliance with closed season regulations.

Compliance with closed season regulations

Results from the interviews showed that 71.9% of the farmers claimed to have had slashed cotton stalks in their fields while 28.1% admitted not to have slashed cotton stalks (Figure 2). However visits to the field by the survey team revealed that only 34.4% of growers had slashed while 65.6% had not slashed. Therefore the actual level of compliance with the above legislation by the time the survey was conducted was 34.4%. Under consideration also was whether the grower had slashed cotton stalks by the legislated date and not by the date of the interview.

Dependance on insecticides in the study areas

Cotton pest management in the study areas is dominated by use of insecticides.

A total of eight insecticides and two acaricides namely Mitac (Amitraz) and Tedion (Tetradifon) were recorded as having been used by cotton growers in the study area.

CONSOLIDATED GRAPH COMBINING CHECK AND SPRAY, INSECTICIDES @ 7 DAYS, INSECTICIDES, PYRETHROIDS, AND SCOUT SPRAY

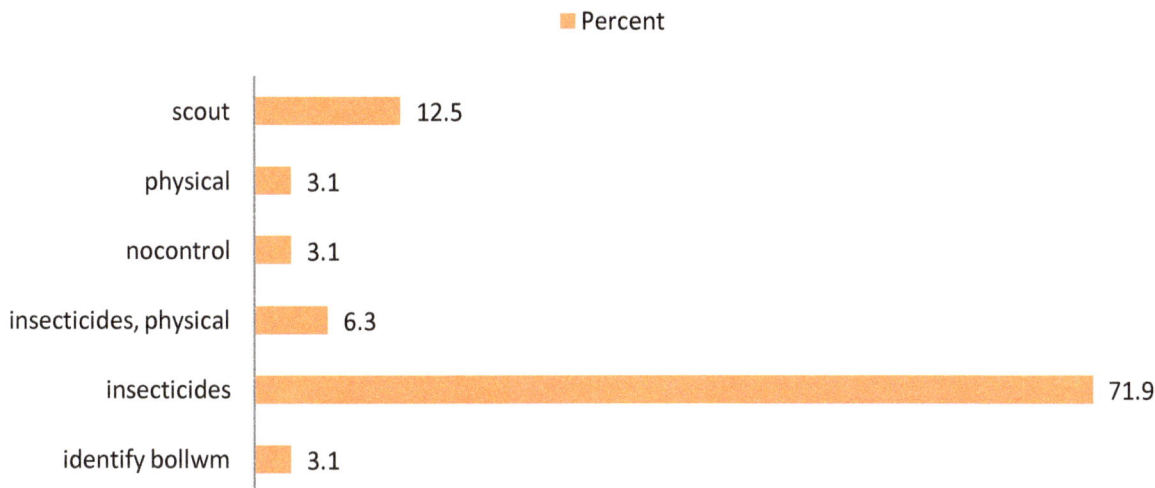

■ Percent

scout	12.5
physical	3.1
nocontrol	3.1
insecticides, physical	6.3
insecticides	71.9
identify bollwm	3.1

Figure 3. Showing the extent of reliance on chemicals for pest control by cotton growers.

SPECTRUM OF INSECTICIDE USE BY GROWERS IN STUDY AREAS

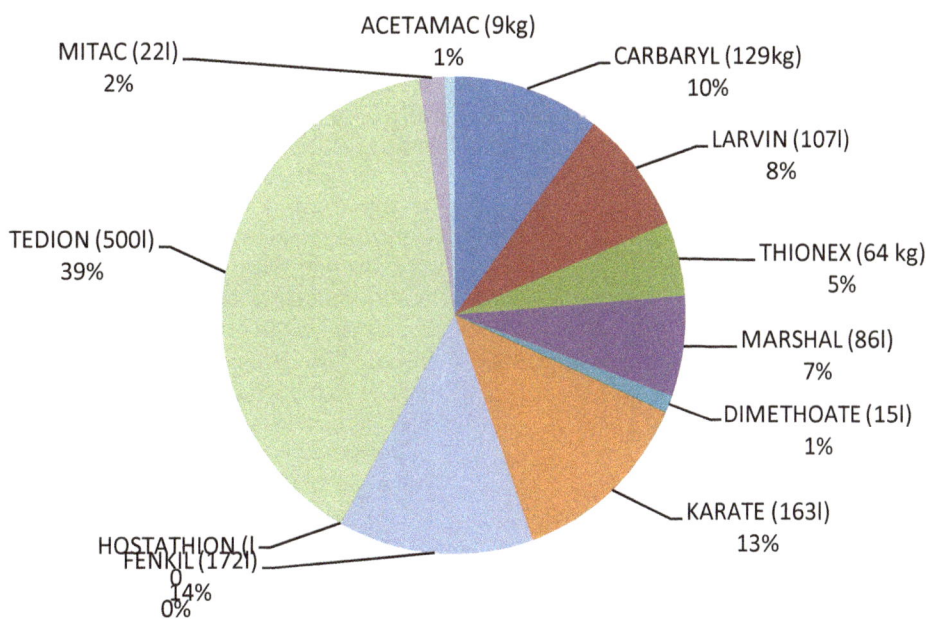

ACETAMAC (9kg) 1%
MITAC (22l) 2%
CARBARYL (129kg) 10%
LARVIN (107l) 8%
TEDION (500l) 39%
THIONEX (64 kg) 5%
MARSHAL (86l) 7%
DIMETHOATE (15l) 1%
HOSTATHION (l) 0%
FENKIL (172l) 14%
KARATE (163l) 13%

Figure 4. Showing a range of chemicals farmers said they used.

Insecticides accounted for 71.9% of responses given by cotton growers in the study areas (Figure 3).

Karate (lambda) and Fenkil (Fenvelarate) Carbaryl, Larvin (Thiodicarb 37.5 FW), and Thionex (Endosulphan 35 EC) were being used mainly against bollworms (Figure 4). Together the chemicals accounted for 50% of all chemicals used in the study areas. That could imply considerable bollworm pressure. Monocrotophos was also in use at Checheche. Such a highly poisonous product is not recommended for application using hand held equipment

ITEMS OF PROTECTIVE CLOTHING USED BY COTTON GROWERS

■ Percent

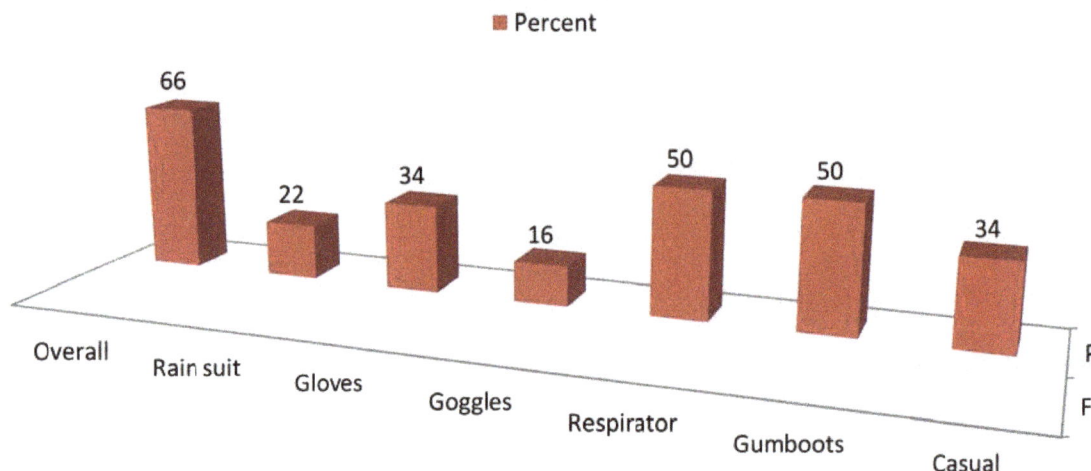

Figure 5. Graphical presentation of items of protective clothing used by growers in study areas combined.

as was the case in Checheche.

Assessment of protective clothing for use during handling of insecticides

Sixty six percent of respondents applied chemicals without any form of protective clothing while 66% handled chemicals with bare hands and 50 % without respirators (Figure 5). An important observation was that in some cases what respondents regarded as respirators were actually dust masks.

DISCUSSION

Factors affecting grower compliance with plant pests and diseases act chapter 19:08.

Level of compliance

The level of compliance with the closed season regulation by the time the survey was conducted was 34.4%. Under consideration also was whether the grower had slashed cotton stalks by the legislated date and not by the date of the interview. In the Low veld, Checheche included, the closed season started on 1st August and ends on 5th October each year. The survey was conducted from 6 to 8 September 2010 and most farmers had not slashed cotton stalks by that time; that was five weeks into the closed season. In the Middle veld, where Nemangwe, Sanyati, and Panmure are located, the closed season starts on 15 August and ends on 20 October each year. Most cotton growers had not slashed cotton stalks by 16-18 September 2010 when interviews were conducted in

Nemangwe and Sanyati. That was a full month into the closed season.

Factors affecting compliance

Production of seed cotton from ratoon was cited by most growers as the major cause for not destroying cotton stalks. Ratoon is becoming popular because it ensures an early cotton crop which normally matures before a crop established from seed. The ratoon crop establishes quickly with the first rains of the season. The ratoon grows from a well-established root system hence can better survive through mid-season droughts. Farmers are able to sell their seed cotton and earn money earlier in the harvest period. Production of ratoon crop is cheap considering the current price of US \$1.00 per kg of cotton seed. The recommended seed rate for cotton is 25 kg/hectare implying that the grower would have saved US \$25.00 for every hectare.

Destruction of cotton stalks is becoming unpopular because of the low market price for seed cotton. There is no grower motivation to go back to the fields to slash and destroy cotton stalks after selling the crops at "unviable prices". Destruction of cotton stalks does not offer a direct monetary benefit hence the reluctance to commit labour, the laziness, and, the prioritization of other family welfare issues over the future of the crop. Issues of labour, laziness and family illnesses are linked. When combined they account for 28% frequency.

From the point of view of the plant inspectorate there are several factors that led to complacency towards destruction of cotton stalks by cotton growers chief among them being lack of visibility of inspectors in cotton growing areas due to poor mobility and delayed amendment of the legislation to enable the inspectors to issue United

States dollar (US$) tickets (*Pers. Com*, 2010). At the time of the survey the Plant Pest and Diseases Act still stipulated fines in Z$ but the economy was using US$ it is not possible to punish offenders. The closed season regulation would continue to be ignored for as long as the enforcement agent remains logistically and legally incapacitated.

Pest management practices existing in the study areas

Over 70% of cotton growers in the study areas relied on insecticides alone for pest control. This is consistent with situations where the closed season is not strictly observed in cotton production. The only other method which rarely cited though was physical control, whereby grower pick and kill pests they find as they walk through the field. Scouting for pests before chemical application was mentioned although most growers failed to explain the technique.

The total area put to cotton by the study areas' sampled cotton growers in Checheche, Nemangwe and Sanyati was 137.5 hectares. Karate (Lambda) and Fenkil (Fenvelarate) are systemic pyrethroids used for the control of bollworms. Combined these chemical accounted for 335 L over 137.5 hectares, which is 2.4 L/hectare instead of about 0.8 l/hectare (Cotton Handbook, 1998). Conventional contact insecticides that growers indicated to have been using to target bollworms were Carbaryl, Larvin, and Thionex all of which account for 300 kg over 137.5 hectares, which is 2.2 kg/hectare. The national average cotton area is 360000 hectare/year. Assuming that each grower applies 2.4 L of pyrethroids per hectare and 2.2 kg of conventional insecticides per hectare, then, 877 090 L and 785 454 kg respectively could be sprayed into the environment annually. Such generous applications of insecticides to control crop damage by pests increases the direct risk of environmental pollution and kill non target pests. The environment is suffering. Growers are suffering too. Their standard of protection when handling insecticides is low.

The main ecologically appropriate cotton pest control tool is observance of host-free period. The low level of compliance with the closed season is linked to general "indiscipline" in the whole pest management regime at the expense of the environment. It was observed that even the acaricide rotation scheme is not being observed. In 2009/2010, Tedion (Tetradifon) was supposed to be used for red spider mite control in Region II only. Tedion is a sulphur compound with a long residual action (Mabveni, 2000). Cotton growers in Region III also used the same thereby increasing the risk RSM developing resistance to acaricides. Some 500 L of tedion was applied on 137.5 hectares that is 3.5 l/hectare against a recommended rate of 1.2 l/hectare (Cotton Handbook, 1998). Assuming that each grower in Zimbabwe applied 3.5 L, then, 1 2 million L of tedion alone could be sprayed

into the environment annually. Of interest was that two growers in Checheche were using nuvacron (monocrotophos), a highly poisonous organophosphate. They got it from neighbouring Mozambique.

The latest global trends in pest control in cotton show that insecticide use is on the decline in most countries and cotton producers are rapidly moving toward minimal insecticide dependent cotton production systems (ICAC, 2007). It appears that Zimbabwean farmers are going in the opposite and wrong direction. There is need to rigorously promote environmentally friendly sustainable pest control systems. The closed season is environment friendly and will undoubtedly reduce insecticide use when strictly observed. While the total elimination of insecticides may not be feasible everywhere it is certainly possible to drastically reduce their use.

Conclusions

Cotton bollworm management in Zimbabwe was largely insecticidal. Cotton growers had a high risk of contamination by insecticides due to poor protection during handling. Integrated pest management was not popular among cotton growers.

Recommendations

Recommendations for future research and pest management practices are given below:

1) The legislation governing cotton closed season and destruction of cotton stalks should be enforced by Plant Quarantine Services as a matter of national priority. That could have the effect of suppressing pest population and of cutting down on the level of insecticide application into the environment.
2) Cotton growers should be trained in the application of integrated pest management techniques most of which are environment friendly and economically sustainable. Rigorous extension is essential in order to increase the level of social and environmental responsibility of cotton production.
3) Finally, legislation alone cannot bring about cooperation. Cotton merchants have to address growers'' grievances regarding producer prices and cost of input.

Conflict of Interests

The author(s) have not declared any conflict of interests.

REFERENCES

Cotton Handbook Zimbabwe (1998). Commercial cotton growers association. Government Printers, Harare.
Cummins R (2003). Biodevastation. Clothes for a change.Organic

Consumers Association. USA.

Esterhuizen D (2009). Zimbabwe Commodity Report-Cotton and Products Annual. Office of Global Analysis, Pretoria.

ICAC (2007).Technical Committee Section.XXV (1). March 2007.

Laura K (2010). Cotton: Dirtiest Crop, Organic Authority, USA.

Mabveni ARS (2000). Crop Physiology and Crop Protection: Entomology, Modul 2, Zimbabwe Open University.

Mudimu GD, Chigume S, Chikanda M (1995). Pesticide use and policies in Zimbabwe–Current perspectives and emerging issues for research: Pesticide Policy Project Publications, No 2, pp. 61-65 University of Hanover, Germany.

Qaim M, Cap EJ, de Janvry A (2003). Agronomics and sustainability of transgenic cotton in Argentina. Ag Bio Forum 6(1&2):41-47.

Vitale J, Boyer T, Uaiena R, Sanders JH (2007). The economic impacts if introducing Bt technology in smallholder cotton production systems of West Africa: A case study from Mali. AgBioForum 10 (2):71-84.

25

Institutionalizing environmental hazards for 'public needs': Destruction of forest for drinking water supply in Kerala, India

S. Mohammed Irshad

Jamsetji Tata Centre for Disaster Management, Tata Institute of Social Science,
Malatil and Jal A.D. Naoroji (New) Campus, Sion-Trombay Road, P. O. Box 8313, Deonar, Mumbai-400 088, India.

Natural resource exploitation is increasingly being considered as a technical issue with the assumption that it can be compensated for. The public concern shifts towards such destruction only when it affects the normal course of day to day life. Immediate needs often undermine the process of institutionalizing knowledge to ensure conservation of natural resources. The question of immediate needs often acts as the determining factor in decision making. This paper is focused on such an environmentally-legal issue in ensuring water availability through the destruction of rain forest. This paper discusses this issue in detail and raises the question of failure of institutionalizing knowledge.

Key words: Sustainability, institutionalizing disasters, beneficiary group, neo-liberalism.

INTRODUCTION

This paper is about the socio-environmental relationship in contemporary ecology management. Diversion of rain forest for drinking water supply scheme is discussed as a contradiction in environmental governance.

Pipe water supply services are largely depending on perennial sources of water, especially surface flowing water. Fresh water eco-systems are integral parts of surface water and flowing water. The dependencies on perennial sources are also rising. Thus, drinking water supply providers are forced to pay less attention on ecology and put more efforts on ensuring water supply. Water polices are become incapable to protect the aquatic biodiversity and other vital resources. The ecological impact of freshwater ecosystems is undervalued across the world. This crises put in place certain recommendations, such as equitable market, realistic pricing and protection of ecosystems (Johnson et al., 2001).

The cost of supplying water supply is generally calculated on the basis of neo-classical economics, which gives thrust on 'full cost pricing' based on the 'user pays' principle. This cost criterion failed to assess the environmental cost. Environmental and social costs are having multiple dimensions and implications are also different (Abeysuriya K et al 2008). Environmental costs are directly connected with the deterioration of natural assets due to economic activities (United Nations, 1997).

The natural resources such as streams, rivers, lakes,

enhances its service to society (Forestry Commission: getting widened across the world. Climate change, population and unregulated consumption of freshwater will lead to freshwater crisis in the coming century. Sustainable utilisation of freshwater is depends on the changing culture of water management (Robert B et al 2002). Environmental conservation programmes faces multiple challenged in the developing world due to the dependency on resources (Brooks and Roumasset 2002).

The market process of pure demand management do not just between objects (commodities or inputs) and objective functions (demand, supply, utility, profit), but rather reflect relations between living human beings (Zafirovski, 2000). Such relation is often challenging the institutions formed to meet the demand and supply. However, The sustainability of institutions is getting public support. Institutional understanding of ecology has been de-limited into the sustainability of collective needs rather than ecological equity. This paper discusses this issue with reference to forest vis a via drinking water provision in Kerala, India.

The paper examines the role of the State in institutionalizing ecological damage in the neo-liberal development governance. This paper attempts to examine the institutional nature of ecological destruction and its consequences on governing environment. The neoclassical economic understanding of ecology and natural resource conservation do not reflect the social, economic and environmental realities of the world. The strongly integrated utilitarian approach to development critically destroy the value system based on ecology and alternatives to protect (Anthony M 1992).

The paper focuses on the bureaucratic understanding of ecology and the problems of identifying alternatives. This paper is the result of an enquiry on a foreign funded drinking water supply project in Kerala, India. The project is called Japan Bank of International Cooperation (JBIC) Funded Urban Drinking water supply augmentation programme.

One component of the project is to increase the amount of pipe water supply to the urban consumers of Kerala Water Authority (KWA) in Thiruvanthapuram District of Kerala, India. However, after completing all the infrastructure development, KWA had faced an unparalleled legal hurdle to carry over the project. The hurdle was the availability of water in reservoir situated in one of the rain forests to meet the demand. Increasing the dam's height at the cost of rain forest was the only solution left to augment the water supply. Since the reservoir is in the Wild Life sanctuary across the Karamana River, Government of Kerala had to take the prior consent of the Ministry of Environment and Forest, Government of India to increase the dam's height. This issue has not met with any public criticisms and resistance. This is the motivation of this paper, and the paper tries to raise the following issues.

2011). The gap between demand and supply for water

a) How does the project influence the environmental governance of the State?
b) How are the larger alternative ideas discussed in these types of projects?

Secondary sources from Kerala Water Authority, discussions with the forest department staffs, Kerala Water Authority staff and members of Kani Tribal community have been used.

SOURCES OF WATER AND CHALLENGES

Kerala is well known for its water resources and the state is experiencing severe water scarcity. Protections of river resources are highly problematic. The annual rainfall availability is estimated to be about 3000 mm (Induchoodan, 1996). However, there is significant variation and a shortage of safe drinking water in many places of the state, besides the availability of average rainfall of 3,055 mm (Indian Meteorological Divisions).

Ground water availability in Kerala

Kerala has an annual replenishable ground water resource of 7,900 million cubic meters (MCM). The Central Ground Water Resource Board, Kerala Division has estimated that the net ground water availability of the State is 6229.04 MCM. According to Central Ground Water Board, only 48 per cent of the ground water sources in Kerala has been exploited (State Planning Board, 2003). Open wells are major as well as the traditional source of drinking water in Kerala; in fact the whole concept of drinking water is still attached to open wells. Centre For Water Resources Development And Management in 1989 revealed that there were three million wells in the state, of which 20 lakhs were private wells. The density of open wells is also very high in Kerala, with density around 250 well per sq.km in the coastal belt; 150 in the midlands and 25 in the highlands. Table 1 explains this in detail.

Surface water availability in Kerala

Kerala has 44 rivers, out of which 41 are west flowing and 3 east flowing. These rivers are characterized as ephemeral; so monsoon rainfall is the main source of survival (James, 2003) (Table 2).

The annual utilizable yield from 31 rivers is 49,199 MCM (63% of the total), with the state's share of 87 per cent (42,672). But it has been estimated that the state is utilizing only 25 percent of the annual utilizable yield (State Planning Board, 2003).

Though Kerala has huge water sources potential, the natural water conservation receives less public attention

Table 1. Ground Water Resource of Kerala as on March 2003 km²/Year.

Provision for domestic and industrial and other uses	1.31
Available ground water resources for irrigation	6.59
Net draft	1.46
Balance ground water Resources for future use	5.13
Level of ground water development (%)	22.17

Source: Central Ground Water Board (2005).

Table 2. Medium river basins of Kerala.

River basins	Length (km)	Catchment area (sq. km)
Chaliar or Baypore	169	2788
Periyar Sivajini Hills	244	5398
Pamba Devarmalai	176	290

Source: Central Water Commission (1999) and Ministry of Water Resources (2004).

Table 3. Destruction of Forest Areas for Developmental Projects in Kerala (1980-2003)

State	Approved cases during 1980-2003	
	Number of cases	Area diverted (in hectare)
Kerala	182	40729.082
India	10358	872791.991

Source: Department of Forests & Wildlife, Government of Kerala.

Table 4. Types of forests in Kerala (Lakh ha.).

Forest type	Area	Total area (%)
Tropical wet evergreen forests	3.48	37.02
Tropical moist deciduous forests	4.1	43.62
Tropical dry deciduous forests	0.094	1
Mountain sub tropical	0.188	2
Plantations	1.538	16.36
Total	9.4	100

Source: Department of Forests & Wildlife, Government of Kerala.

and forest protection as well. Tables 3 and 4 explain the nature of the forest as well as its destruction. In Kerala, big cash crop plantation sector like rubber is considered as forest area.

Thus, the forest management policy of the state is heavily hinged on the interest of plantation lobby. Forest degradation is becoming an order of the day in the

state; for instance from 1905 to 1965 forest degradation had been about 0.27% of total geographical area per year; in 1965 to 1973, 1%, from 1972 to 1975, 8611 sq. km, and from 1980 to 1982 is estimated as 7370 sq.km. 3.17% per annum (national average during the period = 2.79%).

PEPPARA WILDLIFE SANCTUARY

Peppara Forest area is one of the 18 'Biological hot spots' of the world. Peppara Sanctuary has great floral and faunal significance; now the number of hot spots is raised to 34. The sanctuary spreads over 53 sq.km of forest which forms the catchment of Peppara Reservoir. It was declared as a sanctuary in 1983. There are more than 4500 species of flowering plants found in Kerala. The sanctuary consists of part of Palode Reserve (24 sq.kms) and part of Kottoor Reserve (29sq.kms). The total area of the sanctuary is 53 sq.km. The total water spread of the reservoir is 5.82 sq.km. The sanctuary is located at about 50kms north east of Thiruvananthapuram city in Nedumangad taluk (Kerala State, India); it is between longitude 76°40' and 77°17' east and latitude 80°7'and 8°53' north. The records of Kerala Forest Research Institute indicate that there are 145 species of mammals of which 14 species are endemic to Western Ghats; 169 species of fresh water species; 93 amphibian species of which 40 species are endemic and 486 species of birds, with 16 endemic to Western Ghats, in addition to innumerable micro-flora and fauna.

General topography of the area is hilly with elevation varying from 100 to 1717 m. Rainfall and other climate factors are similar to that of Neyyar Wildlife Sanctuary. There are 13 tribal settlements in the sanctuary. Eleven are in Athirumala section and two are in Thodayar section. Peppara wild life sanctuary is a part of Agasthyamala forest. It is in Thiruvananthapuram District of the Kerala State, India. The forest types in the area are west coast tropical evergreen, west coast semi-evergreen, southern hill-top tropical evergreen, southern wet temperature, Southern moist mixed deciduous and southern montane grasslands. The biological wealth of these forests is not fully explored. The presence of medicinal herbs is a positive aspect. One of the 7 Medicinal Plants Conservation areas in Kerala Forests is in the sanctuary.

The water supply to Thiruvananthapuram City and adjoining sub-urban areas depended exclusively on the perennial streams sprouting from the rain forest of this sanctuary. See Table 6 for details of the dam.

PROBLEMS AND CHALLENGES

The major source of pipe water in the area is the Karamana River. The total production and demand and

Table 5. Demand and supply of drinking water in Thiruvananathapuram.

Year	2006	2011	2021	2036
Demand ML	246	261	294	331
Supply ML	151	151	151	151
Deficiency	95	110	143	180

Source: Demand and supply assessment report of Thiruvananathapuram. Kerala Water Authority 2005.

\ supply gap of drinking water supply in the city are as follows:

Total production – from major schemes - 190 ML
Miner schemes - 28 ML
Total - 218 ML
Loss due to leakages - 67 ML
Supply - 151 ML
Existing demand - 246 ML
Demand supply gap - 95 ML

Apart from that, Table 5 tells us the demand- supply forecast of the city.

The aforesaid Figure 1 show the increasing demand for augmented drinking water supply in the Thiruvananthapuram urban and semi-urban areas. Hence increasing the dam's storage capacity was vital for ensuring adequate supply of drinking water to Thiruvananthapuram in the coming years.

IMPLICATIONS OF AUGMENTATION OF WATER SUPPLY ON FOREST

Attaining the target is not just technical in nature; in order to reach this target, the storage capacity of the Peppara Dam would augment, which was the sole source of drinking water for Thiruvananthapuram. The Full Reservoir Level (FSL) at Peppara would go up from104.5 to 110.5 meters. The total water that can be stored in the dam would then go up from 40 million metric cubes to 70 million metric cubes. However, the issue arising here is that the augmentation of the dam's capacity would result in submerge of 267 hectares of forest land, for which Kerala Water Authority has estimated that compensatory forestation would compensate the rain forest of Peppara and it would require $6 Million. (This is from the reply for my request under Right to Information Act from Kerala Water Authority).

Legal issue

In 1972, the Government of India had declared The Wildlife (Protection) Act, according to which "every specified plant or part or derivative thereof shall be the

Table 6. Details of the dam.

Details of the dam	
Length of dam	438 m
Height of dam	36.5 m
Top width	4.0 m
Gross 70.70 Mm3	
Dead storage 1.6 Mm3	
Live storage 68.5 Mm3	

Hydrology	
Catchment area	60 Sq km
Average Rain Fall	481.00 cm
Annual Run off at Dam site	312.30 Million m^3
Peak Design Flood	860 m^3/second

Reservoir	
Full reservoir level	+ 110.50 m
Minimum Draw Down Level	+ 85.25 m
Gross Storage	70.00 Million m^3
Dead Storage	1.60 Million m^3
Live Storage at F R L	68.40 Million m^3
Water Spread Area at F R L	849.60 Hectares

Dam	
Type	Concrete straight gravity
General Bed level of river at dam site	+ 75.50 m
Length of Dam Top	423 m
Height of Dam Above River Bed	36.50 m
Top Width	4.00 m
Length of Spill way	49.00 m
Elevation of crest of spill way	+ 104.50 m
Spill way crest gate	4 nos 10 m x 6 m size
Outlet	1 No of size 1.50 dia
Quantity of Concrete	1,80,000 m^3

Source: Peppara Dam site

property of the State Government if there is any offence committed against this Act or any rule or order made there-of, and, where such plant or part or derivative thereof has been collected or acquired from a sanctuary or national park declared by the Central Government,

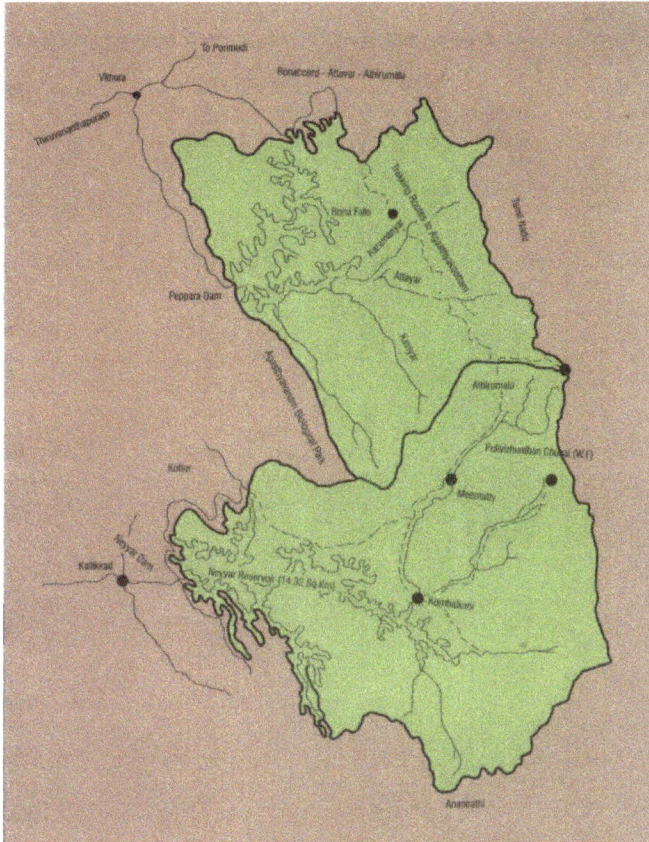

Figure 1. Area of Peppara Dam. Source: Broacher of Peppara Wild Life Sanctuary.

such plant or part or derivative thereof shall be the property of the Central Government". In 2002 the Government of India had introduced The Wildlife Protection Amendment Bill, which did not make any substantial changes in the previous bill but extended the purview of the wildlife protection.

New legal task

The Government of Kerala is entrusted to seek the Central Government's help to amend the 1972 and 2002 Act to augment the dam's capacity. The State Government had made all efforts to get the act amended. Delay in getting permission would create a further economic crisis owing to the huge aid from JBIC. The project is a package of five water supply schemes approved for loan assistance by the Overseas Economic Cooperation Fund (OECF) of Japan (now the Japan Bank for International Cooperation) in 1996. The objectives of the project are to supplement and rehabilitate water supply systems of two urban regions namely Thiruvanathapuram and Kozhikode and to construct water supply systems for three rural regions namely, Meenad, Cherthala and Pattuvam including their

adjoining villages. It is the largest ever urban Water Supply Project in the state. The total cost of the project is $ 389 million; Government of Kerala contributes $ 58.5 million and loan component is $ 331.45 million[1]. The Ministry of Environment and Forest has permitted the Kerala Water Authority to increase the dam's height in order to augment the water supply.

DISPLACEMENT OF TRIBAL

About 100 Kani Tribal families were displaced in 1981 as part of Peppara Dam construction and they were offered 5 acres of land as compensation. There has been no attempt by the Government of Kerala and KWA to distribute the land. The community settled themselves near the area close to the dam. There are about 13 Kani settlements in the reservoir area, of which 7 are within the close vicinity of the reservoir and facing the second phase of displacement. The displacement and livelihood laws of the Kani community have not been critically looked into while taking decisions on augmenting the drinking water supply.

The study raises the following issue for further discussion. The question is how do we define the 'water need' vis a vis forest/ecology? If we pose this question to the society, the likely answer would be in favour of water and not the forest. Thus an ecology hardly becomes an issue. Providers of water supply (Government) and beneficiaries (consumers) are equally responsible for this. Thus the following are the likely implications of the issue.

1. The possibility of institutionalising ecological hazards: The forest submerges have been introduced to the public not as an institutional failure but as part of a government's programme. This new perspective eventually institutionalizes the ecological hazards.
2. Lack of public protest: The urban consumers are not raising voice in favour of forest and the Kani community's displacement has not been considered as a social issue.
3. Challenges in the long term ecological sustainability: The crucial impact of these types of institutionalizations is that it would undermine the need of long term ecological management in particular.
4. Compensatory afforestation: The money offered by KWA to forest department for compensatory forestation needs to be critically looked at. Replacing the rain forest in Peppara is in fact an ambiguous statement, and indeed there are hardly any models available to justify this argument[2].

---

[1] Loan Agreement between Government of India and JBIC for Kerala
[2] The chief forest officer said that KWA would give Rs 30 Crore for compensation and forest department would spend the money through social forestry department.

5. Lack of search for alternatives: Search for alternatives is completely set aside in the project. KWA and JBIC units are paying little interest in alternatives such as rainwater harvesting, leakage detection of KWA supply lines. Of course this may not be sufficient enough to meet the growing demand for water; however, these alternatives are able to meet the partial demand for household water requirements. Nevertheless, the larger impact of this type of project is that it would never allow such alternative to get institutionalised. Such institutions get less importance in environmental governance (Goldsmith: 1992).

ACKNOWLEDGEMENT

Ms. Meena S Nair and Ms. Usha for supporting the field work and details of tribal displacement.

REFERENCES

Abeysuriya KCM, Juliet W (2008). Expanding Economic Perspectives For Sustainability In Urban Water And Sanitation. Development 51:23-29.

Anthony M. (1992) Economics, Ecology and Sustainable Development: Are They Compatible? Environ. Values 1(2):157-170.

Brooks K, James R (2002). Valuing indirect ecosystem services: the case of tropical watersheds. Environ. Dev. Econom. 7(4):701-714.

Forestry Commission (2011). Edinburgh Guidelines Fourth Edition.

Goldsmith AA (1992). Institutions and planned socioeconomic change: Four Approaches. Public Admin. Rev. 52(6):582-587.

Induchoodan NC (1996). Ecological studies on the sacred groves of Kerala, research thesis of Pondicherry University.

James CJ (2003). Water Management in Sooryamoorthy R and Antony (eds.). Managing Water and Water Users: Experience from Kerala's Maryland: University Press of America.

Johnson N, Carmen R, Jaime E (2001). Managing water for people and nature science, new series. Am. Assoc. Adv. Sci. 292(5519):1071.

Kerala Water Authority's Demand Supply Analysis of Thiruvanathapuram (2005). Prepared by KWA Engineers. Kerala Water Authority. p. 1.

Robert BJ, Stephen RC, Clifford ND, Diane MMR (2002). Water In A Changing World" Issues in Ecology 9 pp.

State Planning Board (2003). Economic Review, Government of Kerala, Thiruvanathapuram. Kerala. India.

Wild Life Protection Act, 1972 amended in 2002. Ministry of Environment and Forest, Government of India p. 1.

Zafirovski MZ (2000). An Alternative Sociological Perspective on Economic Value: Price Formation as a Social Process' International J. Pol. Cult. Soc. 14(2):265-295.

26

Promoting tertiary education through ecotourism development

Thomas Yeboah

Department of Hospitality and Tourism Management, Sunyani Polytechnic, School of Applied Sciences and Technology, P.O Box 206, Sunyani, Brong Ahafo, Ghana.

In general, this study was to encourage students to pursue, environmental science, biology and tourism programmes at tertiary level by embarking on educational tour with tertiary students to the study area, while at the same time, the paper performs its functions such as finding out how residents were empowered in funding, capacity building and conflict resolution skills in tourism development. Specifically, the study was to analyse areas in which residents were empowered to involve in ecotourism development. The sample of the study was 281 respondents including 14 key informants. Data were collected using household surveys, made up of questionnaires and interviews. The findings show that the residents were empowered through funding, capacity building and conflict resolution skills. In general, there was no significant difference in methods of empowerment among the socio-demographic characteristics of residents in the projects. Resident's commitment to ecotourism development in their communities is commendable. It was recommended that the government and the NGO's committed to the development of the projects in the local communities should integrate the local people fully and empower them as partners in the management of the projects by not only asking for their views when making decisions, but also, putting their ideas into action for the benefit of the projects. Again, since effective management of the projects is essential, residents should be empowered through training to enable them to participate fully in the projects.

Key words: Residents, empowerment, participation, ecotourism, development.

INTRODUCTION

Empowerment is a means and a goal to obtain basic human needs, education, skills and the power to attain a certain quality of life (Parpart et al., 2002). Obviously 'empowerment' is more than participation in decision-making; it must also include the processes that lead people to perceive themselves as able and entitled to make decisions (Rowlands, 1997). Empowerment may facilitate involvement in agreed-upon activities or alternatively, it can mean exclusion from activities that elements

of the community may not wish to engage in (Ramos and Prideaux, 2014). This implies that the local people should be encouraged to enable them to have direct involvement in and control over what happens in their lives (Bahaire and Elliot-White, 1999).

The ability for community members to participate in ecotourism development projects is however limited by the extent to which ecotourism is accepted as replacement for traditional activities. Where there is an agreement for participation in ecotourism projects, the ultimate success of such projects depends to a large extent on the level of involvement of external stakeholders including tour operators, government agencies and wholesalers (Nault and Stapleton, 2011).

Empowerment may also be seen as the development of skills and abilities of residents to manage existing development projects better and have a say in whatever is done in their community. The Food and Agriculture Organization (FAO, 1990) used the term 'empowerment' to describe any development process or activity such as skill training, management techniques and capacity building which could have some impact on people's ability to deal with different political and administrative systems, and influence decision making.

According to Whitford and Ruhanen (2010), almost all policies for indigenous tourism lacked the vigour and depth required to achieve sustainable ecotourism development. Farrelly (2011) identified a lack of formal education and perceptions of weak leadership from residents which contributed to an inability for local communities to make fully informed decision in community-based ecotourism, leaving them politically disempowered. Rogerson (2004) also found that the main obstacle to meeting government objectives for promoting economic empowerment of the owners of small tourism firms was lack of training in marketing tourism products. In an examination of empowerment in tourism destination, it has been found out that power struggles in local communities continue to affect the most disadvantaged groups such as ethnic and racial minorities, women and the poor (Timothy, 2007) .

Participation in development projects however, reinforces empowerment through an individual's inclusion in an organization and its organizational decision-making (Rocha, 1997). To apply the concept of empowerment to ecotourism development, it would mean that tourist destination communities, rather than governments or the multinational business sector, hold the authority and resources to make decisions, take action and control ecotourism development (Timothy, 2007). Consequently, in order to achieve sustainable ecotourism, the empowerment of communities affected by ecotourism development is attached to the importance of political and socio-economic justice (Sofield, 2003). As a way to achieving public participation and empowerment, Reid (2003) stresses the necessity of communities' awareness raising and transformative learning processes in understanding their situation and the need to handle problems themselves. Ecotourism resources in Ghana and in particular Brong-Ahafo Region include national parks, nature reserves, waterfalls, cultural and historical attractions and tropical flora and fauna. Community involvement in the development of these natural resources into tourist attractions may offer the necessary antidote for sustainability in the development of the tourist attractions in the region (GTB, 2008). Yet there has been little work undertaken by researchers into issues such as funding, capacity building and conflict resolution skills related to empowerment of the local

people in the study communities.

METHODOLOGY

The study area

Brong-Ahafo region is the second largest region of Ghana, after the Northern region, with a territorial size of about 39,557 km². Geographically, it is located at the centre of Ghana. It has a tropical climate with high temperatures of between 23 and 39°C, with a maximum rainfall of 450 mm in the northern parts, and up to 650 mm in the south of the region (Ghana Tourist Board (GTB), 2008). There are two main types of vegetation; the moist semi deciduous forest and the guinea savannah woodland.

The region has tourism facilities such as hotels, restaurants and fast food outlets found mainly in Sunyani and some of the district capitals. Some tourist's attractions in the region include Digya National Park, Bui National Park, Buoyem Caves and Bats Colony, Tanoboase Sacred Grove, Boabeng-Fiema Monkey Sanctuary, Hani Archaeological Site, Bono Manso Slave Market, Kintampo and Fuller Waterfalls (GTB, 2008). The study was however, conducted at Tanoboase, Boabeng and Fiema which form part of the communities selected for the implementation of the community-based ecotourism projects in the Brong-Ahafo region. Furthermore, the sites selected for this study were the earliest to be established in the region as CBEP sites (Zeppel, 2006) and as a result, were due for evaluation.

Tano sacred grove

Tanoboase is located 15 km north of Techiman, along Techiman-Kintampo road. The community began the development of Tano Sacred Grove as an ecotourism site in 1996 with the help of Ghana Association for the Conservation of Nature (GACON), which assisted the community in activities such as construction of green fire belt, tour guide training, wildlife conservation and bushfire prevention education at the initial stages of the project.

In 2001, Tanoboase was selected among the 14 communities to be developed under the Community-based Ecotourism Projects (CBEPs) in Ghana (GTB, 2008). Even though USAID assisted the project financially, its implementation was a collaborative effort among the major stakeholders such as the GTB, NCRC, United States Peace Corps Volunteer, SNV and the local community.

The aim of the project was to develop community-owned and operated ecotourism activities, which will conserve the ecosystems and also serve as income generating opportunities for the local people (GTB, 2008). A tourism management team made up of local community members was set up to manage the project at the local level. Development activities were based on community input, local workmanship and communal labour.

The community has a semi-deciduous forest which covers about 300 acres of land, a distance of about 1 km away from the village. The forest contains bats, baboons, antelopes, and a historic Bono Shrine. It also encloses a cluster of striking sandstone rock formations. The grove is believed to be the cradle of Bono civilization, and it served as a hideout for the Bono people during the slave trade and the Ashanti-Bono wars. Other tourist activities being promoted in the community are a visit to Tano Shrine and a 'village life' tour which includes a visit to local farms, homes and schools. This gives a visitor the opportunity to view local food preparation, village industries and listen to traditional songs and stories.

Boabeng-Fiema monkey sanctuary

Boabeng-Fiema is located 22 km north of Nkoranza. Two communities (Boabeng and Fiema) began the development of the monkey sanctuary as an ecotourism site in the early part of the

1970s with the help of officers from Ghana Wildlife Department (GWD), who protected the sanctuary from encroachers. The sanctuary, which is home to the black and white Colobus and the Mona monkeys that are used to interaction with human beings was opened to tourists in 1997 (Zeppel, 2006). The aim of the project was to develop community-owned and operated ecotourism project. It was also to serve as prospects for generating income by conserving local ecosystems and protecting the monkeys, which are generally regarded by the local people as sacred.

Stakeholders include the local communities, the Nkoranza Traditional Council, the Nkoranza North District Assembly, and NGOs such as NCRC, the United Nations Development Programme (UNDP), the European Union, the United States Peace Corps Volunteer and SNV (Netherlands Development Organization). Initially, USAID funded the project whilst Ghana Tourism Authority (GTA), NCRC and other NGOs supported it through training.

At the moment, the monkeys have spread to the surrounding communities, and based on the advice received from UNDP, the local people have set up a tourism management committee (TMC) made up of residents from all the nine communities which surround the sanctuary to direct the project at the local level. These communities are Boabeng, Fiema, Akrudwa Number 1, Akrudwa Number 2, Busunya, Bonte, Bomini, Senya and Kokorompe. Activities involving the development of the project are based on community input, local workmanship and communal labour.

Fieldwork

The study was conducted between 25th May, 2009 and 11th June, 2009. Four field assistants (two tour guides and two senior high school leavers) from Tanoboase, Boabeng and Fiema were given one day's training in English and Twi to assist the researcher in the distribution and administration of the questionnaires.

All the in-depth interviews were conducted at places of choice by the interviewees in the various communities. Though a total of 281 questionnaires were administered, 268 responses were obtained. This indicated a total response rate of 95.4%. The returned questionnaires were made up of 122 (43.4%), 50 (17.8%) and 96 (34.2%) respondents from Tanoboase, Boabeng and Fiema respect-tively.

Target population and sample size

The target population for the study was household heads or their representatives aged 18 years and above in the selected communities. This age group of people was targeted because people in this group were among the economically active population in the study area (Ghana statistical service - GSS, 2005). A list of household heads was compiled and used as a sampling frame for the selection of the respondents. The unit of data collection was individual household heads in the communities.

Those selected for the in-depth interview were the key informants or the opinion leaders in the study area. They were made up of fourteen representatives of the local people including TMC members, traditional authorities, service providers, assemblymen and unit committee members from Tanoboase and Boabeng-Fiema project sites.

Since it was not practically possible to observe all the elements in the target population, a sample was selected for the survey. The size of the sample required for the study depended on the purpose of the study and the availability of resources. In order to determine the sample size for the study, it was estimated that about 79% of the economically active population in the study area were aware of visitors' interest in the communities' tourism projects (GSS, 2005). This is because the region abounds in a wide range of tourist attractions. The sample size was therefore determined using Fisher's formula of determining samples (Chandam et al., 2004).

The calculated sample size indicated that at least 255 respondents had to be selected from Tanoboase, Boabeng and Fiema to get a representative population. 10% was however, added to make room for non-response. In total, 281 members of the communities took part in the study.

Sampling procedure

The study utilized a multi-stage sampling procedure to select respondents. The first phase centered on the listing of household heads in each of the communities. As part of this exercise, field assistants were tasked to list and identify the number of people in each house and also give identification marks to each of the household heads. Household refers to a person or group of persons related or unrelated who live together in the same house or compound, share the same housekeeping arrangement and are catered for as one unit (GSS, 2005).

The second phase dealt with the proportional allocation of the sample size of 281 among the three selected communities (Boabeng, Fiema and Tanoboase). To ensure fair representation, this exercise was based on the population of the communities instead of the household list. With this approach, the community with more people had more household heads participating in the study than its counterpart with less people. Therefore, using the list of household heads as a sampling frame, these sample sizes; 51, 103 and 127 were allocated to Boabeng, Fiema and Tanoboase respectively.

At the third phase, simple random sampling (without replacement) was used in selecting the individuals from the list of heads of households. Using simple random sampling, one household head was selected from the sampling frame to complete a questionnaire.

Additionally, 14 in-depth interviews were conducted with the opinion leaders or the key informants in the study area using an interview guide. Ten representatives of TMC members (including assemblymen and unit committee members), two elders representing traditional authorities and two service providers were purposively selected. It was the researcher's hope that the individuals selected would have knowledge, experience or information that would be useful to know about.

Research instruments

An interview schedule was the main instrument used for the study. The questionnaires were verbally administered in Twi. This approach was adopted because of the low literacy rate in the study area. The GSS (2005), reports that the effective literacy level for the study area is 48%, which is lower than the national average of 54.5%. Additionally, Twi was used because it is the lingua franca of the people involved in the study. Respondents were asked to respond to a series of close-ended and open-ended questions.

Data processing and analyses

The data were analysed by using the Statistical Product and Service Solution (SPSS) version 16. The quantitative responses were categorised, analysed, and examined based on various respondent groups such as sex, age and place of residence. Percentages and frequencies were also used in the analyses.

Qualitative data arising from open-ended questions that respondents answered using their own words, were coded into a set of categories developed from identified commonalities, that is, repeated themes were recorded together and categories of themes identified as they emerged. All the qualitative data were para-phrased while remaining faithful to the original meaning as it was given by the respondents during the in-depth interviews. It is also important to note that all the qualitative data had to be translated from Twi to English.

Table 1. Socio - Demographic Characteristics of Respondents.

Individual characteristics	Frequency	Percent
Sex		
Male	171	63.8
Female	97	36.2
Total	268	100.0
Residential status		
Resident	254	94.8
Non-resident	14	5.2
Total	268	100.0
Age		
30 – 39	109	40.7
40 – 49	118	44.0
50 and above	41	15.3
Total	268	100.0

RESULTS AND DISCUSSION

Socio-demographic characteristics of respondents

The socio-demographic characteristics of respondents were sex, residential status, and age. The highlights of the findings are as shown in Table 1.

Development of ecotourism tends to be sex-selective, thereby altering the composition of the population as well as its size in the destination area (Pearce, 1992). Mason and Cheyne (2000) observe that sex affects the needs, aspirations and attitude of people to issues and events.

Place of residence in relation to area of tourism concentration is known to affect people's perception and attitude towards tourism development. The impacts of tourism on urban areas or on people residing in tourism concentrated areas are found to be potentially so great that some method to reassure local residents has assumed prime importance (Bahaire and Elliot-White, 1999).

Age is known to determine individuals' needs, attitudes and perceptions towards tourism development in a community. Gilbert and Clarke (1997) notice that young and middle aged had a strong support for ecotourism development. The importance of the people found between 30 and 50 years age categories in the study area was that their ideas and grievances were generally heard and felt by the larger community. Consequently, these were the age groups which could influence certain decisions about the tourism projects. Most of the people in these age groups were breadwinners at home.

Residents' empowerment in ecotourism development projects

Among the key aspects of community participation in tourism development is the empowerment of the people to enable them to participate in decisions that affect their community and their lives. Community development involves empowerment of residents by providing them with the skills they need to make changes in their own lives and their communities (Korten, 1990). With specific reference to Ghana, CBEPs introduced support mechanisms to empower residents to participate effectively in the programme. Such support mechanisms were to improve the capacity of residents to plan and manage ecotourism development projects at the community level.

About 63.0 % of respondents agreed that funds were provided to residents to help them participate effectively in tourism development (Table 2). It was observed that, NCRC provided community leaders with money which enabled them to attend in-service training in bush fire prevention and sanitation. It was also revealed that funds were given to some of the Tourism Management Committee (TMC) members and individuals in the communities by UNDP to attend training workshops in tree planting and also, funds to buy mango seedlings for cultivation. One of the interviewees said: "*NCRC helps in capacity building, organise workshops to equip us with financial management and also offer technical advice. UNDP were giving us money to attend workshops but at the moment, it is not active. Landowners, at the end of every quarter of the year, are given part of the revenue from the project. Again Hotel, Catering and Tourism Training Centre (HOTCATT) in Ghana gave us training on how to receive visitors and how to present our local dishes to meet the taste of visitors.*"

At every quarter of the year, proceeds from the projects at Boabeng-Fiema for example, were shared among communities that surround the sanctuary. However, according to respondents, the money received from the project at a quarter of the year was woefully inadequate for any meaningful development in the communities. This made it necessary for management of the projects to halt 'quarterly sharing' of proceeds till the end of the year before they would decide whether to use the money accrued to develop at least one of the communities or use it to buy a bus that will carry tourists to and from the nearby towns. Meanwhile, at Tanoboase, proceeds from the tourism project were used to fund needy children's education, repair football field and the street lights in the community. Affected land owners in the study area were also given a share of the proceeds from the project.

Closely related to financial support to residents is the source of funding for the tourism projects. It was observed during the study that, funds were provided for the development of the ecotourism projects through levies. Moreover, the projects were funded through penalties or fines from those arrested for encroaching the forest reserves. Other sources of funding were through annual harvests and donations from individuals, groups and organizations.

Capacity building was identified as one of the modes through which residents were empowered to participate

Table 2. Ways in which residents were empowered to participate in ecotourism projects

Techniques	% in agreement	% not sure	% in disagreement
Provision of funds	63.0	2.0	35.0
Capacity building	62.0	8.0	30.0
Conflict resolution	87.0	2.0	11.0

in the CBEPs in the study area. This is a process and means through which a country, its people and organisation develop skills necessary to manage their resources in a sustainable manner (Gubbles and Koss, 2000). The purpose of the capacity building as a component of the project was to strengthen the institutional structures in the communities to deal with the task of tourism development. At the institutional level, it was meant to promote decentralised management of tourism.

About 62.0% of respondents agreed that, the programme offered capacity building to local people especially residents who were desirous of venturing into ecotourism development (Table 2). The reason is that, series of workshops were organized for residents in the communities. For instance, NCRC organized workshops for the Tourism Management Committee members in financial accounting. That is, how to save money and what to do with the money accrued from the tourism projects. More so, courses were organised for residents on how to manage the resources in the forest reserves, and again, to train tour guides for the ecotourism projects in the communities. Workshops were also organized by UNDP to train residents in tree planting. Likewise, Netherlands Development Organization trained residents to plant trees like *mangifera indica,* popularly known as 'mango tree' and *terminalia glaucescens,* which is locally called *framo,* to serve as food for the animals in the forest reserves.

Skill training is one of the means of meeting the human resource capacity needs of any organization. This can be used to build the skills and knowledge as well as attitudes of local people in ecotourism development. As a result, it was initiated by the project to increase the quality of service provision and also raise ecotourism awareness. It was observed during the study that, in order to improve service quality, Hotel, Catering and Tourism Training Centre (HOTCATT) initiated skills development programme for community members on standards for service provision (how to receive visitors and also package the local dishes for tourists) and the protection of the communities' natural assets. This was also confirmed by a resident at Tanoboase during the in-depth interview. He said: *"NCRC gives in-service training to us on how to manage the projects. I have been trained in data reporting, first aid, management and governance, private sector community partnership and tourism development plan. I sometimes give talks to the community on the benefits of tourism"*

This finding is consistent with that of Holland et al. (2003) study of heritage trails in the Czech Republic, where beneficiary communities were involved in capacity building through training in tourism skills. Paul (1987) observes that local people who participate in tourism projects need training and support to facilitate the development of the projects. This is also in line with Friedmann's (1992) observation that empowerment of local people to participate in development projects could lead to both their economic and socio-political well-being. The empowerment of community members helps them to assume key roles and responsibilities in the management of ecotourism projects.

Eighty-seven percent of respondents were in agreement that residents, especially the community leaders, were taught how to resolve conflicts relating to tourism projects in the communities (Table 2). The in-depth interview conducted also revealed that most of the project's management committee members attended workshops on conflict management. Efforts to provide community members with conflict resolution skills may have been informed by lessons from other projects in other countries. Abraham and Plateau (2001) have reported on the time consumed by community leaders in Kibera, a Nairobi slum, in protracted mediation to settle interpersonal conflicts. This was linked to the fact that community leaders lacked training in conflict resolution. Conflicts in the communities were often settled by the traditional authorities or the community leaders and in some few cases, judiciary. These were confirmed during the in-depth interview with some of the opinion leaders in the communities; *"Conflicts in this community may include herbs taken from the forest by local people, destruction of farms by monkeys in the forest and embezzlement of money by some leaders of the project. To resolve it, offenders are asked to refund the money embezzled. The Chief sometimes settles conflicts through re-allocation of land to the affected people whose lands have been taken by the projects."*

The in-depth interview asserts that, the communities involved in the projects were empowered through funding, capacity building and conflict resolution skills. As confirmed by this study, empowerment involves getting rid of the barriers that work against the local communities and building their capacity, providing them with funds and conflict resolution skills to engage effectively in tourism development (Arnstein, 1969; Fariborz and Ma'of, 2008;

Table 3. Means of Empowerment by Socio-Demographic Characteristics of Respondents.

Individual Characteristics	Funding				Capacity Building			Conflict Resolution		
	N	Mean	Test Stats.	p-value	Mean	Test Stats.	p-value	Mean	Test Stats.	p-value
Sex										
Male	171	2.61	t-test	0.000*	2.63	t-test	0.000*	1.98	t-test	0.699
Female	97	2.18	2.017		2.27	2.080		1.97	0.057	
Age										
<34	109	2.37	Anova	0.567	2.52	Anova	0.355	2.05	Anova	0.205
34-54	118	2.47	0.746		2.39	1.308		1.93	1.492	
<54	41	2.40			2.45			1.71		
Community										
Tanoboase	122	2.55^a	Anova	0.029*	2.54	Anova	0.084	1.93	Anova	0.848
Boabeng	50	1.88^{ab}	3.604		2.12	2.504		2.00	0.165	
Fiema	96	2.63^b			2.64			1.97		

N = 268. *The mean difference is significant at the 0.05 level; a and b indicate the difference in mean scores of the dependent variables across the individual characteristics.

Pretty, 1995; Tosun, 2000; ZhaoandRitchie, 2007).

Means of empowerment by respondents' socio-demographic characteristics

The mean responses of the ways in which residents were empowered by sex, age and community are presented in Table 3. Both t-test and one-way analysis of variance (ANOVA) were performed in order to assess the differences in the manner in which residents were empowered to participate in the projects. T-test statistical technique was employed on socio-demographic variable that was measured along a dichotomous scale such as sex (1 = male, 2 = female) of respondents. Other characteristics of respondents like age and community of residents which were measured along interval scale differences, were tested using one-way analysis of variance. It was hypothesized that; there is no significant difference in methods of empowerment among the socio-demographic characteristics (sex, age and community) of residents in the projects.

Pearce (1992) observes that development of tourism tends to be sex-selective, thereby altering the composition of the population as well as its size in the destination area. The t-test results (Table 3) show that there was a significant statistical difference between sex of respondents and funding (p-value 0.000) of tourism projects, and capacity building (p-value 0.000) of residents in the destination communities. There was however no significant difference between sex of respondents and conflict resolution (p-value 0.699) in the study area. Female respondents expressed high levels of agreement (funding:

mean = 2.18, capacity building: mean = 2.27), whilst their male counterparts expressed their doubts (funding: mean = 2.61, capacity building: mean = 2.63) as to whether residents were empowered through funding and capacity building. The involvement of residents, especially women, in productive enterprises could lead to both their economic and socio-political well-being and empowerment (Friedmann, 1992). Responses from both males (mean = 1.98) and females (mean = 1.97) confirm that residents, were trained in how to resolve ecotourism-related conflicts in their various communities as shown in Table 3.

Gilbert and Clarke (1997) observe that young and the middle aged are in favour of ecotourism development in their communities. The one-way ANOVA revealed that there was no significant difference in the provision of funds (p-value 0.567), capacity building (p-value 0.355), and conflict resolution skills (p-value 0.205) with respect to respondents' age in the study area.

Responses from those aged ≤ 39 years indicated that residents benefited from community empowerment programmes such as funding (mean = 2.37) and conflict resolution skills (mean = 2.05) but expressed their doubts as to whether residents received training in capacity building (mean = 2.52). The reason being that many of the people aged ≤ 39 years might not have been around at the time the training was going on, or were in the communities but did not see that people were being trained. However, respondents found within 40 - 49 years age brackets agreed that residents were empowered through funding (mean = 2.47), capacity building (mean = 2.39) and conflict resolution skills (mean = 1.93). This was confirmed by respondents aged 50 years and above as shown in Table 3.

Place of residence in relation to area of ecotourism concentration is known to affect people's perception and attitude towards ecotourism development. The impacts of ecotourism on people residing in ecotourism concentrated areas are found to be so great that, some methods need to be taken to reassure the safety of local residents in destination communities (Bahaire and Elliot-White, 1999).

The one-way ANOVA revealed that there was a significant difference among the communities and funding (p-value 0.029) of ecotourism projects. There was however no significant difference among the communities and capacity building (p-value 0.084), and conflict resolution (p-value 0.848).

The results show that whilst respondents at Boabeng (community directly affected by the ecotourism project) were in agreement that residents were compensated by providing them with funds (mean = 1.88) and training in capacity building (mean = 2.12), their counterparts at Tanoboase and Fiema were divided as to whether people in their communities were provided with funds (Tanoboase: mean = 2.55, Fiema: mean = 2.63) and training in capacity building (Tanoboase: mean = 2.54, Fiema: mean = 2.64). The reason for the divided opinion among the communities may be due to the fact that the negative impact of the ecotourism project is felt by the people at Boabeng more than the rest of the communities. This finding confirms Bahaire and Elliot-White (1999) report that place of residence in relation to area of tourism concentration affects the local people.

It was observed that at Boabeng, residents were staying with the monkeys in their homes. Therefore, it was not surprising when the local people at Boabeng-Fiema Monkey Sanctuary project site received funds to build shrines, boreholes, visitor centres and internet café. But it is envisaged that, sooner or later, if measures are not put in place to compensate the local people adequately, they will run out of patience, looking at the inconveniences created by the monkeys to them and the inability of the management to provide the basic needs of the communities like senior high school, health centres, good roads and jobs to the youths.

Conclusions

The findings of the study led to the conclusion that residents at both project sites (Tano Sacred Grove at Tanoboase and Boabeng-Fiema Monkey Sanctuary at Boabeng and Fiema) were provided with funds, capacity building and conflict resolution skills all of which have influence on local people's participation in the projects. Effective management of the projects is very crucial if the communities and indeed all stakeholders are given the necessary training to enable them participate fully in the CBEPs. The statistical analysis done supports the null hypothesis set. That is, there is no significant difference in ways of empowerment among the socio-demographic characteristics of residents in the projects. The implication is that socio-demographic characteristics of residents did not significantly have impact on the way they were empowered.

The authorities need to integrate the communities fully and recognize them as partners in the management of the projects by not only asking for their views when making decisions but also putting their ideas into action for the benefit of the projects.

For communities to have more knowledge in ecotourism development, they should seek assistance of experts from organisations and institutions like Ghana Wildlife Department, GTA, and NGOs. Similarly, to rekindle students' interest in ecotourism and tertiary education, the authorities in tertiary institutions such as the universities and the polytechnics around the tourist sites, should encourage their students to embark on educa-tional tour to these sites at least once every academic year. The communities could join resources in the protection of attractions, training of human resources for ecotourism development, construction of roads, joint promotion, and research which relates to impact assess-ment and monitoring of communities' attitude towards ecotourism development. Collaboration would enable the communities to enjoy economies of scale as well as gaining recognition and support from government and international donors.

Since local people can be empowered through access to credit, efforts should be made to address it. Most of the community members would like to sell food and drinks to the visitors. Unfortunately, they do not have the initial capital for such establishments. In order to empower them to engage in this business, government through the rural banks in the area should initiate a special tourism-related micro-finance scheme for the communities. The interest rate on the loan facility should be affordable to make it attractive to ordinary people in the communities.

Development of every economy relies on its infrastruc-tural base. As a result, government should come to the aid of the communities to help improve security, drinking water, sanitation, roads, education and health care facilities in the area.

Conflict of interest

The author did not declare any conflict of interest.

REFERENCES

Abraham A, Plateau JP (2001). Participatory Development in the Presence of Endogenous Community Imperfections, World Bank, Washington DC.

Arnstein SR (1969). A Ladder of Citizen Participation. J. Am. Statute of Planners. 35(4):216-224.

Bahaire T, Elliot-White M (1999). Community Participation in Tourism Planning and Development in the Historic City of York, England. Lincoln: University of Lincolnshire and Humberside.

Chandam JS, Singh J, Khanna KK (2004). Business Statistics (2nd ed), Vikas Publishing House Ltd., New Delhi.

Food and Agriculture Organisation (FAO) (1990). Participation in

Practice: Lessons from the FAO People's Participation Programme, Rome. FAO.

Farrally TA (2011). Indigenous and democratic decision-making: Issues from community-based ecotourism in the Bouma National Heritage Park. Fiji. J. Sust. Tour. 19(7):817-835

Fariborz A, Ma'of BR (2008). Barriers to Community Leadership towards Tourism Development in Shiraz, Iran. Eur. J. Soc. Sci. 7:172 - 178.

Friedman J (1992). Empowerment: The Politics of Alternative development. Blackwell, Oxford.

Ghana Statistical Service (2005). Analysis of District Data and Implications for Planning in Brong Ahafo Region: Ghana Population and Housing Census Report for 2000, GSS, Accra.

Ghana Tourist Board (2008). Official Listing of Hotels, Restaurants, Travel and Tour Agencies, Car Hire Services. Directory (31), GTB, Accra.

Gilbert D, Clarke M (1997). An Exploratory Examination of Urban Tourism Impact, With Reference to Residents' Attitudes in the Cities of Canterbury and Guildford. J. Urban Policy and Planning-CITIES. 14(16):343-352.

Gubbles P, Koss C (2000). From the Root Up: Strengthening Organisational Capacity through Guided Self-Assessment. World Neighbours, Oklahoma, USA.

Holland J, Burian M, Dixey L (2003). Tourism in Poor Rural Areas: Diversifying the Product and Expanding the Benefits in Rural Uganda and the Czech Republic. PPT Working Paper, No 12.

Korten D (1990). Getting to the 21st Century, Kumarian Press, Connecticut.

Mason P, Cheyne J (2000). Residents' Attitude to Proposed Tourism Development. Annals of Tourism Research, 27:391-411.

Nault S, Stapleton, P. (2011). The community participation process in ecotourism development. A case study of community of Sogoog. Bayan- Ulgii, Mongolia. J. Sust. Tour. 19(6):695-712.

Parpart JL, Rai SM, Staudt K (2002) Rethinking empowerment, gender and development: An introduction. In J.L. Parpart, S.M. Rai and K. Staudt (eds) Rethinking Empowerment: Gender and Development in a Global/Local World (pp. 3-21). Routledge; London and New York.

Paul S (1987). Community Participation in Development Projects: The World Bank Experience. Discussion Paper 6. The World Bank, Washington D.C.

Pearce DA (1992). Tourist Development (2nd ed). England: Longman.

Pretty J (1995). The Many Interpretations of Participation. In Focus 16:4-5.

Reid DG (2003) Tourism, Globalization and Development: Responsible Tourism Planning. Pluto Press, London.

Rocha EM (1997) A ladder of empowerment. J. Planning Educ. Res. 17:31-44.

Ramos AM, Prideaux B (2014). Indigenous ecotourism in the Manyan rainforest of Palenque: empowerment issues in sustainable development, J. Sust. Tour. 22(3):461-479.

Rogerson C (2004). Transforming the South African tourism industry: The emerged black-owned bed and breakfast economy, Geojournal. 60:273-281.

Rowlands J (1997) Questioning Empowerment: Working with Women in Honduras. Oxfam Publications, Oxford.

Sofield THB (2003) Empowerment for Sustainable Tourism Development. Oxford: Pergamon, Elsevier Sci.

Timothy DJ (2007) Empowerment and stakeholder participation in tourism destination communities. In A. Church and T. Coles (eds) Tourism, Power and Space (pp. 199-216). Routledge, London and New York.

Tosun C (2000). Limits to Community Participation in the Tourism Development Process in Developing Countries. Tourism Management. 21:613-633.

Whitford M, Ruhanen L (2010). Australian indigenous tourism policy: practical and sustainable policies? J. Sust. Tour. 18(4):475-476

Zeppel H (2006). Indigenous Ecotourism: Sustainable Development and Management, Cromwell Press, Trowbridge.

Zhao W, Ritchie JR (2007). Tourism and Poverty Alleviation: An Integrative Research Framework, Current Issues in Tourism, 10(3) 119-143.

27

Visible near infra-red (VisNIR) spectroscopy for predicting soil organic carbon in Ethiopia

Abebe Shiferaw[1,2] and Christian Hergarten[2]

[1]International Livestock Research Institute (ILRI), Addis Abeba, Ethiopia.
[2]University of Bern, Hochschulstrasse 4, 3012 Bern, Switzerland.

Over the past few decades, the advantages of the visible-near infra-red (VisNIR) diffuse reflectance spectrometer (DRS) method have enabled prediction of soil organic carbon (SOC). In this study, SOC was predicted using regression models for samples taken from three sites (Gununo, Maybar and Anjeni) in Ethiopia. SOC was characterized in laboratory using conventional wet chemistry and VisNIR-DRS methods. Principal component analysis (PCA), principal component regression (PCR) and partial least square regression (PLS) models were developed using Unscrambler X 10.2. PCA results show that the first two components accounted for a minimum of 96% variation which increased for individual sites and with data treatments. Correlation (r), coefficient of determination (R^2) and residual prediction deviation (RPD) were used to rate four models built. PLS model (r, R^2, RPD) values for Anjeni were 0.9, 0.9 and 3.6; for Gununo values 0.6, 0.3 and 1.2; for Maybar values 0.6, 0.3 and 0.9, and for the three sites values 0.7, 0.6 and 1.5, respectively. PCR model values (r, R^2, RPD) for Anjeni were 0.9, 0.8 and 2.7; for Gununo values 0.5, 0.3 and 1; for Maybar values 0.5, 0.1 and 0.7, and for the three sites values 0.7, 0.5 and 1.2, respectively. Comparison and testing of models shows superior performance of PLS to PCR. Models were rated as very poor (Maybar), poor (Gununo and three sites) and excellent (Anjeni). A robust model, Anjeni, is recommended for prediction of SOC in Ethiopia.

Key words: Prediction, soil organic carbon, visible near infra-red, spectrometer, Ethiopia.

INTRODUCTION

Concerns about global warming have resulted in an international agreement on reducing the emission of greenhouse gases (Kandel et al., 2011). The concern created a renewed interest in determination of soil organic carbon (SOC) content (Brunet et al., 2007). SOC represents one of the major pools in the global C cycle. Therefore, small changes in SOC stocks cause an important CO_2 fluxes between terrestrial ecosystems and the atmosphere (Stevens et al., 2006). Determination of SOC content is an important part of research to examine the fluxes. Current technologies to determine SOC depend on two categories of technologies often described as "intensive" and "non-intensive" (McCarty et al., 2002).

To quantify SOC, "intensive technology", uses several different techniques of fractionation and chemical extractions procedures. The intensive technologies include dry combustion for total carbon, calcimeter method for inorganic carbon and wet oxidation for SOC (Janik et al., 1998; Sankey et al., 2008; Walkley and Black, 1934). "Intensive technologies" are conventional and standard procedures but are time-consuming, laborious and expensive. The existence of several deviations in analytical procedures among the standard methods makes them more complex (McCarty et al., 2010).

In recent years, the "non-intensive technology" method is used as an alternative method because of its multiple advantages. Attention is given for such an alternative method as Visible near infrared reflectance (VisNIR) using diffuse reflectance spectroscopy (DRS) (Brunet et al., 2007). VisNIR-DRS methods are new, rapid, simple, non-destructive, reproducible, cost effective and some times more accurate than conventional analytical methods (Chang et al., 2001; Brown et al., 2005; Gomez et al., 2008; Cecillon et al., 2009; McCarty et al., 2010).

It is well-known fact that infrared predicted data can never be better than the original laboratory values. VisNIR-DRS method is less accurate than conventional laboratory methods such as wet oxidation and dry combustion (Stevens et al., 2006). If the sources of laboratory error can be identified, however; the VisNIR method may in fact be a better tool for interpretation than the 'appropriate' chemical analysis (Janik et al., 1998). A comprehensive review on advantages and disadvantages of VisNIR Spectrometer exist in Blanco and Villarroya (2002). VisNIR Spectrometer methods have also a limitation associated with instrumentation, data transferability, variation in study scale (Mouazen et al., 2010). In spite of these limitations, progress has shown the potential of Visible-Near Infra-Red Reflectance (VisNIR) for soil analysis (Janik et al., 1998).

In predicting SOC various types of spectrometers (DRS) are used (Blanco and Villarroya, 2002). The most common types of spectrometers are described as diffuse reflectance (DR), Mid Infrared (MIR) and Near Infrared (VisNIR). In this study, VisNIR spectrometer was used with range from 700 to 2,500 nm wavelength (Viscarra Rossel et al., 2006; Viscarra Rossel and McBratney, 2008). DRS has been used in soil science research since the 1950s (Viscarra Rossel and McBratney, 2008), however, characterizing soil using VisNIR-DRS dates back to the 1960s (Brown et al., 2005). Over the past 40 years, VisNIR-DRS methods have been developed as tool to predict SOC (Kang, 2006). Today the wide application of VisNIR-DRS methods has resulted in a modern technique for landscape modeling (Brown et al., 2005) precision agriculture (He and Song, 2006; Brown et al., 2005) digital soil mapping (Viscarra Rossel and McBratney, 2008) and soil C monitoring (Brown et al., 2005; Ge et al, 2011) for use in carbon sequestration studies and carbon finance.

VisNIR-DRS method involves analytical correlation of spectral data for predicting soil physical and chemical properties (He and Song, 2006; Chang et al., 2001; Genot et al., 2011) including SOC (Brown et al., 2005; Brown et al., 2005; Kang, 2006; Reeves et al., 2006; Gomez et al., 2008; Ge et al., 2011). The method has been reported as an accurate way of predicting SOC in laboratory (Gomez et al., 2008; McCarty et al., 2002; Stevens et al., 2006). Existing challenges limiting use of

VisNIR-DRS includes finding suitable data treatment and calibration strategies (Chang et al., 2001). As soil organic matter is complex, spectra results are not directly informative (Brunet et al., 2007). There is complexity of spectra and overlapping bands associated with its soil organic matter component (Kang, 2006; Sankey et al., 2008). The VisNIR spectra for SOC have not been well described so far, perhaps due to the complexity of material (Brown et al., 2005). Moreover, soil constituents various materials other than organic matter, which interact in a complex way to produce a given spectrum. So, direct quantitative prediction of soil characteristics is impossible (Cecillon et al., 2009; Chang et al., 2001). It is good to note that soils are more diverse in composition compared with traditional VisNIR products like grains or forages (Ge et al., 2011). It is therefore rather possible to calibrate model to predict soil organic carbon.

Simple equations involving pedo-transfer functions are used for predicting soil properties (Janik et al., 1998). Likewise, over the past decades, both physical and chemical properties of soils have been predicted from soils spectral data using multivariate equations (Kang, 2006; Cecillon et al., 2009). The prediction is successful for soil organic carbon. Multivariate analysis is used to construct models capable of accurately predicting properties of unknown samples. Multivariate calibration methods such as multiple linear-regression (MLR), principal components regression (PCR), Boosted Regression Trees (BRT), Artificial Neutral Networks (ANN), Locally Weighted Regression (LWR) and partial least squares regression (PLSR) has been applied to all spectroscopic studies (quantitative analysis) with variable degrees of success (Kang, 2006; Chang et al., 2001; Genot et al.,2011). PLS, PCR, MLR are good where there is linear relationship while ANN and others can be used where there is no linear relationship (Blanco and Villarroya, 2002). None of the above models are universally accepted and there are variously proposed calibration techniques (Chang et al., 2001; Genot et al., 2011).

Regression techniques involve relating the soil spectral data measured using VisNIR-DRS to laboratory measured soil properties (Ge et al., 2011). In this study, spectral data was related with SOC determined using analytical (Walkley and Black) method using multivariate regression models. Models built are tested using full prediction method and checked for accuracy using statistical parameters (Chang et al., 2001; Kandel et al., 2011).

This study makes use of three models: PCA, PLS and PCR. These models were selected for three reasons. First, they are full spectrum data compression techniques (Viscarra Rossel and McBratney, 2008; Naes et al., 2002). Second, the models can handle co-linearity. Third, they are most widely used and successful in SOC predictions (Blanco and Villarroya, 2002; Ge et al., 2011). As reviewed by Stevens et al. (2006), PLS and PCR are more frequently used than other models. MLR model was not used in this study because of its limitation in leverage correction and handling co-linearity (Stevens et al., 2006; CAMO, 2012).

As reviewed by Brown et al. (2005), soil properties were predicted using VisNIR Spectrometer in a wide range of scale representing soil variability from local, regional to global libraries. Regional libraries refer to a greater geographic extent than local libraries while global libraries are based on major soil taxa from multiple continents (Sankey et al., 2008; Brown et al., 2005). A comparison of results by Sankey et al. (2008) and review by Chang et al., (2001) and Stevens et al., (2006) shows that local libraries have better calibration accuracy compared with regional and global libraries. This study attempts to build four models (for individual 3 sites and all three sites) and recommends the most robust model for prediction of SOC in Ethiopia. Until recently, VisNIR-DRS has not been used as a tool to predict soil properties in Ethiopia. The paper specifically attempts to show the effect of data treatment on models, model testing and selection.

MATERIALS AND METHODS

The study area

The study areas are located in the Ethio-Swiss Soil Conservation Program (SCRP) sites established in 1980s. The sites are Gununo in South, Maybar in North-Eastern and Anjeni in North-Western Ethiopia. Gununo site is situated in Wolayita Zone, at 16 km WNW of Sodo town at 37° 38' E /6° 56 'N (SCRP, 2000, b) in Damote-Sore district. Maybar site is situated in South Wello Zone, 14 km SSE of Desse town at 39° 40' E /11 00 'N (SCRP, 2000d) in Albulko district. Anjeni site is situated in West Gojam Zone,Dembecha district at 15 Km North of Demecha at 37° 31' E /10° 40 'N(SCRP, 2000c) (Figure 1).

Methods

An equivalent mass depth soil sampling method was used as suggested for soil carbon study by Stolbovoy et al. (2002). Soil samples were taken from 64 soil profiles in three sites. Although the study sites are small in size, there are different types of soil types in the areas (Table 1) resulted in an intensive sampling. Depending on profile depth, samples were taken from 0-10, 10-30, 30-50, 50-100 cm depths. Although SOC distribution decrease with soil depth, its concenteration is visible up to 1 meter (Allen et al., 2010). Thus, deep sampling protocol is suggested for SOC study (Baker et al., 2007). Total soil samples are 96 from Gununo, 98 from Anjeni and 81 from Maybar. As recommended by Brunet et al. (2007) and Knadel et al. (2011) soil samples were grinded and sieved through 0.2 mm for better carbon prediction as used in this study.

A field spectroscopy (VisNIR-DRS) by Analytical Spectral Device (ASD) Incorporation was used for measurement of 275 samples taken from three sites. SOC was measured in laboratory using standard procedure for wet oxidation method as described in Walkley and Black (1934). Scanning procedures are as described in Brown et al. (2005) with detail protocols as indicated in Viscarra Rossel (2009). Reflectance spectra were measured on petri dishes, twice for each sample using a mug light. Spectra wavelength ranges from 350 to 2500 nm. Data reduction methods are needed in VisNIR Spectrometer study (Blanco and Villarroya, 2002). Following spectra data transposing for pre-processing, data was

reduced using average (for replicate sample spectra measurement). Then every 10th of the wavelength was selected.

There also seems to be lack of clarity on pre-processing to optimize spectral data (Brunet et al., 2007). Proper data pre-treatment help develop accurate calibration (Reeves et al., 2006; Blanco and Villarroya, 2002). Having tested various data pre-treatment procedures, Multiplicative scatter correction (MSC) and Detrending (DT) were selected to get best calibration and validation result. Steps used in developing multivariate models are as described in Blanco and Villarroya (2002) and CAMO (2012).

Unscrambler X 10.2 (CAMO Software, Analytical Spectral Device {ASD}, Oslo, Norway) (CAMO, 2012) was used for data pre-treatment, model calibration, validation and testing. Using test set validation method; principal component analysis (PCA) was used to examine hidden structure of data, to visualize relationship (similarity and difference) between soil samples and spectral wavelength (variables). PCA was used mainly to describe sample effect on models. PCA was used as descriptive tool while PCR and PLS were used as predictive tool. SOC content was regressed against soil spectra using PLS and PCR.

All model calibration involves selecting 10 components (factors), testing regression coefficients at *$P < 0.05\%$ significance level with test set validation. A total of 4 models were built for three individual sites independently and for all the three sites (altogether). To develop model for the three sites, data (n=275) was divided in to validation (30%, n=82) and calibration (70%, n=193) set. In developing each site models, validation and calibration samples are 28 and 68 for Gununo, 29 and 69 for Anjeni and 24 and 57 for Maybar, respectively.

The regression models were compared to examine accuracy and predictive ability using correlation coefficient (r), slope, coefficient of determination (R^2), root mean error of calibration (RMEC) and prediction (RMEP). Ratings of the models in this study were based on combining two parameters. The first parameter was based on R^2 values rate as suggested by Viscarra Rossel and McBratney (2008). The second parameter was based on RPD value rate as suggested by Mouazen et al. (2010). The accuracy of developed models were tested using full prediction by examining (predicted and reference plot) which shows the difference between measured and predicted values.

RESULTS AND DISCUSSION

Soil organic carbon (SOC) analytic result

The soil of the study sites were described and classified by the Ethio-Swiss Soil Conservation Program (SCRP) (Kejela, 1995; Weigel, 1986,a, Weigel, 1986,b). Altitude of the study area varies from 1982 to 2858 meter above sea level (m.a.s.l). Traditional agro-ecology of the sites varies from Moist WeynaDega to Wet WeynaDega.

SOC samples of the three sites (n= 275) have 2.5 mode and 1.9(g/Kg) median. SOC data is skewed positively (0.8, standard error of skewness = 0.14) with first quartile (Q1) = 1.0 and third quartile (Q3) = 2.6 values.

Previous soil studies in the area, SOC was also determined using Walkley and Black method (though sampling procedure varies). Anjeni was described as soils with low organic carbon (Zeleke, 2000; SCRP, 2000, c). Kejela (1995) found OC variation with maximum values with Phaeozem surface layers with 4.6% and minimmum with sub soils of (Gleysol-Fluvisol) with 0.05. SOC % in Zeleke (2000) and SCRP (2000c) varied from 1.1

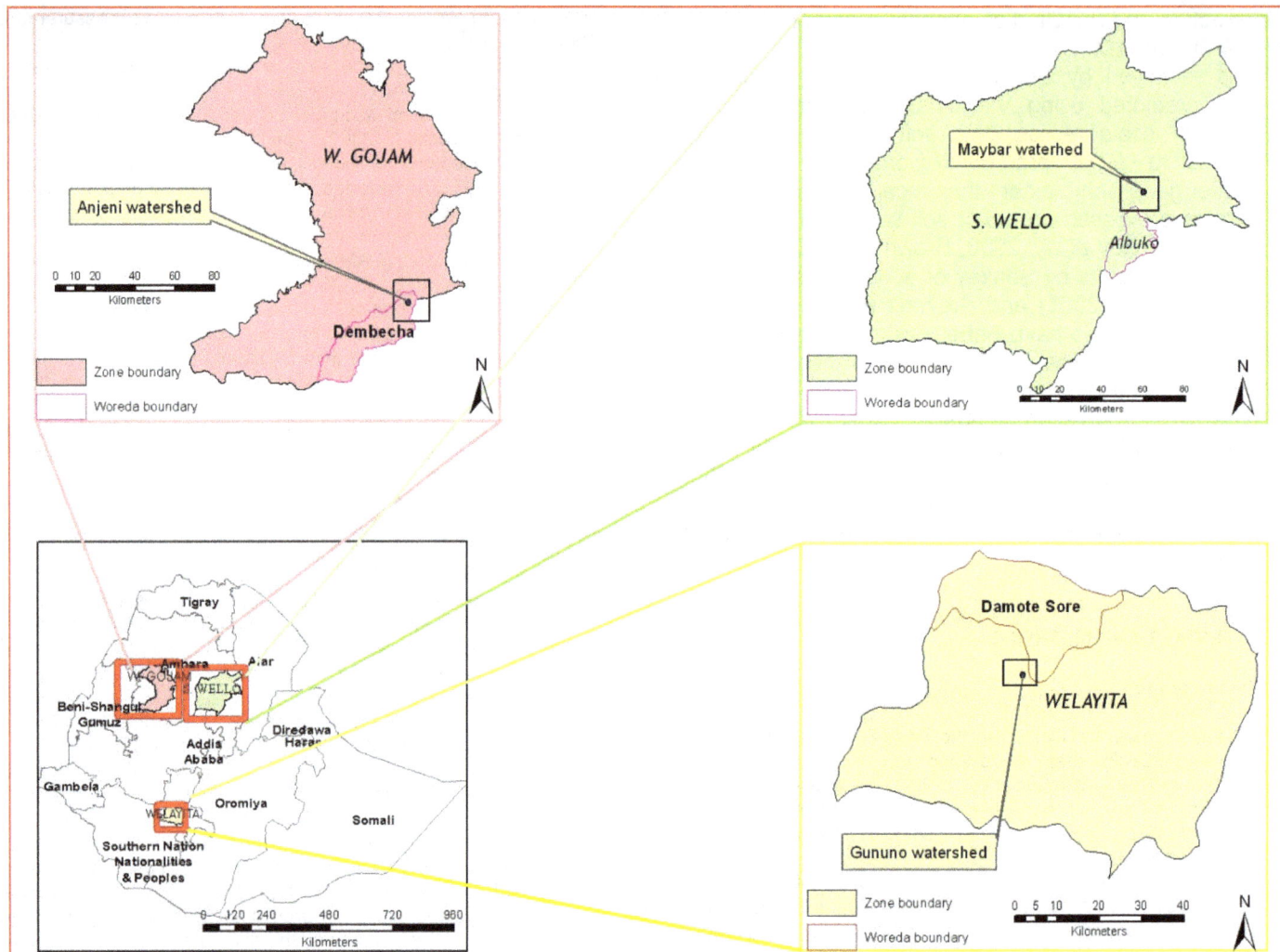

Figure 1. Location of study sites in Ethiopia.

Table 1. Description of soils of the study sites.

Name of research site	Gununo (GUN)	Maybar (MAY)	Anjeni (ANJ)
Climate (Thornthwaite classification) *±	Temperate , humid	Temperate , Sub-humid	Temperate , Sub-humid
Parent materials *,±	Trapp series of tertiary volcanic eruptions, ignimbrites,rhyolite , trachites and tuffs	Volcanic Trapp series with alkali-olivine basalts	Basaltic Trapp series of the tertiary volcanic eruption, tuff
Major soil Types (FAO-UNESCO)	Nitosols, Acrisols, Phaeozems, Fluvisols	Phaeozems , Lithosols, Gleysols	Alisols, NitosolsCambisols
Size of study area (ha)	166.8*	519.7*	918.4*

*Based on SCRP, 2000a; SCRP, 2000b; SCRP, 2000c; SCRP, 2000d; ± Kejela (1995), Weigel (1986a), Weigel (1986b).

to 3.9% mainly because survey area was smaller compared with Kejela (1995). Weigel (1986a) indicated that high percentage of OC is available in Gununo with some soil units of Humic Acrisols and Nitisols. Organic

Table 2. Soil organic carbon (SOC, g/kg)) descriptive statistics.

Site	Sample number (n)	Min	Max	Mean	Std. Deviation	Variance
*MAY (North - West- Ethiopia)	81	0.26	6.7	2.8	1.5	2.2
*GUN (South - Ethiopia)	96	0.20	3.3	1.8	0.8	0.7
*ANJ (North - East Ethiopia)	98	0.05	3.7	1.4	1.1	1.0
3 sites (all sites)	275	0.05	6.7	2.0	1.2	1.6

MAY*=maybar, ANJ*=Anjeni, GUN*=Gununo.

Table 3. SOC % variation accounted by first components with raw spectra.

Raw spectra[+]	Maximum components*	% Variation accounted by components (PC$^{\pm}$)		
		PC1	PC2	PC3
Gununo(GUN)	10	78	20	2
Anjeni(ANJ)	10	82	16	2
Maybar(MAY)	10	89	10	1
3 SITES (all)	10	71	25	3

PC$^{\pm}$= major principal component (1, 2, 3) *Optimum components = 3, [+]No treatment.

Matter (OM) variation shows that some layers of Humic Acrisols has a maximum of 6.2% while Eutric Nitosols has a minimum of 1.2% (% OM = O.C% X 1.72).Weigel (1986, b) characterized SOC variation of Maybar with maximum values at depths of Phaeozem soil profiles with 5.9% OM and minimum value of 1.5 % OM at some depth. Comparison of variation of SOC (g/Kg) across the sites shows that the minimum values were recorded in Anjeni and higher values in Maybar (Table 2)

Principal component analysis (PCA)

PCA shows that the first two principal components accounted for a minimum of 96% of the variance (raw spectra for all the three sites). Percent variance increased for specific sites (Table 3) and with data treatment. For example, for the three sites, with De-trending the first two components accounts for 99% of the variance.

PCA is used to find out outliers in a data set (Tobler, 2011). Maybar samples have 4% potential outliers (Figure 2). Under normal situation, 5% of the samples may lie outside the ellipse (CAMO, 2012). Samples far from center have high leverage (potentially influential) (Naes et al., 2002; CAMO, 2012). If leverage values for samples are above 0.4, it is "bothering" (CAMO, 2012). Maybar sample has 9% highest and worse absolute leverage values with 4% potential outliers which have reduced model quality.

The result explains why Maybar model has least predictive ability as reflected in values of correlation (r), coefficient of determination (R^2) and residual prediction deviation (RPD) in both PLS and PCR models (Figures 3 and 4). Samples, which appear as potential outliners,

were not removed in this study because they contain real soil information measured under laboratory condition. Comparison of variances showed the closeness of calibrated and validated curves, which reflected that models were true representativeness and there is absence of threat from outliers. A further data treatment with Multiplicative Scatter Correction (MSC) and De trending (DT) also developed better PCA with fewer components.

Principal component regression (PCR)

PCR is a multivariate regression analysis technique. PCR is used in predicting SOC using VisNIR-DRS. PCR and PLS provide similar results, though PLS usually converges in less factors than PCR. Although there seems to be confusion on data pre-processing to optimize spectral features for SOC prediction, Chang et al. (2001) points out that finding suitable data treatment is main challenge in VisNIR-DRS study.

Some authors prefer derivatives (Brunet et al., 2007) but in this study, results using first and second order derivatives were even worse than the raw spectral data. Various data treatment methods (moving average, baseline, standard normal variant (SNV) were tested before selecting MSC and Detrending (DT). The various data treatment procedures (baseline effect, moving average) have improved the models a little compared with raw spectral data.

Partial least square regression (PLS)

Review shows that the most frequently used regression

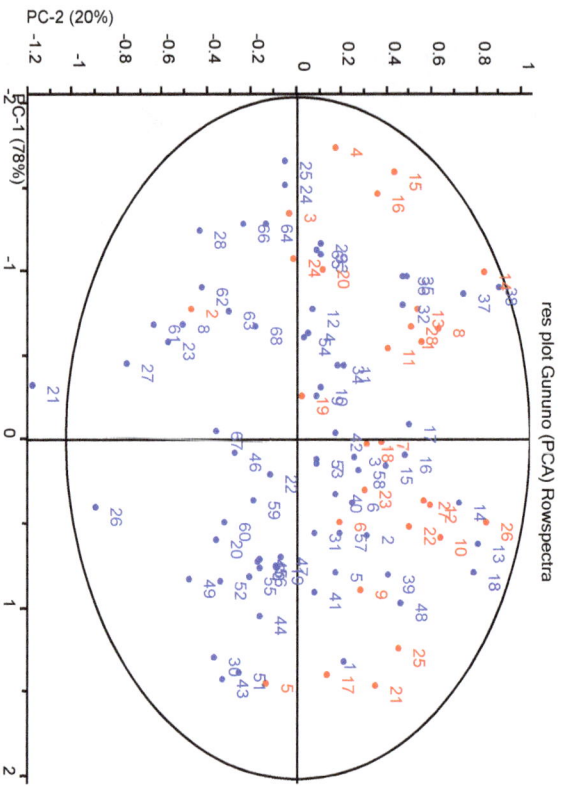

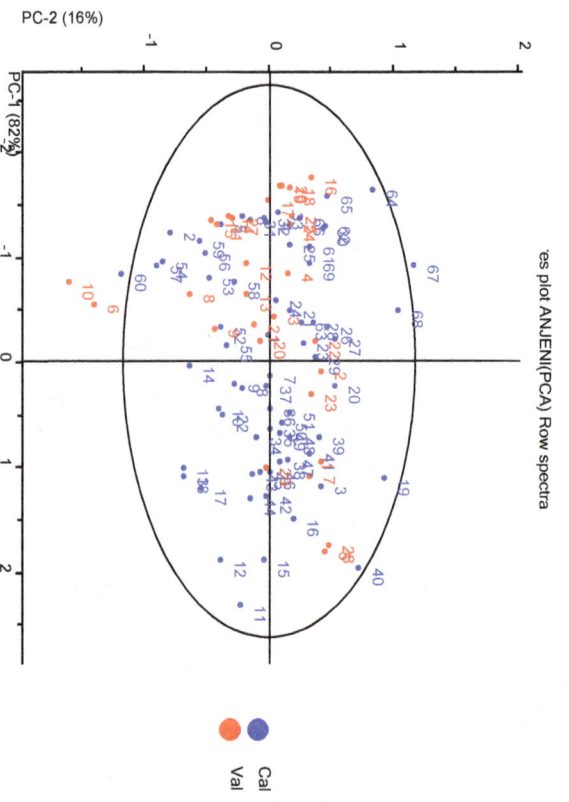

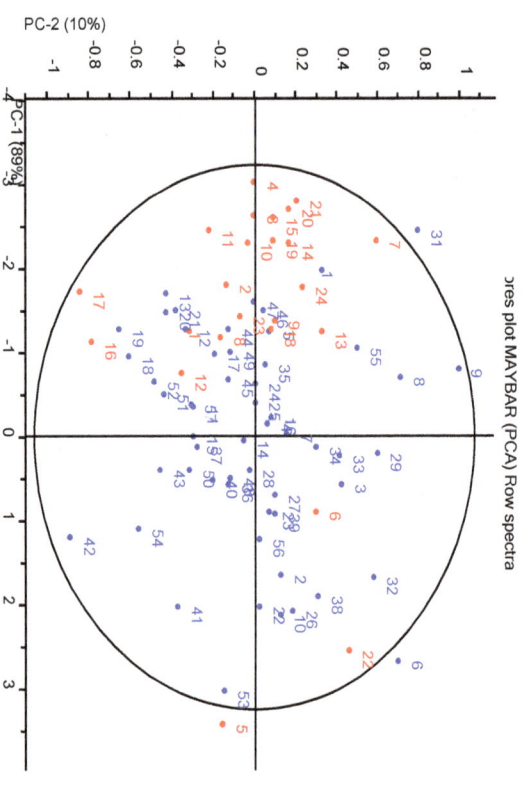

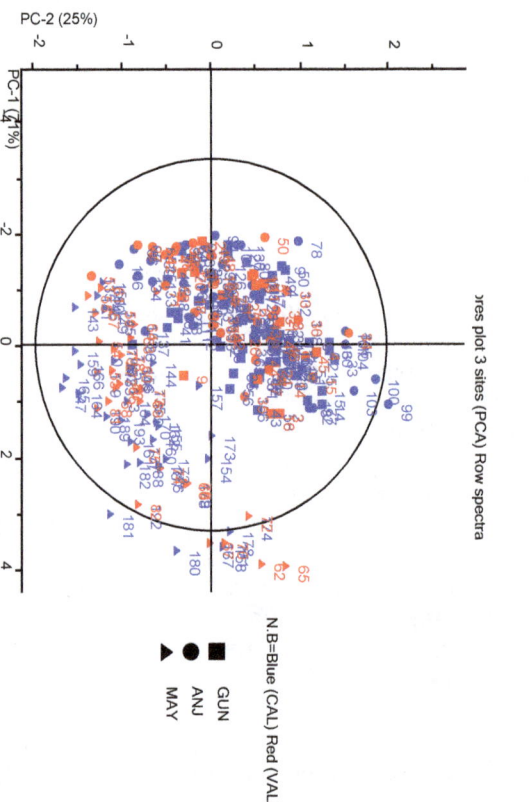

Figure 2. Score plot for first principal components (PC1, PC2) for each and 3 sites altogether.

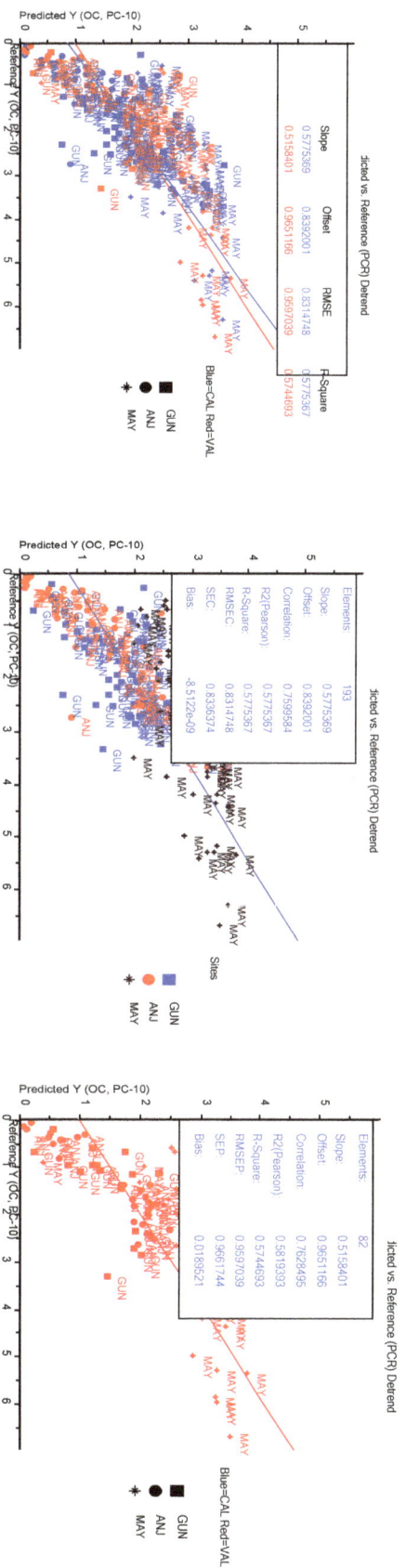

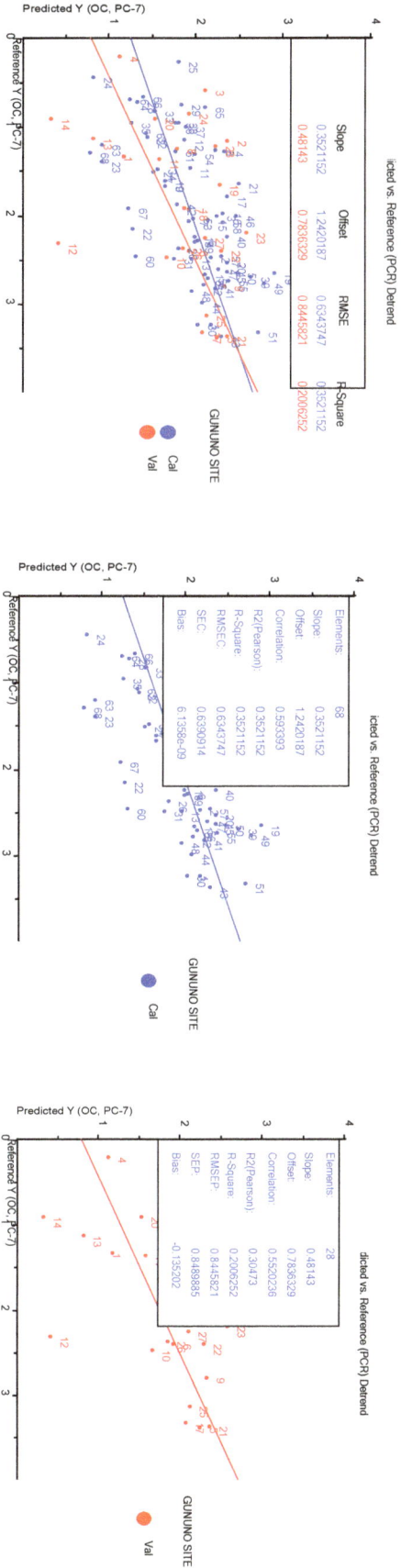

Figure 3. PCR models for individual sites and three sites altogether (validation and calibration). Offset = intercept, SEC= standard error of calibration, SEP = standard error of performance/prediction, R-Square (R²) = coefficient of determination, Correlation (r) = correlation, RMSEP = root mean square of error of prediction, RMSEC = root mean square of error of calibration MSC = multiplicative signal correction, Deterend = De trending, PCR = principal component regression, PLS = partial least square regression, SEC = standard error of calibration SEP = standard error of performance/prediction, NB = The % SOC predicted values (y) are based on spectral measurement while the measured values (x) are measured using Walkley and Black method.

PCR Model Calibration and Validation (Anjeni) (MSC, PC10)

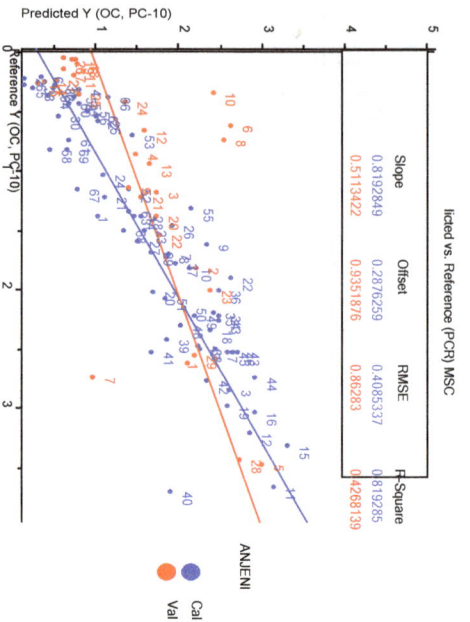

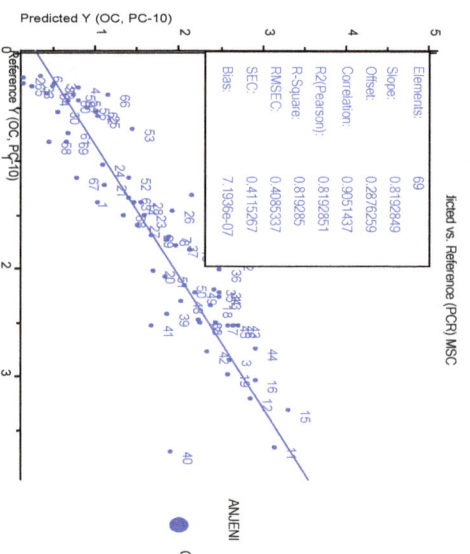

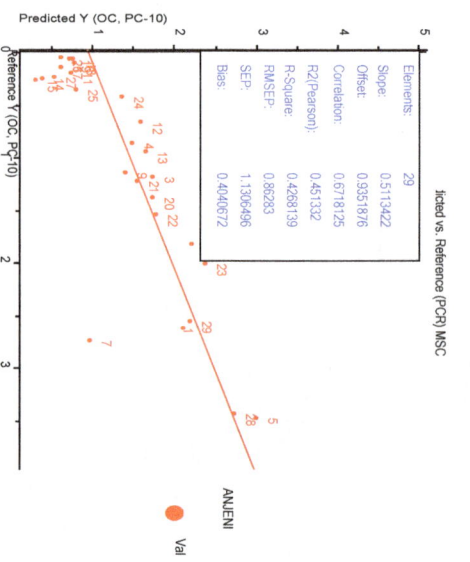

PCR Model Calibration and Validation (Maybar) (MSC, PC10)

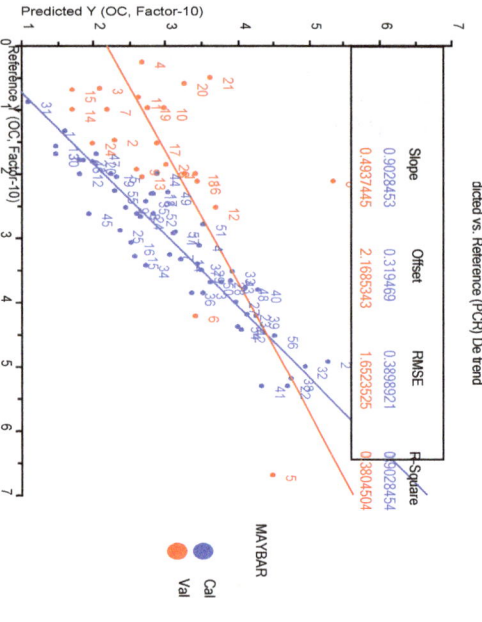

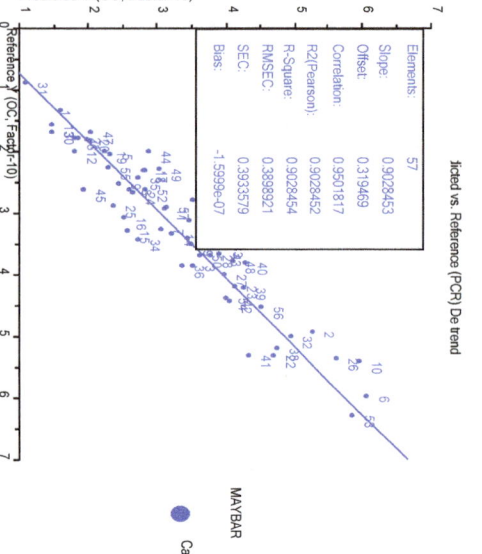

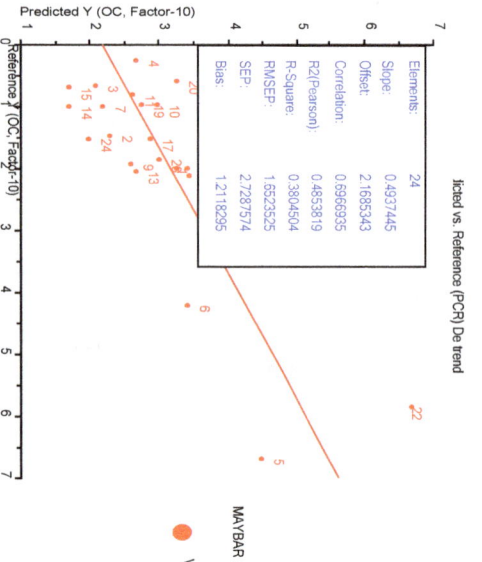

Figure 3. Contd.

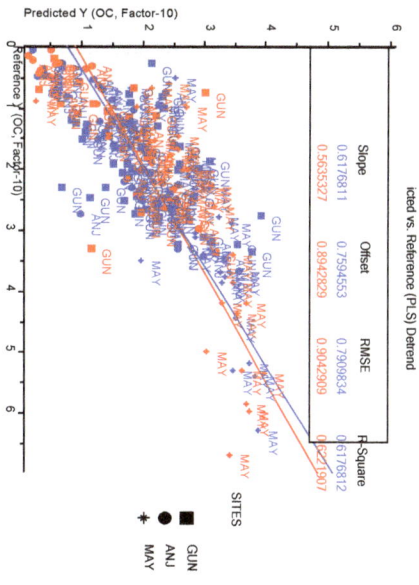

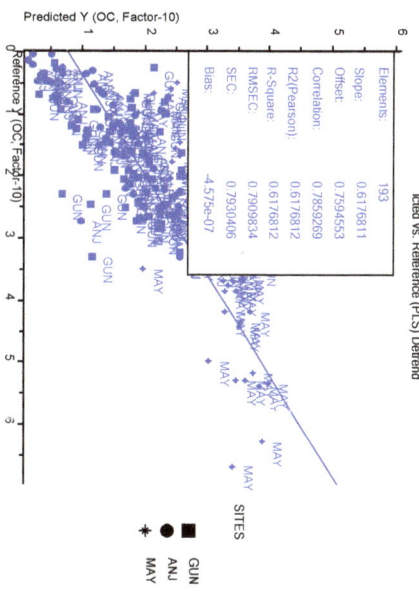

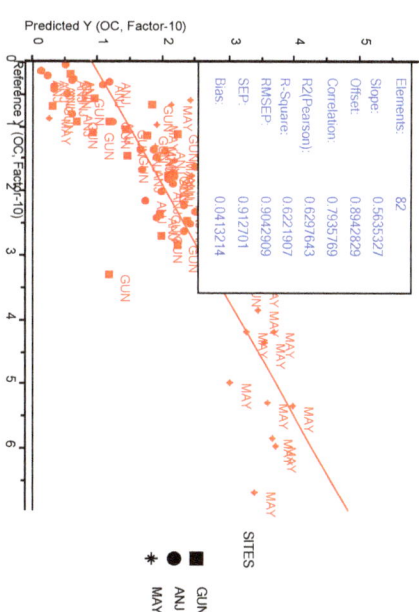

PLS Model for three sites (PC10) (deterend)

PLS Model for Gununo PC4 (deterend)

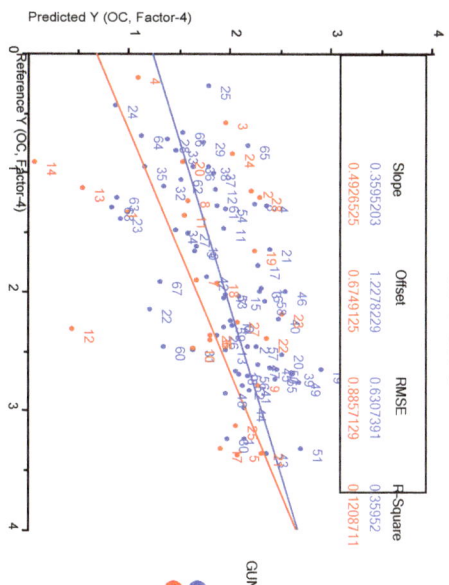

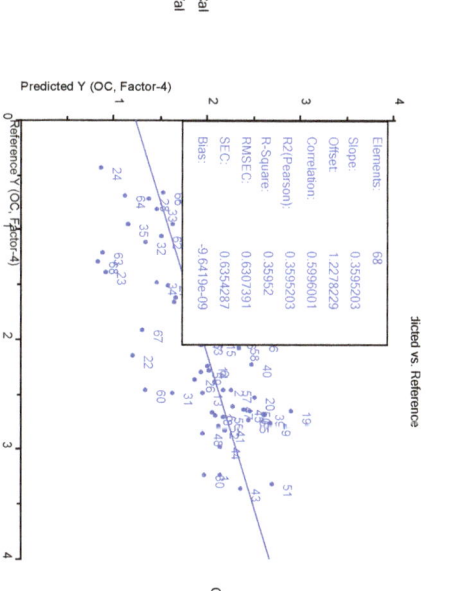

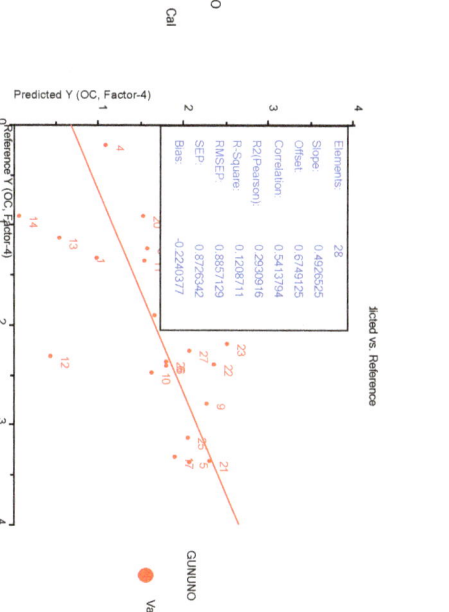

Figure 4. PLS Models for individual sites and three sites altogether (validation and calibration). Offset = intercept, SEC = standard error of calibration, SEP = standard error of performance/prediction, R-square (R²) = coefficient of determination, correlation (r) = correlation, RMSE = root mean square of error of prediction, RMSEC = root mean square of error of calibration MSC= multiplicative signal correction, Deterend = De trending, PCR = principal component regression, PLS = partial least square regression, SEC = standard error of calibration SEP = standard error of performance/prediction, NB = the % SOC predicted values (y) are based on spectral measurement while the measured values (x) are measured using Walkley and Black method.

Figure 4. Contd.

PLS Model for Anjeni PC10 (MSC)

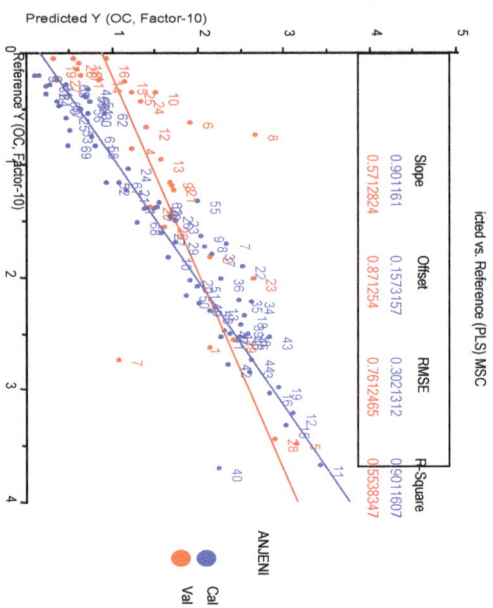

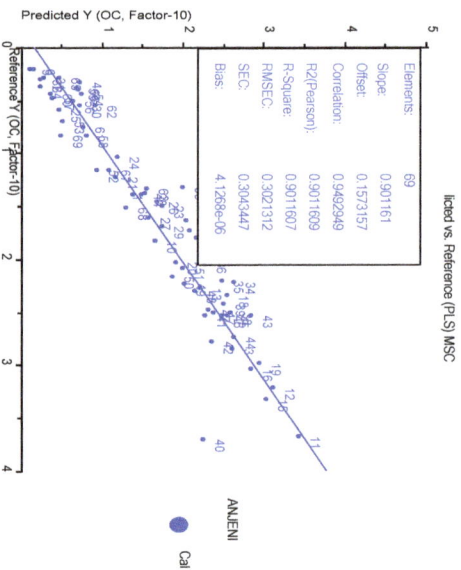

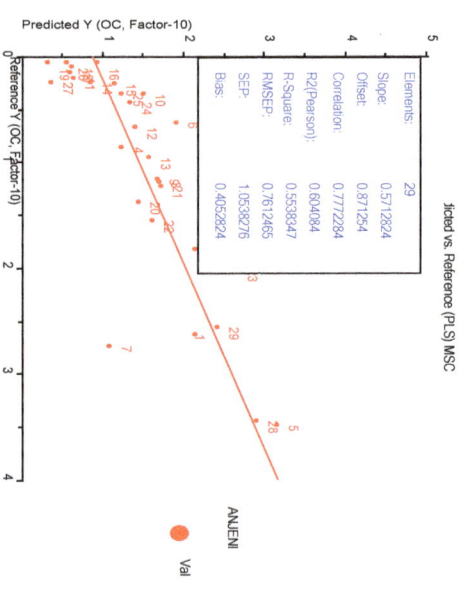

PLS Model for Maybar PC10 (deterend)

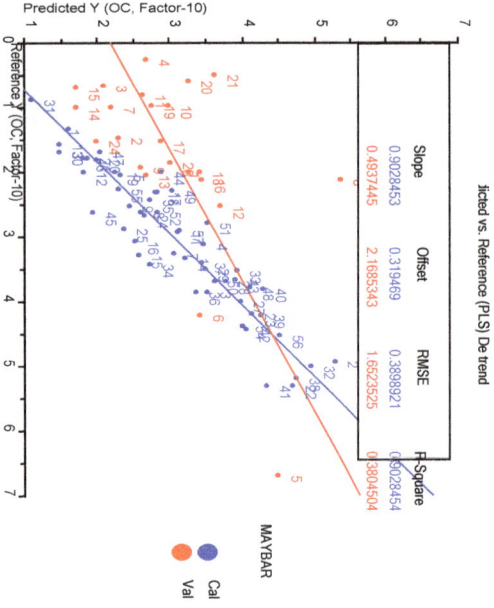

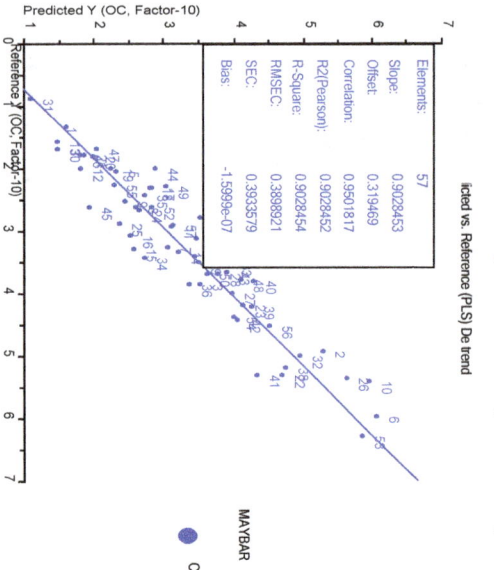

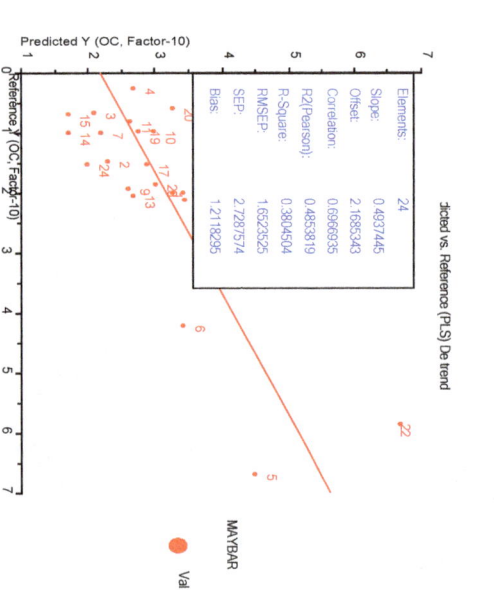

Partial least square regression (PLS)

Review shows that the most frequently used regression models in VisNIR-DRS are PCR and PLS (Blanco and Villarroya, 2002; Viscarra Rossel et al., 2006). Both PCR and PLS can cope with data containing large numbers of predictor variables that are highly collinear (Viscarra Rossel and McBratney, 2008). PLS is the most preferred and popular method to predict SOC (Kang, 2006; Viscarra Rossel et al., 2006; Viscarra Rossel and McBratney, 2008). PLS is used for accurate prediction of site-specific data sets to establish local spectral library (Sankey et al., 2008). SOC measured with Walkley-Black method have been predicted from local to global spectral level using PLS in VisNIR-DRS. Review of past studies on SOC by He and Song (2006) found correlation of 0.9 for soil organic matter (n= 30) RMSEP = 0.12, RMSEC=0.058. Brown et al. (2005) predicted SOC (n= 3793) with correlation of 0.82, Slope =0.76, RMSD=0.9% (with first derivative, D1). McCarty et al., (2002) predicted SOC for different set of sample (n=177- 257) with correlation of 0.82-0.98, RMSD =5.5-7.9. Kang (2006) found correlation of 0.9 for soil samples (n=26) to predict SOC using PLS regression model (r = 0.9) with RMSEC = 0.07 and RMSEP = 0.12.

Testing and comparison of models for SOC prediction

Using full prediction test, the minimum and maximum deviation values were compared for PLS and PCR models. PLS model for Anjeni is the best while PCR model for Maybar is the worst. PLS as a whole has better performance compared with PCR (Table 6). This agrees with findings of Kang (2006), Viscarra Rossel et al. (2006) and Viscarra Rossel and McBratney (2008).

To compare models, accuracy indices are used (Chang et al., 2001; Brunet et al., 2007; He and Song 2006; Ge et al., 2011, Kandel et al., 2011; Stevens et al., 2006). These indices are statistical parameters based on high value (close to 1) correlation coefficient (r^2), coefficient of determination (R^2) and slope values. Moreover, values of residual predictive deviation (RPD), root mean square error (RMSE), standard of error of calibration (SEC) and standard error of performance or prediction (SEP) also assesses model quality (Chang et al., 2001; Brunet et al., 2007; He and Song, 2006; Mouazen et al., 2010; Ge et al., 2011; CAMO, 2012). In this study (Tables 4 and 5) accuracy indices are better for PLS than PCR.

Root mean square error of predication (RMSEP) is expressed in the same units than the variable of analyses (soil organic carbon, g kg^{-1}). Standard error of prediction/performance (SEP) assesses the ability of the model to predict SOC. Standard error of calibration (SEC) is the standard deviations of all the points from the reference values in the calibration set (Stevens et al.,

2006). Best model has lowest SEP. That means, SEP indicates variation in the precision of predictions (Mouazen et al., 2010; CAMO, 2012). In this study (Table 4 and 5) SEP values are better for PLS than PCR.

R^2 values for prediction of soil properties are rated as very good (>0.81), good (0.61-0.8), fair (0.41-0.6) and poor (<0.4) (Viscarra Rossel and McBratney, 2008). The value of R^2 varies from 0.1 (Maybar) which is rated as poor to 0.9 (Anjeni) which is rated as very good (Tables 4 and 5). R^2 values reflect that Anjeni has good predictive ability for SOC while the three site model has is fair. But, Maybar and Gununo models are too poor to be used for prediction.

Ratio of standard deviation to RMSEP or RMSEC is RPD (Chang et al., 2001; Stevens et al., 2006; Mouazen et al., 2010; Kandel et al., 2011; Ge et al., 2011). RPD is used as indicator of predictive ability of models. Genot et al., (2011) indicated that RPD is used to compare samples from diverse variability. Rating shows that RPD< 1 is very poor model, RPD from 1 to 1.4 is poor model, RPD from 1.4 to 1.8 is fair model, RPD from 1.8 to 2 is good model, RPD from 2 to 2.5 is very good model and PRD >2.5 is excellent model (Mouazen et al., 2010). The value of RPD in this study varies from 0.7 (Maybar) to 3.6 (Anjeni).

Values of r^2, R^2, slope and RPD (Tables 4 and 5) shows that PLS has better predictive capacity compared with PCR. Finding in this study agrees with PLS better performance over PCR as indicated by Mouazen et al. (2010) and Viscarra Rossel et al. (2006).PCR and PLS are related techniques and in most situations prediction errors will be similar (Viscarra Rossel and McBratney, 2008), though PLS has comparatively lower predication error. As a whole, taking in to account the two rating methods based on R^2 values as suggested by Viscarra Rossel and McBratney (2008) and RPD value as suggested by Mouazen et al., (2010), Anjeni model is excellent while Gununo and Maybar models are poor. Maybar model has least predictive capacity and rated as very poor based on the above two rating parameters.

Conclusions

Visible-near infrared reflectance (VisNIR) diffuse reflectance spectrometer (DRS) method was used to predict SOC in Ethiopia. Analytical data shows that SOC (g/Kg) from three sites (n=275) has a mean value of 2.0 with 1.2 standard deviation. Most frequent value of SOC is 2.5 g/Kg with a minimum of 0.05 and maximum of 6.7.

PCA score plot shows first two components accounts for a minimum of 96% variation. The closeness of the samples in score plot shows samples similarity with respect to the first principal components.

Although performance of PLS is superior to PCR, in both cases Anjeni model is the best while Maybar the worst. The poor performance of Maybar model might be

Table 4. PCR model calibration and validation results.

Site	Spectra treatment	Process	n (samples)	PCs	Correlation (r)	Slope	Offset	Final		RPD
								R²	RMSEC/RMSEP	
3 sites	Raw spectral	CAL	193	10	0.71	0.51	0.96	0.51	0.89	1.3
		VAL	82	10	0.68	0.42	1.19	0.46	1.07	1.2
	De-trend	CAL	193	10	0.75	0.57	0.83	0.57	0.83	1.4
	Figure 3a	VAL	82	10	0.76	0.51	1.96	0.57	0.95	1.2
Gununo	Raw spectral	CAL	68	7	0.52	0.40	0.95	0.22	0.84	0.9
		VAL	28	7	0.52	0.40	0.95	0.22	0.83	0.9
	De-trend	CAL	68	7	0.59	0.35	1.24	0.35	0.66	1.2
	Figure 3b	VAL	28	7	0.55	0.48	0.78	0.20	0.80	1
Anjeni	Raw spectral	CAL	69	10	0.90	0.81	0.29	0.81	0.41	2.
		VAL	29	10	0.65	0.52	0.78	0.46	0.83	1.3
	MSC Figure 3c	CAL	69	10	0.90	0.81	0.28	0.81	0.40	2.7
		VAL	29	10	0.67	0.51	0.93	0.42	0.86	1.2
Maybar	Raw spectral	CAL	57	10	0.70	0.63	1.19	0.63	0.75	2
		VAL	24	10	0.45	0.23	2.73	0.18	1.89	0.7
	De-trend	CAL	57	10	0.78	0.61	1.28	0.61	0.78	1.9
	Figure 3d	VAL	24	10	0.50	0.21	2.90	0.12	1.96	0.7

CAL = Calibration, VAL = validation, MSC = multiplicative signal correction offset = intercept, R-square (R^2) = coefficient of determination, correlation (r) = correlation, RMSEP = root mean square of error of prediction, RMSEC = root mean square of error of calibration n = sample number RPD = residual prediction deviation PCs = principal components (factors).

Table 5. PLS model calibration and validation results.

Site	Spectra treatment	Process	n (sample)	No. of components	Correlation (r)	Slope	Offset	Final		RPD
								R²	RMSEC/RMSEP	
3 sites	Raw spectral	CAL	193	10	0.77	0.59	0.80	0.59	0.81	1.4
		VAL	82	10	0.76	0.53	0.93	0.58	0.94	1.5
	De-trend	CAL	193	10	0.79	0.61	0.75	0.61	0.79	1.5
	Figure 4a	VAL	82	10	0.79	0.56	0.89	0.62	0.90	1.3
Gununo	Raw spectral	CAL	68	6	0.62	0.38	1.17	0.38	0.61	1.1
		VAL	28	6	0.62	0.38	1.17	0.38	0.61	1.1
	De-trend	CAL	68	4	0.59	0.35	1.22	0.35	0.63	1.2
	Figure 4b	VAL	28	4	0.54	0.49	0.67	0.1	0.88	1.0
Anjeni	Raw spectral	CAL	69	10	0.94	0.90	0.15	0.9	0.30	3.6
		VAL	29	10	0.80	0.59	0.55	0.70	0.62	1.7
	MSC	CAL	69	10	0.94	0.90	0.15	0.90	0.30	3.6
	Figure 4c	VAL	29	10	0.77	0.57	0.87	0.55	0.76	1.4
Maybar	Raw spectral	CAL	57	10	0.93	0.82	0.42	0.87	0.44	3.4
		VAL	24	10	0.69	0.52	2.04	0.41	1.60	0.9
	De-trend	CAL	57	10	0.95	0.90	0.31	0.90	0.38	3.9
	Figure 4d	CAL	24	10	0.69	0.49	2.16	0.38	1.65	0.9

CAL= Calibration, VAL = validation, MSC = multiplicative signal correction offset = intercept, R-square (R^2) = coefficient of determination, Correlation (r) = Correlation, RMSEP = root mean square of error of prediction, RMSEC = root mean square of error of calibration n = sample number RPD = residual prediction deviation PCs = principal components (factors).

Table 6. Testing PCR and PLS models using full prediction.

Site	Model	Spectra treatment	n (sample)	PCs	Deviation from reference (n)	
					Min	Max
3 sites	PLS	De-trend	275	10	0.5	1.9
	PCR	De-trend	275	10	0.4	1.6
Gununo	PLS	De-trend	96	4	0.3	1.7
	PCR	De-trend	96	7	0.3	2.4
Anjeni	PLS	MSC	98	10	0.1	0.7
	PCR	MSC	98	10	0.2	0.9
Maybar	PLS	De-trend	81	10	0.7	3.7
	PCR	De-trend	81	10	0.8	3.9

n = Number of soil samples PC = principal component/factors, MSC = multiplicative signal correction, Deterend = De trending, PCR = principal component regression, PLS = partial least square regression.

attributed to the 9% high leverage values and 4% potential outliers. PLS correlation (r), coefficient of determination (R^2) and residual prediction deviation (RPD) were used to compare PLS and PCR models. Models testing showed better performance of PLS compared with PCR. Based on two statistical parameter rating (R^2 and RPD), Maybar, Gununo and three sites models are not recommended for prediction of SOC. Models were rated as very poor (Maybar) and poor (Gununo and three sites). Anjeni model, however, is excellent and can be used for prediction of SOC in Ethiopia. Anjeni model is more applicable to Nitisols, Alisols and Cambisols, soil units (FAO/UNESCO) (according to decreasing order of application).

Although there are standard protocols in soil spectroscopy for spectral measurement, gaps still exist in having clear guideline on data pre-treatment, calibration and validation for SOC prediction. The study recommends developing further predictive models to represent the diverse soil units in Ethiopia.

ACKNOWLEDGEMENTS

This study was funded by U.S. National Science Foundation (NSF) under the grant number GEO-0627893 through International START secretariat as 2010 Grants for GEC Research in Africa. The authors are grateful to Lorenz Ruth for his technical assistance in laboratory spectral measurement. We are also grateful to CAMO software team, Prof. Dr. Hans Hurni, Dr. Bettina Wolfgramm, Dr. Gete Zeleke, Tadele Amare (University of Bern, Switzerland) for their contribution to finalize this study. Special thanks go to Bosena Buzunhe and Nugussue Bekele (Ethiopian Institute of Agricultural Research, Debre Ziet, Ethiopia) for their assistance from field sampling to laboratory analytical measurement.

REFERENCES

Allen DE, Pringle MJ, Page KL and Dalal RC (2010). A review of sampling designs for the measurement of soil organic cabon in Australian grazing lands. Rangeland J. 32(3):227-246.

Baker JM, Ochsner TE, Venterea RT, Griffis TJ (2007). Tillage and soil carbon sequestration - What do we really know? Agric. Ecosyst. Environ. 118:1-5.

Blanco M, Villarroya I (2002).NIR spectroscopy: a rapid-response analytical tool. Trend Anal. Chem. 21(4):240-250.

Brown DJ, Shepherd KD, Walsh MG, Mays MD, Reinsch TG (2005). Global soil characterization with VNIR diffuse reflectance spectroscopy. Geoderma 132(3-4):273-290.

Brunet D, Barthes BG, Chotte JL, Feller C (2007). Determination of carbon and nitrogen contents in Alfisols, Oxisols, and Ultisols and from Africa and Brazil using NIRS analysis: Effects of sample grinding and set heterogeneity. Geoderma 139:106-117.

CAMO (2012). Complete Multivariate Analysis and Experimental Design Software (CAMO Software Release Notes, Unscrambler X version 10.2), CAMO Software Research & Development Team, CAMO Software, NedreVollgate 8, N-0158, Oslo, Norway, retrieved on March 20/2012 from http://www.camo.com/

Cecillon LC, Barthes BG, Gomez C, Ertlen D, Genot V, Hedde M, Stevens A, Burn JJ (2009). Assessment and monitoring of soil quality using Near-Infrared Reflectance Spectroscopy (NIRS). Eur. J. Soil Sci. 60:770-784.

Chang CW, Laird DA, Mausbach MJ, Hurburgh CR (2001). Near-Infrared Reflectance Spectroscopy-Principal Components Regression. Soil Sci. Soc. Am. J. 65:480-490.

Ge Y, Morgan CLS, Grunwald S, Brown DJ, Sarkhot DV (2011). Comparison of soil reflectance spectra and calibration models obtained using multiple spectrometers. Geoderma 161:202-211.

Genot V, Colinet G, Bock L, Vanvyve D, Reusen, Y, Dardenne P (2011).Near Infrared Reflectance Spectroscopy for estimating soil characteristics valuable in the diagnosis of soil fertility. J. Near Infrared Spec.19:117-138.

Gomez C, Viscarra Rossel RA, McBrantney AB (2008), Comparing predictions of soil organic carbon by field Vis-NIR Spectroscopy and hyper spectral remote sensing. Geophys. Res. Abstr.(10)1-2, SRef-

ID:1607-7962/gra/EGU2008-A-00317.

He Y, Song H (2006). Prediction of soil content using near-infrared spectroscopy, SPINE news room. The international Society for Optical Engineering. DOI: 10.1117/2.1200604.0164

Janik LJ, Merry RH, Skjemstand JO(1998).Can mid infrared diffuse reflectance analysis replace soil extractions?. Aust. J Exp. Agric. 38:681-96.

Kang M (2006). Quantification of soil organic carbon using mid and near Diffuse Reflectance Infrared Fourier Transform spectroscopy, M.Sc thesis, Texas A&M University, Department of Geology and Geophysics Department. Accessible at <geoweb.tamu.edu/Faculty/Herbert/docs/02KangMSThesis.pdf>

Kejela K (1995). Soils of the Anjeni Area-Gojam Research Unit, Ethiopia. Soil Conservation Research Program (SCRP), Center for Development and Environment (CDE), University of Bern, Switzerland, Research Report 27.

Knadel M, Thomsen A, Greve MH (2011). Multi-sensor On -The -Go Mapping of Soil Organic Carbon Content. Soil Sci. Soc. Am. J. 75:1799-1806.

McCarty GW, Reeves JB, Follett RF, Kimble JM (2002). Mid-Infrared and Near Infrared Diffuse Reflectance Spectroscopy for Soil Carbon Measurement. Soil Sci. Soc. Am. J. 66:640-646.

McCarty GW, Reeves JB,.Yost R, Doraiswamny PC, Doumbia M (2010).Evaluation of methods for measuring soil organic carbon in West African soils. Afr. J Agric. Res. 5(16):2169-2177.

Mouazen AM, Kuang B, DeBeardemaeker J, Ramon H(2010). Comparison among principal components, partial least square and back propagation neutral network analyses for accuracy of measurement of selected soil properties with visible and near infrared spectroscopy. Geoderma 158:23-31.

Naes T, Isaksson T, Fearn T, Davies T (2002). A User-Friendly Guide to Multivariate Calibration and Classification, NIR publications, Chichester, UK, p. 344.

Reeves JB, Follett RF, McCarty GW, Kimble JM (2006). Can Near or Mid-Infrared Diffuse Reflectance Spectroscopy Be Used to Determine Soil Carbon Pools?. Commun. Soil Sci. Plan. 37:2307-2325.

Sankey JB, Brown DJ, Bernard ML, Lawrence RL (2008). Comparing local vs. global visible Near-Infrared (VisNIR) diffuse reflectance spectroscopy (DRS) calibrations for the prediction of soil clay, organic C and inorganic C. Geoderma 148:149-158.

SCRP (2000a). Concept and Methodology: Long-term Monitoring of the Agricultural Environment in Six Research Stations in Ethiopia. Soil Conservation Research Program (SCRP), Center for Development and Environment (CDE), University of Bern, Switzerland.

SCRP (2000b). Area of Gununo, Sidamo, Ethiopia: Long Term Monitoring of the Agricultural Environment (1981-1994). Soil Conservation Research Program (SCRP), Center for Development and Environment (CDE), University of Bern, Switzerland.

SCRP (2000c). Area of Anjeni, Gojam: Long-term Monitoring of the Agricultural Environment 1984-1994. Soil Erosion and Conservation Database, Soil Conservation Research Program (SCRP), Center for Development and Environment (CDE), University of Bern, Switzerland.

SCRP (2000d). Area of Mayber, Wello: Long-term Monitoring of the Agricultural Environment 1984-1994. Soil Erosion and Conservation Database, Soil Conservation Research Program (SCRP), Center for Development and Environment (CDE), University of Bern, Switzerland.

Stevens A, Wesemael B, Vandenschrick G, Toure S, Tychon B (2006). Detection of Carbon Stock Change in Agricultural Soils Using Spectroscopic Techniques. Soil Sci. Soc. Am. J. 70:844-850.

Stolbovoy V, Montanarella L, Filippi N, Jones A, Gallego J, Grassi G (2002). Soil Sampling Protocol to Certify the Changes of Organic Carbon Stock in Mineral Soil of The European Union, Version 2 , EUR 21576 EN/2, 56 pp. EC(European Commission), Office for official Publication of the European Communities, Institute for Environment and Sustainability, Luxembourg.

Tobler M (2011)Assessment of dominant land-uses systems in the Tajik Pamiron on the basis of a soil spectral library , Master Thesis, Federal Institute of Technology (ETH) Zurich, Center for Development and Environment (CDE) Bern and Swiss Federal Institute for Forest, Snow and Landscape Research (WSL), 90 pp. Available at <www.ehs.unu.edu/palm/file/get/8662>

Viscarra Rossel RA (2009).The Soil Spectroscopy Group and the development of a global soil spectral library. Geophys. Res. Abstr. 11:1-2, EGU2009-14021.

Viscarra Rossel RA, Walvoort DJJ, McBratney AB, Janik LJ, Skjemstad JO (2006).Visible near infrared, mid infrared or combined diffuse reflectance spectroscopy for simultaneous assessment of various soil properties. Geoderma 131:59-75.

Viscarra Rossel RA, McBratney AB (2008). Diffuse Reflectance Spectroscopy as a Tool for Digital Soil Mapping, In: Digital Soil Mapping with Limited Data (eds. Hartemink et.al) Springer Science Business Media B.V. Chapter 13, pp. 165-172.

Walkley A, Black IA (1934). An examination of the degtjareff method for determining soil organic matter and a proposed modification of the chromic acid titration method. Soil Sci.37: 29-38.

Weigel G (1986a). Soils of the Gununo Area-Sidamo Research Unit, Ethiopia. Soil Conservation Research Program (SCRP), Center for Development and Environment (CDE), University of Bern, Switzerland, Research Report 8.

Weigel G (1986b). Soils of the Maybar Area, Wello Area: their potential and constraints for agricultural development, Volume A4., a case study in the Ethiopian Highlands, African Studies Series, Geographica Bernensia, University of Berne, Switzerland.

Zeleke G (2000). Landscape Dynamics and Soil Erosion Process Modeling in the North-western Ethiopian Highlands. African Studies Series 16. Berne: Geographica Bernensia.

Permissions

List of Contributors

Pratibha Arya
Botany Department, Government P. G. College Augustyamuni, Rudraprayag, Garhwal, Uttarakahand, India

S. C. Sati
Botany Department, Government P. G. College Augustyamuni, Rudraprayag, Garhwal, Uttarakahand, India

J. O. ADEFILA
Department of Geography, Ahmadu Bello University, Zaria, Nigeria

Edward D. Wiafe
Department of Environmental and Natural Resources Management, Presbyterian University College, P. O. Box 393, Akropong-Akuapem, Ghana

Richard Amfo-Otu
Department of Environmental and Natural Resources Management, Presbyterian University College, P. O. Box 393, Akropong-Akuapem, Ghana

L. O. Odokuma
University of Port Harcourt, Rivers State, Nigeria

C. J. Ugboma
University of Port Harcourt, Rivers State, Nigeria

Reuben A. Garshong
Department of Animal Biology and Conservation Science, University of Ghana, P. O. Box LG67, Legon-Accra, Ghana

Daniel K. Attuquayefio
Department of Animal Biology and Conservation Science, University of Ghana, P. O. Box LG67, Legon-Accra, Ghana

Lars H. Holbech
Department of Animal Biology and Conservation Science, University of Ghana, P. O. Box LG67, Legon-Accra, Ghana

James K. Adomako
Department of Botany, University of Ghana, P. O. Box LG55, Legon-Accra, Ghana

Tinsae Bahru
Ethiopian Institute of Agricultural Research (EIAR), Forestry Research Center (FRC), Non-Timber Forest Products (NTFPs) Case Team, P. O. Box 30708, Addis Ababa, Ethiopia

Zemede Asfaw
The National Herbarium, Department of Biology, Faculty of Science, Addis Ababa University, P. O. Box 3434, Addis Ababa, Ethiopia

Sebsebe Demissew
The National Herbarium, Department of Biology, Faculty of Science, Addis Ababa University, P. O. Box 3434, Addis Ababa, Ethiopia

M. O. Mohammed
Environment and Natural Resources Research Institute, the National Centre for Research, Khartoum, Sudan

M. E. Ali
Fisheries Research Centre, Agricultural Research Corporation, Khartoum, Sudan

Ram Chandra Kandel
Department of National Parks and Wildlife Conservation, P. O. Box 860, Kathmandu, Nepal

Olaotswe Ernest Kgosikoma
University of Edinburgh, Crew Building, West Mains Road, EH9 3JN, Scotland
Department of Agricultural Research, Private Bag 0033, Gaborone, Botswana

Witness Mojeremane
Botswana College of Agriculture, Private Bag 0027, Gaborone, Botswana

Barbra A. Harvie
University of Edinburgh, Crew Building, West Mains Road, EH9 3JN, Scotland

Javaid Ahmad Shah
Centre of Research for Development (CORD), University of Kashmir, Srinagar-190006, J & K, India

Ashok K. Pandit
Centre of Research for Development (CORD), University of Kashmir, Srinagar-190006, J & K, India

G. Mustafa Shah
Department of Zoology, University of Kashmir, Srinagar-190006, J & K, India

Ezeaku, Peter Ikemefuna
Department of Soil Science, Faculty of Agriculture, University of Nigeria, Nsukka, Nigeria

Dickson Otieno Owiti
Department of Zoology, Maseno University, P.O Box 333-40105, Maseno, Kenya
Department of Fisheries and Natural Resources, Maseno University, PO Box 333-40105, Maseno, Kenya

Raphael Achola Kapiyo
Department of Environmental Science, School of Environment and Earth Science, Maseno University, P.O Box 333-40105, Maseno, Kenya

Esna Kerubo Bosire
Department of Environmental Science, School of Environment and Earth Science, Maseno University, P.O Box 333-40105, Maseno, Kenya

B. K. Tripathy
Department of Botany, Dharmasala Mohavidyalaya, Dharmasala, Jajpur, Odisha, India

R. B. Mohanty
Department of Botany, N. C.Autonomous College, Jajpur, Odisha, India

N. Mishra
Department of Zoology, Chandbali College, Chandbali, Bhadrak-756133, Odisha, India

T. Panda
Department of Botany, Chandbali College, Chandbali, Bhadrak-756133, Odisha, India

Demeke Datiko
Department of Biology, Addis Ababa University, P. O. Box 1176, Addis Ababa, Ethiopia

Afework Bekele
Department of Biology, Addis Ababa University, P. O. Box 1176, Addis Ababa, Ethiopia

John N. Kigomo
Kenya Forestry Research Institute, P.O. Box 20412-00200, Nairobi, Kenya

Gabriel M. Muturi
Kenya Forestry Research Institute, P.O. Box 20412-00200, Nairobi, Kenya

Chansa Chomba
Zambia Wildlife Authority, P/B 1 Chilanga, Zambia

Ramadhani Senzota
Department of Zoology and Wildlife Conservation, University of Dar es Salaam, P.O. Box 35065 Dar es Salaam Tanzania

Harry Chabwela
Department of Biological Sciences, University of Zambia, P. O. Box 32379 Lusaka, Zambia

Jacob Mwitwa
School of Natural Resources, Copperbelt University, P. O. Box 21692 Kitwe, Zambia

Vincent Nyirenda
Office of the Director General, Zambia Wildlife Authority, P/B 1 Chilanga, Zambia

Mushtaq Ahmad Ganie
Department of Zoology, Faculty of Basic Sciences, Bundelkhand University Jhansi-284 128, U. P., India

Mehraj Din Bhat
Department of Zoology, Faculty of Basic Sciences, Bundelkhand University Jhansi-284 128, U. P., India

Mohd Iqbal Khan
Department of Zoology, Faculty of Basic Sciences, Bundelkhand University Jhansi-284 128, U. P., India

Muni Parveen
Department of Zoology, Faculty of Biological Sciences, University of Kashmir, Srinagar-190 006, J & K, India

M. H Balkhi
Faculty of Fisheries, Sher-e-Kashmir University of Agricultural Sciences and Technology of Kashmir, Srinagar, India

Muneer Ahmad Malla
Department of Zoology, Faculty of Basic Sciences, Bundelkhand University Jhansi-284 128, U. P., India

I. A Nweke
Department of Soil Science, Anambra State University, Igbariam Campus Anambra State, Nigeria

Rafia Rashid
Department of Environmental Science, University of Kashmir, Srinagar - 190006, India

Ashok K. Pandit
Department of Environmental Science, University of Kashmir, Srinagar - 190006, India

R.K. Garg
Centre of Excellence in Biotechnology, M.P. Council of Science and Technology (MPCST), Vigyan Bhawan, Nehru Nagar, Bhopal-462003 (M.P.), India

R.J. Rao
School of Studies in Zoology, Jiwaji University, Gwalior-474011 (M.P.), India

D.N. Saksena
School of Studies in Zoology, Jiwaji University, Gwalior-474011 (M.P.), India

Irfan Mustafa
Graduate School of Science and Technology, Kumamoto University, 860-0862, Japan
Biology Department, Faculty of Sciences, Brawijaya University, 65145, Indonesia

Hiroto Ohta
Graduate School of Science and Technology, Kumamoto University, 860-0862, Japan

Takuro Niidome
Graduate School of Science and Technology, Kumamoto University, 860-0862, Japan

Shigeru Morimura
Graduate School of Science and Technology, Kumamoto University, 860-0862, Japan

Ram Prasad Lamsal
Department0 of Environmental Science and Engineering, Kathmandu University, Nepal

Bikash Adhikari
Department0 of Environmental Science and Engineering, Kathmandu University, Nepal

Sanjay Nath Khanal
Department0 of Environmental Science and Engineering, Kathmandu University, Nepal

Khet Raj Dahal
Department of Civil Engineering, Kantipur Engineering College, Tribhuvan University, Nepal

Simon L. A. Mwansasu
Institute of Resource Assessment, University of Dar es Salaam, P O Box 35097, Dar es Salaam, Tanzania
Department of Physical Geography and Quaternary Geology, Stockholm University, SE-106 91 Stockholm, Sweden

Lars Ove Westerberg
Department of Physical Geography and Quaternary Geology, Stockholm University, SE-106 91 Stockholm, Sweden
Bolin Centre for Climate Research, Stockholm University, S-10691 Stockholm, Sweden

W. Mubvekeri
Cotton Research Institute, Kadoma, Department of Research and Specialist Services Zimbabwe

J. Bare
Department of Biological Sciences, Faculty of Science and Technology Midlands State University Gweru, Zimbabwe

Caston Makaka
Department of Biological Sciences, Faculty of Science and Technology Midlands State University Gweru, Zimbabwe

F. Jimu
Cotton Research Institute, Kadoma, Department of Research and Specialist Services Zimbabwe

S. Mohammed Irshad
Jamsetji Tata Centre for Disaster Management, Tata Institute of Social Science, Malatil and Jal A. D. Naoroji (New) Campus, Sion-Trombay Road, P. O. Box 8313, Deonar, Mumbai-400 088, India

Thomas Yeboah
Department of Hospitality and Tourism Management, Sunyani Polytechnic, School of Applied Sciences and Technology, P.O Box 206, Sunyani, Brong Ahafo, Ghana

Abebe Shiferaw
International Livestock Research Institute (ILRI), Addis Abeba, Ethiopia
University of Bern, Hochschulstrasse 4, 3012 Bern, Switzerland

Christian Hergarten
University of Bern, Hochschulstrasse 4, 3012 Bern, Switzerland

www.ingramcontent.com/pod-product-compliance
Lightning Source LLC
Chambersburg PA
CBHW050443200326
41458CB00014B/5053